VOLUME ONE HUNDRED AND SEVENTY EIGHT

PROGRESS IN MOLECULAR BIOLOGY AND TRANSLATIONAL SCIENCE

Advances in CRISPR/Cas and Related Technologies

VOLUME ONE HUNDRED AND SEVENTY EIGHT

PROGRESS IN MOLECULAR BIOLOGY AND TRANSLATIONAL SCIENCE

Advances in CRISPR/Cas and Related Technologies

Edited by

DIPANJAN GHOSH

Department of Natural Products,
National Institute of Pharmaceutical Research and Education (NIPER), Kolkata, India

ELSEVIER

ACADEMIC PRESS
An imprint of Elsevier

Academic Press is an imprint of Elsevier
50 Hampshire Street, 5th Floor, Cambridge, MA 02139, United States
525 B Street, Suite 1650, San Diego, CA 92101, United States
The Boulevard, Langford Lane, Kidlington, Oxford OX5 1GB, United Kingdom
125 London Wall, London EC2Y 5AS, United Kingdom

First edition 2021

Notices

Knowledge and best practice in this field are constantly changing. As new research and experience broaden our understanding, changes in research methods, professional practices, or medical treatment may become necessary.

Practitioners and researchers must always rely on their own experience and knowledge in evaluating and using any information, methods, compounds, or experiments described herein. In using such information or methods they should be mindful of their own safety and the safety of others, including parties for whom they have a professional responsibility.

To the fullest extent of the law, neither the Publisher nor the authors, contributors, or editors, assume any liability for any injury and/or damage to persons or property as a matter of products liability, negligence or otherwise, or from any use or operation of any methods, products, instructions, or ideas contained in the material herein.

ISBN: 978-0-12-821590-6
ISSN: 1877-1173

For information on all Academic Press publications
visit our website at https://www.elsevier.com/books-and-journals

Publisher: Zoe Kruze
Acquisitions Editor: Ashlie M. Jackman
Developmental Editor: Jhon Michael Peñano
Production Project Manager: James Selvam
Cover Designer: Alan Studholme

Typeset by SPi Global, India

Contents

Contributors

Kazi Tawsif Ahmed
Department of Botany, Visva-Bharati, West Bengal, India

Sangram Bagh
Biophysics and Structural Genomics Division, Saha Institute of Nuclear Physics, Homi Bhabha National Institute (HBNI), Kolkata, India

Sulagna Basu
Division of Bacteriology, ICMR-National Institute of Cholera and Enteric Diseases, Scheme XM, Beliaghata, Kolkata, India

Sujit Kumar Bhutia
Department of Life Science, National Institute of Technology, Rourkela, Odisha, India

Swadesh Ranjan Biswas
Department of Botany, Visva-Bharati, Santiniketan, West Bengal, India

Deepro Bonnerjee
Biophysics and Structural Genomics Division, Saha Institute of Nuclear Physics, Homi Bhabha National Institute (HBNI), Kolkata, India

Dhritiman Dey
Department of Natural Products, National Institute of Pharmaceutical Research and Education (NIPER), Kolkata, India

Dipanjan Ghosh
Department of Natural Products, National Institute of Pharmaceutical Research and Education (NIPER), Kolkata, India

Darshana Gupta
Department of Botany, Visva-Bharati, West Bengal, India

Rahul Gupta
Infectious Diseases and Immunology Division, CSIR-Indian Institute of Chemical Biology, Kolkatta, India

Munshi Azad Hossain
Department of Molecular, Cell, and Developmental Biology, University of California, Los Angeles, CA, United States

Anbarasu Kannan
Department of Protein Chemistry and Technology, CSIR-Central Food Technological Research Institute; Academy of Scientific and Innovative Research (AcSIR), CSIR-Central Food Technological Research Institute (CSIR-CFTRI) Campus, Mysuru, India

Tawsif Ahmed Kazi
Department of Botany, Visva-Bharati, Santiniketan, West Bengal, India

Subhadip Mukhopadhyay
Department of Radiation Oncology, Laura and Isaac Perlmutter Cancer Center, NYU Medical School, New York, NY, United States

Sharmi Naha
Division of Bacteriology, ICMR-National Institute of Cholera and Enteric Diseases, Scheme XM, Beliaghata, Kolkata, India

Rajalakshmi Prakash
Department of Protein Chemistry and Technology, CSIR-Central Food Technological Research Institute; Academy of Scientific and Innovative Research (AcSIR), CSIR-Central Food Technological Research Institute (CSIR-CFTRI) Campus, Mysuru, India

Shubhchintan Randhawa
Department of Molecular Medicine & Biotechnology, Sanjay Gandhi Postgraduate Institute of Medical Sciences, Lucknow, India

Ankur Rao
Division of Bacteriology, ICMR-National Institute of Cholera and Enteric Diseases, Scheme XM, Beliaghata, Kolkata, India

V. Ravichandiran
Department of Natural Products, National Institute of Pharmaceutical Research and Education (NIPER), Kolkata, India

Sobhan Roy
Departments of Biochemistry and Molecular Biology, University of Chicago, Chicago, IL, United States

Subhasree Roy
Division of Bacteriology, ICMR-National Institute of Cholera and Enteric Diseases, Scheme XM, Beliaghata, Kolkata, India

Syamal Roy
Department of Natural Products, National Institute of Pharmaceutical Research and Education (NIPER), Kolkata, India

Shatakshi Sengar
Imperial Life Sciences, Gurugram, India

Rajveer Singh
Department of Natural Products, National Institute of Pharmaceutical Research and Education (NIPER), Kolkata, India

Snehasikta Swarnakar
Infectious Diseases and Immunology Division, CSIR-Indian Institute of Chemical Biology, Kolkatta, India

Preface

The CRISPR-Cas9 system has undergone one of the fastest developments in the field of targeted genome editing. From its discovery in 1987, through its establishment as a bacterial adaptive immune system, to it being improved into the most efficient and simple-to-design genome editing tool, observed to be functional in prokaryotic, mammalian and plant systems, CRISPR-Cas9 system has shown the potential to have wide applications ranging from medication using gene therapy to crop improvement. In addition to this, its contribution to the field of research has been enormous whereby it has helped in studying functional parameters at DNA and protein levels and further accelerated research in its own domain by allowing easy modifications at the DNA level. Further research is still in progress to exploit all possible information that can be extracted using this technology. The system serves as yet another inspiration, signifying the diversity of nature and emphasizing the fact that nature holds some of the simplest solutions to some of the most complicated problems.

The book "Advances in CRISPR/Cas and related technologies" deals with a wide variety of research topics related to recent advancement in genome editing techniques. As a genome editing tool, the CRISPR structure consists of two major components: a guide RNA molecule and a Cas9 protein. However, a serious drawback of the genome editing tool is its off-target effect. This particular drawback is a serious hindrance in its therapeutic application. Currently, a lot of work is going on to overcome this hurdle. The series gives in-depth knowledge and updates the reader on the important modified molecular techniques used in regard to the CRISPR/Cas system to offset its off-target activity. This series, while updating the reader of higher efficiency genome editing will also throw light upon unmet challenges like targeted delivery of CRISPR/Cas9 inside mammalian organisms and also application of CRISPR-Cas system for metabolic bioengineering.

CHAPTER ONE

The evolution and history of gene editing technologies

Shubhchintan Randhawa[a,*] and Shatakshi Sengar[b]
[a]Department of Molecular Medicine & Biotechnology, Sanjay Gandhi Postgraduate Institute of Medical Sciences, Lucknow, India
[b]Imperial Life Sciences, Gurugram, India
*Corresponding author: e-mail address: shubirandhawa@yahoo.com

Contents

Abstract

Scientific enquiry must be the driving force of research. This sentiment is manifested as the profound impact gene editing technologies are having in our current world. There exist three main gene editing technologies today: Zinc Finger Nucleases, TALENs and the CRISPR-Cas system. When these systems were being uncovered, none of the scientists set out to design tools to engineer genomes. They were simply trying to understand the mechanisms existing in nature. If it was not for this simple sense of wonder, we probably would not have these breakthrough technologies. In this chapter,

Progress in Molecular Biology and Translational Science, Volume 178
ISSN 1877-1173
https://doi.org/10.1016/bs.pmbts.2021.01.002

we will discuss the history, applications and ethical issues surrounding these technologies, focusing on the now predominant CRISPR-Cas technology. Gene editing technologies, as we know them now, are poised to have an overwhelming impact on our world. However, it is impossible to predict the route they will take in the future or to comprehend the full impact of its repercussions.

1. Introduction

The unprecedented progress in the gene editing technologies in the last decade has ushered in a marvelous time for the field of genetics. The ripples created by the advent of gene editing technologies, particularly CRISPR/Cas, have been felt almost in every field of biological sciences, from model organisms, evolution, agriculture, diagnostics and therapeutics. Traditionally, genetic research has depended upon uncovering and analyzing mutations that occur spontaneously. Mendel, Avery and Morgan relied on these methods and in the mid-20th century. Muller[1] and Auerbach[2] enhanced this rate of mutations using chemical treatment or radiation. Later technologies included transposon-based insertions in some organisms; however, all these methods induce genomic changes randomly. The 1970s and 1980s saw the first attempts at the first targeted changes to the genome in yeast and mice, respectively.[3–6] This targeted gene editing relied upon homologous recombination to produce the desired genetic changes. This method, although extremely precise, was quite inefficient and needed powerful selection procedures for the recovery of the edited products.[7] Due to the low efficiency and lack of mammalian embryonic stem cell lines other than mice, this method of gene editing was not adapted widely to other species. Newer gene editing methods have solved this issue and made direct genomic editing possible in all cell and organism types.[8,9] Technologies that edit the genome identify and change specific sequences, typically do so by introducing double-strand breaks (DSBs), which are then corrected by the cell's endogenous DNA repair mechanisms. Although common knowledge now, the realization that DSBs can stimulate gene editing and local mutagenesis arose from research in DNA damage and repair. In 2011, Youds et al. reported that DSBs stimulate recombination among homologous sequences during meiosis.[10] Furthermore, Latt et al. previously reported that ionizing radiation caused DSBs led to crossovers between sister chromatids.[11] Further, stimulation of the homologous repair pathways was shown in response to DSBs induced by highly specific nucleases.[12–15] In addition to homologous recombination, DSBs can also be repaired using the cells endogenous Non-Homologous End Joining (NHEJ) repair mechanism.[16] NHEJ is error-prone and often introduces insertions or deletions (indels) at the site of the DSB resulting in frequent disruption of the gene and consequently lack of expression.

Currently, we have three main gene editing technologies: Zinc Finger Nucleases (ZFNs), Transcription Activator-Like Effector Nucleases (TALENs) and CRISPR-Cas. Owing to the ease of use, the CRISPR-Cas technology dominates the field of gene editing; however, the other two technologies are also used in several medical and agriculture areas. Interestingly, all three of these technologies have been derived as a result of investigations into basic biological phenomena without the intention of performing genetic editing in animal cells. In this chapter, we will discuss in detail the history, mechanism and applications of these gene editing techniques. Furthermore, we will discuss use of these technologies in human therapeutics and the ethical issues surrounding the application of these technologies to humans. But first, we will start from the very beginning, from where DNA was discovered.

1.1 The story of DNA

The "Griffith's Experiment," conducted in 1928 by English bacteriologist Frederick Griffith described the conversion of a non-pathogenic pneumococcal bacteria to a virulent strain.[17] In this experiment, Griffith mixed the living non-virulent bacteria with a heat inactivated virulent form. He subsequently infected mice with this mixture and much to his surprise, the mice developed pneumonia and died. Furthermore, he was able to isolate colonies of the virulent strain from these mice. Because the original virulent strain was heat inactivated, he concluded that the non-virulent strain had transformed into the virulent type. This phenomenon was observed for the first time but was confirmed a year later by Dawson and Sia who were also able to perform this transformation *in vitro*.[18] The studies Dawson started were continued by James L. Alloway and he took the experiment one step further. He lysed the virulent strain of bacteria and filtered the intracellular substance to obtain a cell-free extract. He observed that even this extract transformed the non-virulent strain to a virulent one and he hypothesized that there was something in the cell-free extract that caused this conversion. He named this "something" the "transforming principle".[19] His studies after that focused on isolating that "something" from the cell-free extract. In 1933, he observed that it precipitates out of solution with addition of alcohol,[20] but it wasn't until 1944 that Avery and McCarty were able to demonstrate that the transformation was caused by a substance known as the deoxyribonucleic acid (DNA).[21] This was a time when common consensus among scientists, including Avery, was that genes are made up of proteins. The discovery of DNA caused a dramatic shift in the fundamental understanding of the molecular core of life and after that, DNA became a focus of intense examination and research.

1.2 The story of DNA repair and homologous recombination

Replacement of exogenous sequences in the genome with a sequence from an exogenous donor template through homologous recombination (HR) has been recognized since decades. Smithies et al. conducted pioneering studies that showed that homologous DNA strands recombine and insert at specific genomic locations.[5,22] These experiments were critical for envisioning and designing gene targeting experiments in murine embryonic stem cells for which Mario Cappecchi, Oliver Smithies and Martin Evans were awarded the Nobel Prize in Physiology or Medicine in 2007.[23] An account of the development of assays and experiments leading up to generation of mice with genes altered through HR has been outlined by Oliver Smithies himself in the article—"Forty years with homologous recombination."[24] Although in the initial experiments using HR, the efficiency of integration of the DNA template in cells was quite low,[5] it was still about 10 times more efficient than the microinjection method used by Brinster et al.[25] Brinster introduced DNA into fertilized murine eggs through direct pro-nuclear microinjection. However, soon it became obvious that inducing a Double Stranded Break (DSB) near the homology region significantly improved the frequency of HR occurrence.[26] A finding that was later confirmed by Allan Bradley and colleagues, solidly establishing the concept that induction of DSBs enhanced the frequency of HR.[27] This phenomenon played a fundamental part in the use of HR in modern gene editing approaches and is the central mechanism employed by all the gene-editing technologies currently in use.

As mentioned before, there are two main types of DNA repair mechanisms that take place when a DSB occurs in the genome: Homology directed repair (HDR) that utilizes HR or Non-Homologous End Joining, a mechanism that takes place in absence of a donor template. Both these mechanisms are used by the cell to seal a DSB in the genome. Depending upon the repair pathway employed by the cell, the DSB can either result in a gene disruption because of nucleotide insertion or deletion *via* NHEJ or precise gene editing associated with HDR. Interestingly, the HR events increase by at least a 1000-fold when these nucleases are used as compared to experiments conducted previously without them.

Genetic manipulation using these tools can be used in a variety of ways to study and analyze the genome. The INDELS induced by the NHEJ mechanism can be an effective method to examine functionality of a gene. RNA interference has been previously used to probe gene functionality. However, it downregulates but never fully disrupts expression of the gene of interest. Gene editing achieved by using NHEJ completely abrogates the expression of the targeted gene and effectively inactivates all the alleles. Although RNAi has proven to be a valuable tool in the analysis of gene function, RNAi might not fully display phenotypes caused by the absence of a gene since there is residual expression of the target gene.[28–30] Additionally, the presence of off-target effects could mean that the observed phenotype could also be due to manipulation of a gene unrelated to the current experiment. Nevertheless, the problem of off-target effects also exists with all the newer gene editing technologies. Another downside of the NHEJ based gene manipulation is that although some cell types display efficiency as high as 95%,[31] it can be markedly lower depending upon factors such as cell type, target gene, and method of delivery. Furthermore, NHEJ based gene manipulation is not homogenous and cannot be controlled. Consequently, it generates a heterogenous population of cells with a variety of INDELs, the effects of which might be different on the expressed protein and the cell. Therefore, expansion of a selected cell clone must be done for many applications. This could inherently introduce a bias in the experiment, since that particular selected INDEL could bestow the cell with unique characteristics. For example, in my own experiments (unpublished) conducted as a part of my doctoral thesis, I observed that CRISPR/Cas mediated deletion of the CXCR4 using the NHEJ pathway made one particular clone of the B-Cell Acute Lymphoblastic Leukemia cell line Tanoue more resistant to cytotoxic effects of dexamethasone and vincristine. Pharmacological inhibition of CXCR4 did not affect the cytotoxicity of these drugs indicating that either the indel created in the CXCR4 gene or another off-target INDEL created by the CRISPR/Cas system could have caused this. However, one advantage of clonal expansion is that it enables creation of a series of mutations in the target gene in an isogenic background. By contrast, HR mediated genome editing creates precise changes in the sequence at the target locus. Therefore, unlike NHEJ based editing, each cell carries the same edited sequence. However, for

HR to take place, the cells also need to be provided a template sequence with homology arms for it to be inserted at the target site. This method can be used to insert long DNA sequences or to generate Single Nucleotide Polymorphisms in a particular genomic location. Additionally, "safe harbor" regions of the genome can be used to insert expression cassettes of the full gene for expression of transgenes.[32] Full-length Open Reading Frames or transgenic cDNA can be inserted precisely next to the start codon of an endogenously expressed gene, thereby putting the transgene under regulation of the target genes endogenous promoter.[33–35] Thus, this method provides an advanced toolbox for highly precise and efficient genetic editing. HR can also be used in place of NHEJ in gene disruption assays, by insertion of a reporter gene that disrupts the endogenous gene. This has the added advantage of providing a way of tracking the cells that contain the desired edit and the reporter gene expression. Furthermore, these gene editing systems can be multiplexed and combined with reporter genes, providing a powerful tool for creation and tracking of cells with complicated genotypes.[36,37] Another non-canonical HR mediated pathway has now been used to make small changes in the genomic sequence. This variant of HDR pathway employs a synthetic, small, single-stranded oligodeoxyribonucleotides (ssODNs) as a template to be used for repair of the DSB. These small changes created through the ssODN, fall broadly under the category of HDR. This mechanism has been named as Single-Stranded Template Repair (SSTR) and does not operate though the canonical HDR pathway. However, the mechanism of this particular method of gene editing has yet to be delineated and may not represent a normally used DNA repair pathway. The advantage of this type of editing is that it can perform gene correction at the level of a single nucleotide. This property of ssODNs make it highly relevant to gene-function studies and even more so for the purposes of gene therapy.[38] Although, ssODNs are very simple to synthesize making it an accessible technique, but its range of genetic edits is limited as compared to the standard gene editing pathways and vectors. Furthermore, the efficiency of ssODNs might be lower than those of standard mechanisms employed for gene editing.

2. Gene editing technologies

2.1 I-Sce1 meganuclease

One of the first nuclease-based gene editing technologies to be developed used I-*Sce*I meganuclease from yeast. It is a rare-cutter endonuclease and it recognizes an 18 bp sequence originally identified as responsible for homing of introns in mitochondria of yeast.[39] A decade after their discovery, I-*Sce*I meganucleases were used to stimulate HR in the mammalian genome. The frequency of I-*Sce*I induced HR events was higher by two orders of magnitude when compared to spontaneously occurring HR.[13] In one of the first studies attempting gene editing using the I-*Sce*I, the authors showed that specific DSBs induced were efficiently repaired through HDR in the presence of a donor DNA molecule with flanking homology regions. Subsequently, the I-*Sce*I restriction site was introduced into the murine Embryonic Stem (ES) cells through HR. After that, the meganuclease was expressed in the presence of the donor template with flanking sequences homologous to regions surrounding the I-*Sce*I restriction site. This experiment resulted in a 100-fold increase in efficiency of I-*Sce*I mediated genome targeting.[40] For obvious reasons, the experiment was lengthy and laborious, since it requires HR at two successive steps and two selection markers. For these reasons, the approach was impractical because the integration of the gene of interest was dependent upon the presence of the I-*Sce*I restriction site at the locus of interest. Moreover, later studies have indicated that in mammalian cells, HR-mediated gene editing using meganucleases might be locus dependent.[41] Consequently, efficient gene editing by the I-*Sce*I system has been reported to be much easily exploited in species like Xenopus[42] and Medaka fish embryos,[43] but has proven to more restrictive in mammalian cells. To overcome this limitation, modified meganucleases that contain a nuclear localization sequence have been developed. These NLS containing I-*Sce*I meganucleases can be effectively used to create efficient germline transgenes in porcine and murine embryos.[44] Engineering the I-*Sce*I meganuclease and customizing recognition sitesare effective ways to negate the limitations posed by lack of suitable target sequences. One such engineered version of the meganuclease that was derived from I-CreI was used to alter the SCID gene at high frequency.[45] Additionally, another engineered I-CreI that cleaves a 22 bp sequence in the Rag1 gene was used to effectively edit target sequences in mouse and rat embryos.[46] The fact that meganucleases were used for the purpose of gene editing before the other more celebrated nucleases [Transcription activator-like effector nucleases (TALENs), Zinc Finger Nucleases (ZFNs) and Clustered Regularly Interspaced Short Palindromic Repeats (CRISPR) and associated proteins] is commonly not adequately acknowledged.

2.2 Zinc finger nucleases

Although the yeast derived meganucleases were the first to be applied for the purpose of editing mammalian genes, a fundamental shift in paradigm took place in 2009 that transformed the way transgenic animals are generated. In a critical study, Guerts et al. utilized ZFNs to generate the first knockout rats ever created.[47] The authors fused the endonuclease domain of the bacterial restriction enzyme Fok1 with the DNA binding Zinc Finger domains, generating a protein that has both the ability to bind to DNA at a specific locus and then cleave it. This method of generating chimeric proteins by fusion of DNA binding zinc finger and Fok1 nuclease domain had been developed a few years earlier when these proteins had been shown to effectively cleave and stimulate HR events at specific chromosomal locations in embryos of *Xenopus laevis*.[48] The methodology was then improved after some years and ZFNs were used in correction of the SCID mutation using HR with an extra-chromosomal donor sequence. The improved method showed an enhanced efficiency of about 18%.[49] The efficient cleavage of the target site to generate a DSB required dimerization of the Fok1 nuclease domain, necessitating that two ZFN proteins come together for the methodology to work effectively.[50] The mechanism was used for development of many approaches for ZFN mediated gene editing and the technology was rapidly adapted by several scientists. Before the ZFN technology, it had not been possible to use the classical gene targeting on rats due to lack of cultured rat ES cells. Therefore, the use of ZFNs to create knockout mice was a breakthrough that stimulated the rapid assimilation of the technique by various laboratories.[51] As a result, ZFN was critical in several gene-editing experiments in mice also.[52,53] Additionally, one of the biggest impacts of the ZFN technology was in livestock in which, like rats, cultured ES cells were unavailable. This had conventionally restricted genomic editing attempts in cattle. The ZFN technology ushered in an era where genome editing of cattle was made possible, for example, a variety of domestic pigs carrying a variant of the RELA gene was created,[54] making them resistant to the African swine fever. Another example is the use of ZFNS to generate cattle that displays increased resistance to mastitis.[55] Apart from *in vivo* studies, ZFN has been vastly applied to *in vitro* studies as well, for which adeno-associated virus particles were used to deliver the components to the cell.

Although the ZFN technology created quite a stir in the field of genetics, some undesired consequences were also discovered.[56] For example, the specificity with which ZFNs bind to their target sequences is not strict and the ZFN protein can bind to other places on the genome where sequences are similar to the target site. This is a limitation that spans across all current gene editing platforms and will be discussed in a separate section later in this chapter. In spite of this drawback, some pre-clinical studies that used ZFN in human cells have proceeded to clinical trials. For example, a trial for disruption of CCR5 (the HIV co-receptor) using ZFN has been approved by regulatory authorities in the United States.[57,58] However, in spite of looking promising the ZFN methodology was not ubiquitously adapted by the research field. This was due to the laborious and complicated procedure for generation of the Zn finger combinations that target specific sequences in the genome. Each Zn Finger binds to a sequence of 3 bp, so combinations of Zn finger domains are used to target particular sequences in the genome. Therefore, not only do we have to engineer the protein to be fused with a nuclease, we also need to engineer the Zn finger domains themselves making the procedure highly complicated and requiring a high level of expertise. The technology is also expensive and time consuming. Consequently, the ZFN technology was not accessible to most standard molecular biology laboratories. Indeed, open platforms were developed for design of the ZFN domains[59] and the chimeric proteins are also commercially available;[60] however, the efficiency of the technology still varies greatly across cell types.

2.3 TALENs

Construction of TALENs were inspired by existing genetic engineering tool Zinc Finger motifs and naturally occurring Transcription Activator Like Effectors (TALE) derived from pathogenic bacteria found in plants. Discovery and identification of TALE protein from pathogenic bacteria of genus Xanthomonas was a critical achievement in the history of genetic engineering. Xanthomanas are a family of gram-negative bacteria known to infect a wide range of plant species and its capacitance to invade host cells is largely mediated by the TALE injected during the onset of infection.[61]

The first TAL effector proteins identified from *Xanthomonas campestris* pv. *vesicatoria* (a pathogen of pepper) was AvrBs3. It conferred resistance to plant cells by activating plant immune response through the activation of the Bs3 gene. First genetic characterization of AvrBs3 on a self transmissible plasmid revealed two proteins of 82 and 125 kDa (each from both the strands of DNA, respectively). Interestingly "mirror" reading frame comparison lead to "a remarkable feature of both ORFs is the presence of

17 direct 102 bp repeats which [within each ORF] share 91%–100% homology with each other."[62] Succeeding studies revealed amino acid sequence for AvrBs3 by concluding the open reading frame for the same on the first DNA strand.

The possibility to confer resistance by TALE attracted scientific community in hopes to identify novel disease resistance genes as scientists were aiming to construct disease resistance plants by understanding its underlying mechanism. Subsequent studies on various Xanthomas species further enlarged the role of TALE, as it may vary from providing enhanced resistance to increasing the susceptibility of the plant cells for bacterial attack. Furthermore, it was revealed that in some plants species, its role will be in accordance with the presence of resistance gene in plant itself.

The conceptualization of TALENs was initiated by the studies identifying and confirming the presence of nuclear localization of TAL effector proteins. Initial studies identified the presence of nuclear localization signal of TAL effector in onion epidermal layer and these initial findings were further verified by other conclusive studies that demonstrated the nuclear localization of AvrBs3 after a pathogenic attack. The revelation of TALE and its intricate role in eliciting a host immune response in pathogenic infections inspired other groups to unravel the structural and functional subunits of these proteins. C-terminal acidic domain was the first identified TALE subunit along with its functional relevance by series of mutagenic studies on various eukaryotic system, from yeast activation system to rice plant.[63,64] Further discovery of DNA binding protein subunits laid the fundamental grounds for TALENs as it provided the acronym "TAL," which stands for "transcription activator-like."[65,66] Other mutational studies unveiled the possible point of interaction between plant and TAL effector proteins, by using mutated gamma subunit of Transcription factor II which further provided resistance to Xanthomanas infection.[66]

It was also discovered that TALE was marked by the presence of conserved repeat regions that range from 5 to 30 nucleotides.[47] They correspond to an array of 33–35 amino acids and are characterized by the presence of variable amino acids only on two positions of 12 and 13, respectively, that have been named as Repeat Variable Di-residue (RVD).[67,68]

In 2009, two independent research groups published papers illustrating the key role of RVDs in determining the length and nucleotide sequence of the target DNA. They observed the target site length was in accordance with the number of repeats in an array, which helped them in unwinding a simple correlation between hypervariable residues and the base attached to an individual repeat.[67,68] For example, "NG" RVD within a repeat corresponded to identification of a cytosine, whereas an "ND" or "HG" RVD corresponds to adenine. This particular understanding led to the possibility of using this recognition mechanism as genetic code for DNA binding sites of TALE and laid the architectural framework for customizable TALE array repeat fit for binding the desired target sequence. Subsequent studies further provided conclusive evidence on successful creation of engineered TALE repeats with desired specificities.[69–71]

Simple rearrangement of modular repeats enabled the facilitation of designed effector proteins to suite the specific DNA binding requirement of target molecule. Having realized that the untapped potential of TALEs; some other scientists attempted and fused it with a FOK I nuclease which marked the completion of the construction of comprehensive gene editing tool "TALENs." Taking an account of this significant development in the area of genetic engineering, *Nature Methods* phrased it "Method of the year" for 2011.

Construct of TALENs essentially requires the C-terminal sequence (nuclear localization), DNA binding domain comprising array of identical repeats with RVD and *Fok*I catalytic domain in a vector backbone same as adding beads onto a string leading to necklace formation. The choice of the vectors and assembling process have varied among scientific groups as per their need and innovation but what remains as challenging task is the construction of domain of almost identical repeats. Subsequently, most of details mentioned below encircles around the process and the progress in developing TALE repeat units from standard cloning method to automated techniques and so on.

A number of methods have been proposed till date for the construction of DNA-binding engineered TALE domain. The earlier methods were arduous and time-consuming; however, the newer automated methods are quicker and more convenient. In the present-day scenario, construction of TALENs no longer remains as a tedious task owing to the automation in technology and availability of pre-existing kits. Most of the recent studies generally make use of Platinum TALENs for the commencement of desired editing. In the time governed by predefined kits, one should also ensure the final architecture of amino acids encoded by the complete TALEN construct is apt for creating desired editing in the specific organism.

TALENs have had a great impact on a variety of areas both *in vitro* and *in vivo,* efficiently allowing the attempts to alter sequence in many model organisms which could not be modified earlier. Edited organisms include taxonomically simpler phylum of microalgae to advanced mammals such as microalga Nannochloropsis,[72] fruit fly,[73] roundworm,[74] zebrafish,[75] rat,[76] pig.[77] Additionally, they were also utilized for modifying endogenous genes in silkworm,[78] and crickets.[79] Traditionally most of the studies only used single TALEN pair for generating NHEJ-induced knockout mutations; apart from few other studies which also reported the use of two TALEN pairs for targeting same chromosome with the aim of creating deletions as well as inversions in larger chromosomal segments. Other reports even used a short single-stranded DNA oligodeoxynucleotide donor along with TALENs to enhance genomic insertions. The efforts for creating mutations in the model organism resulted in specific disease related animal model which in turn paved the way for better understanding of human disease.

Setting aside the conventional image of TALENs as an agent for developing disease specific models, their potential has also been explored in other areas: locomotory behavior of marine fish,[80] role of thyroid hormones and epigenetic regulation in the developmental process of Xenopus.[81] Additionally, they were also deployed to construct mutated Aneophlies model which could unfold various physiological aspects of lifecycle of this vector mosquito.[82,83]

Recently, they have also been utilized for the construction of insects model to upscale the downstream production of commercial product as in case of silkworm and alteration of consumable food item by improving the nutrient content,[84] attempts to reduce allergens in dairy products,[85] targeting and deliberate mutation of secondary metabolic pathways from miniature Nanochloropsis (microalgae) and sugarcane for enhanced production of biofuel, creating cleaner alternative future energy source.

The discovery of TALE revolutionized the field of genetic engineering. TALENs not only confer advantage over Zinc finger by providing ease of customization but also broadened the variability and capacity of gene editing reagents. TAL effector proteins are also capable of binding of gene activators and receptors apart from nuclease.[86–88] These advanced gene editing reagents constructed with the aid of TALE surpassed existing editing tools including zinc finger nucleases and homing endonucleases (or "meganucleases").

TALENs were much easier to design in comparison to ZFN. Recognition of RVD on repeat regions of TALE facilitated the DNA binding specificity on various targets. Along with the improvised DNA binding efficiency, TALENs also reduced the time constrain. TALENs could be engineered in a short span of 2 days. In addition to the shorter time requirement for a construct; it also facilitates construction of hundreds of TALENs at the same time, paving the way for library of TALENs which could target all the genes in a genome. TALENs also posed an upper hand on ZFNs owing to the extendable capacity of repeat regions. TALENs were generally constructed for binding to 18 bp or longer target sequences in the light to achieve higher degree of specificity and affinity in accordance to theoretic concepts, in contrast to ZFN binding to shorter sequences ranging from 9 to 12 bp. Another potential advantage of TALENs over ZFN includes the higher degree of freedom for possible site selection, which increases the chances of finding potential sites for every 100 bp of the genome. Although methylations on target sequence seemed to hamper the binding affinity of TALENs[89]; nevertheless manipulated RVD code could even contain methylated DNA bases.[90,91]

Though off-target mutations are a major concern for all gene editing technologies till date and its degree highly varies with the selection of specific sequence and target organism, still slight reduction in offsite mutation frequency could significantly affect success story of genome engineering technologies. Much data is not available for comparing frequencies of offsite mutation between TALENs and ZFNs but one such study demonstrates lesser offsite mutation on a similar site CCR2 gene when both TALENs and ZFNs were used to target the same CCR5 gene.[92]

Despite of being an easier technology of gene editing, larger size requirement of usually 3 kb adds to another important demerit of TALENs making it less attractive tool than ZFNs which utilizes a smaller construct of around 1 kb. Larger size constructs offer several complications with regards to packaging in vector molecule as well as reduced expression efficiency in host organism.

2.4 CRISPR: The celebrity gene editing technology

Function: The CRISPR-Cas system is an adaptive immune mechanism employed by various single-celled organisms (mainly archaea and bacteria) to fight off invading viruses. In essence, it is the documentation of past genetic attacks on the cell in order to recognize any similar attacks in the future. The bacterial cell incorporates fragments of the invading viral DNA in its genome. Several of these viral fragments (spacers) are separated by the same short palindromic sequence (repeats) that forms a hairpin loop structure when expressed. This immune response staged against viruses by the CRISPR-Cas system contains mainly of three stages: (i) Adaptation, (ii) Expression, and (iii) Interference. The first stage (adaptation) involves acquisition of the spacers from the invading virus, as a memory for future infections. In this stage, a distinctive complex of Cas proteins identifies a short motif called the Protospacer Adjacent Motif (PAM) on the target DNA to bind and cleave a portion of the target DNA. The repeat at the 5′ end of the CRISPR array is then duplicated after which the adaptation complex introduces the protospacer DNA in the array, consequently making it a spacer. A few CRISPR-Cas systems employ other mechanisms to introduce spacers into their CRISPR arrays. The acquire spacers from RNA through reverse transcription encoded within the CRISPR locus.

In the expression phase, the CRISPR gene is transcribed into a single RNA molecule, the pre-CRISPR RNA (pre-crRNA). The pre-crRNA is then processed into the mature form of the CRISPR RNA (crRNA), that contains the spacer sequence and the partial flanking repeats. The generation of the mature crRNA differs among various types of CRISPR-Cas systems, such that some CRISPR-Cas variants use a complex of Cas proteins, some use a single multi-domain Cas protein and some use the host RNAses.

In the third stage (interference), the crRNA (usually bound to the Cas protein complex) acts as a guide for recognition of the protospacer in the target or invading genome, which is cleaved by the Cas proteins. The Cas proteins that cause the cleavage and inactivation can either be present in the Cas-crRNA complex already or be recruited at the time of interference.

The above mentioned mechanisms are a brief, oversimplified summary of the functionality of various gene editing technologies that are illustrated in Fig. 1 and compared in Table 1. There are certainly many details left out due to the scope and the space restrictions of this chapter. The reader is urged to explore further into the mechanistic aspects of these phenomenon and is directed to previously published reviews on the topic[93–98] (Fig. 1).

Classification: As is the case with other bacterial and archaeal biological defense mechanisms, CRISPR-Cas systems show significant diversity in the sequence of the Cas proteins.[97,99–103] The CRISPR-Cas system is in a ceaseless arms race with viruses resulting in evolution at a rapid pace. Furthermore, the architectures of the genomic locus and compositions of individual Cas genes also show remarkable variety. Our current knowledge of this diverse CRISPR-Cas universe continues to expand at lightning speed because of screening of the continuously growing genomic and metagenomic tools and databases. Our classification criteria need to be robust in order to keep pace with such rapid expansion. In addition, they need to be based upon the evolutionary relationships of the CRISPR-Cas systems to stand the test of time. However, generation of such a classification system

Table 1 Comparison of existing gene editing technologies.

	Zn finger motif	TALENs	CRISPR-Cas
DNA Recognition component	Protein	Protein	RNA
Essential components	Zinc finger and FokI domain	TALE and FokI protein	Cas protein and guide RNA
Length of target DNA molecule	9–18 bp	30–40 bp	22 bp + PAM region
Cleavage Mechanism	Double-strand break by *Fok*I	Double-strand break by FokI	Single/Double-strand break by Cas protein
Targeting limitations	Difficulty in targeting non G-rich sites	Requirement of 5′T in the targeted sequence although in some recent works this limitation has been overcome	Target sequence must contain PAM sequence
Design	– Complicated construction. – Predefined target specific Zn finger motifs are available – Array assembly could alter binding specificity of adjacent ZFN motif	– Simpler than ZFN – Established target specific TALEs are available – Relatively complex designing as compared to CRISPR	– Easy to construct – Complimentary sgRNA and Cas9 plasmids for desired target available in pre-defined kits
Cloning	Needs additional requirement of linkage of Zn finger motifs	No requirement of linkage as separate individual motifs can be developed by Golden Gate assembly method	Cas9 expression vectors are established and widely used
Ease of delivery	Relatively easier as small size of ZFN expression system greatly facilitates its applicability in numerous viral vectors	Difficult as the large size of the functional system offers packaging constrains	Moderate as Cas9 is also large in size, however, newer smaller orthologs also exist
Efficiency	Variable	High but variable	High but variable
Cost effectiveness	Expensive	Cheaper than ZFN but Not cost effective	Cost effective
Multiplexing Ability	No	No	Yes

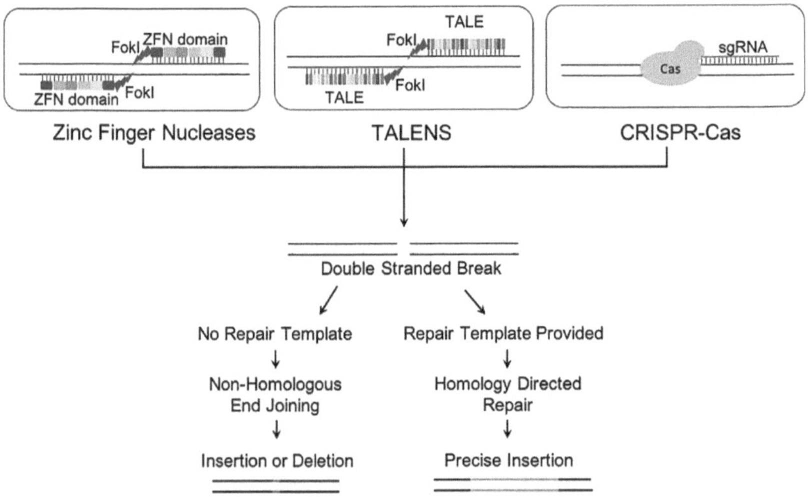

Fig. 1 Genome editing using ZFN, TALEN and CRISPR-Cas. The schematic represents the three main gene editing techniques widely used. The three technologies induce double stranded breaks which are repaired using the cells endogenous DNA repair mechanisms. If there is no repair template provided, the cell uses its Non-Homologous End Joining repair mechanism that often introduces an insertion or deletion, leading to lack of expression due to a frameshift mutation. If, however, a repair template is provided the DSB is repaired using the homology directed repair and the template is inserted at the target site.

presents daunting challenges due to absence of universal markers and rapid evolution of the CRISPR locus.[104] Kira Marakova, Eugene Koonin and colleagues have been the world authority in the classification and nomenclature of different CRISPR-Cas systems and have published classification and subsequent updates over the years.[95,99,105–112] Since the sequence and composition of the CRISPR-Cas system shows wide diversity, a straightforward approach based on these characteristics is not possible. Consequently, the authors use a multi-pronged approach that uses a combination of gene composition comparison, architecture of the locus, clustering based on sequence similarity and phylogenetic analysis of Cas proteins that were conserved. The classification that Marakova et al. published in 2015 listed 5 types and 16 sub-types of the CRISPR-Cas system.[109] The paper also introduced for the first time the division of the systems broadly into two distinct classes (Fig. 2). The effector modules in the class 1 systems are comprised of multiple Cas proteins which might form the crRNA binding complexes that facilitate processing of pre-crRNA and ultimately interference. On the other hand, the CRISPR-Cas systems in class 2 possess only one

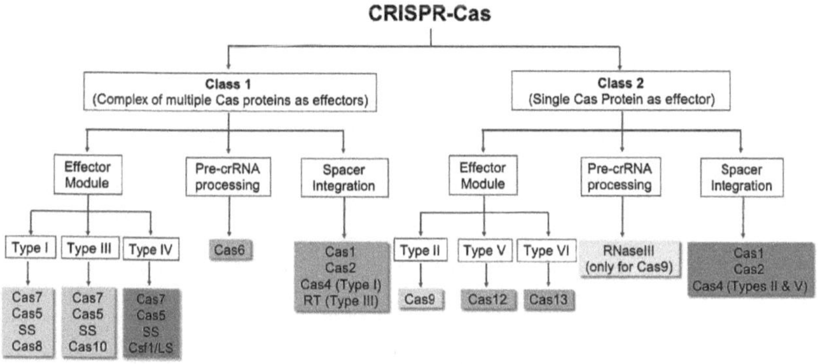

Fig. 2 Classification of CRISPR-Cas systems: CRISPR-Cas systems are broadly divided into two classes differentiated by the effector proteins involved. The interference and target cleavage is carried out by a complex of Cas proteins in the Class 1 systems, whereas, only a single Cas protein performs these functions in Class 2 systems. Several Cas proteins like Cas7 and Cas5 are involved in effector complexes of multiple sub-types of the Class 1 system, although each type has its own unique Cas proteins too. Cas6 is involved in Pre-crRNA processing in several class 1 types. In class 2 systems, the Pre-crRNA processing is performed by the effector Cas molecule except Cas9, which requires RNaseIII for processing of the Pre-crRNA. Spacer integration is performed by Cas1, Cas2 and Cas4 in both class 1 and class 2 systems.

multidomain crRNA binding protein (for example, the Cas9 protein in type II subtype). The domains in this protein perform all the functions needed for interference and, in some varieties pre-crRNA processing. Within the last 5 years, there has been tremendous progress in the discovery of new CRISPR-Cas systems and although it is easy to identify the origin of some of these newer systems, their assignment to a particular type/sub-type in the CRISPR-Cas classification remains challenging. Interestingly, newer CRISPR-Cas systems have been identified whose functions are other than the traditional interference related function of the Cas proteins. For example, some CRISPR-Cas systems have been identified to be implicated in signal transduction and regulation.[113] Such rapid progress in the field makes it necessary for the classification criteria and the classification itself be updated periodically. Furthermore, it makes the classification increasingly challenging. The 2020 classification published by Makarova et al. lists three types each in class 1 and 2 CRISPR-Cas systems with several sub-types in each.[112]

Since the most frequently occurring CRISPR-Cas types and subtypes have now been uncovered, the broad classification currently used will probably remain the same for several years to come. However, newer

CRISPR-Cas systems are continuously being discovered, therefore the classification will continue to evolve and develop. Overall, the basic foundation of the CRISPR-Cas classification has now been laid and has entered the era of refinement and consolidation. One of the limitations of this classification system is that it is based upon predictions generated using computational analysis and experimental validation and characterization has still not been performed. The current classification will assist in experimentation in this field, and in turn experimental studies will facilitate in refinement of the classification.

History: More often than not, stories of sweeping success that seem to have happened overnight involve hard work and toiling over an extended period of time away from plain sight. Only the success and the glitter are conspicuous, making it seemingly instantaneous. Similarly, the stories of most groundbreaking scientific discoveries rarely have a eureka moment. They consist of a series of small discoveries happening over a period of decades, that seem insignificant on their own. However, when the discoveries reach their grand culmination, the resulting breakthrough is sometimes greater than what each of those studies could do individually. The narrative of the development of the CRISPR technology embodies this phenomenon. The story begins in 1987, when Yoshizumi Ishino was studying the gene implicated in alkaline phosphatase conversion in *E.coli*.[114] To that end, Ishino et al. sequenced the *iap* (isozyme of alkaline phosphatase) gene contained within a 1.7 kb fragment from the *E.coli* genome. During the process, the authors realized that a peculiar sequence appeared several times in different clones downstream of the translation termination of the *iap* gene. Furthermore, these fragments made secondary structures, a property that made sequencing much harder. Back then, the process of sequencing these fragments had been quite laborious and time consuming. It is indeed wonderful that today, using Polymerase Chain Reaction (PCR) and current sequencing technologies, the same genetic region can be sequenced in 1 day. So unexpected and mysterious was the appearance of these sequences that although the authors mentioned it in the discussion of their paper, they could not possibly fathom the function of these bizarre sequences. Incidentally, similar sequences had been reported in two prior studies,[115,116] although they were repetitive and palindromic, they did not have any sequence similarities with the ones discovered downstream of the *iap* gene. The prior studies proposed that these sequences could be related to mRNA stability. Soon after the 1987 study, the strange sequences were detected in other strains and organisms,[117–119] although little interest was garnered for them. All these observations were the first encounter of CRISPR.

It was not until 1993[120] that these sequences got the attention they deserved, from Francisco Mojica, who was pursuing his doctorate at the University of Alicante in Spain. When he began his Ph.D. in 1989, his advisor was interested in how restriction enzymes behaved differently in the high salt environment where *Haloferax mediterranei* was found. The very first DNA fragment he analyzed; he noticed several copies of a short 30 bp palindromic sequence separated by 36 nucleotides each. These sequences did not resemble any other repeats found in microbes. Mojica was fascinated with these mysterious repeats and spent the next decade examining them. Soon, he uncovered these repeats in other organisms and also came across the 1987 paper by Ishino that reported similar structures. Although there was no sequence similarity between the 1987 study and his sequences, the structure was similar and Mojica realized that presence of similar genetic structures across such distant microbes indicated toward an important function associated with them. At the time he hypothesized that these repeats might be associated with replicon partitioning. After a short stint as post-doctoral fellow at Oxford, he returned to University of Alicante as a faculty member and due to a lack of start-up funds at the university he diverted his attention to studying these repeats using bioinformatics tools. By 2000, he had uncovered these sequences in over 20 microbes and by 2002 the number of those microbes had doubled than that and Jansen et al. proposed the name CRISPR[121] at Mojicas' suggestion. However, the function of these sequences still proved elusive and several hypotheses were generated only to be proven wrong one by one. By 2003, Mojica had re-focused from the repeats themselves to the spacers using a bioinformatics approach. He had previously tried to match these sequences to all known sequences without any success, however, since the databases were continually updating his latest attempt bore fruit. One of the spacer sequences he had recently analyzed from *E. coli* matched a sequence from a P1 phage that was known to infect several strains. Interestingly, the strain containing the P1 viral spacer sequence was itself resistant to infection from the P1 phage. This curious finding spurred him on and within a matter of a week he had checked sequence similarity for about 4500 spacer sequences. Eighty-eight of these spacers showed sequence similarity to other known sequences, of these two-thirds matched with viruses. Although Mojica recognized the value of this discovery, his enthusiasm was only one sided. After being rejected from several prestigious journals he submitted his paper to Journal of Molecular Evolution where he had to go through 12 months of revision and review after which his paper proposing CRISPRs likely function as an adaptive bacterial immune system was published in

February 2005.[122] In the same year, another crucial paper was published by Pourcel et al. that stated that the CRISPR spacers were acquired by bacterial strains in a polarized fashion in the front end of the CRISPR locus.[123] Moreover, the acquired spacers were similar to a sequence of prophage in the strain demonstrating that the spacers were of viral origin. This study met with the same amount of resistance as Mojica and was published in *Microbiology* in March 2005 after several attempts in different journals. A third seminal paper about the extrachromosomal origins of the CRISPR spacers was published that same year in September 2005.[124] Not surprisingly, this paper was also rejected in another journal before it was accepted in *Microbiology*. This report by Bolotin et al. was the first to propose a mechanism of anti-sense RNA inhibition like RNAi by which the CRISPR bestowed immunity to microbial strains. The speculation seemed reasonable at the time; however, it would be proven wrong later. Direct experimental evidence of CRISPR provided microbial immunity came from a study conducted by Phillipe Horvath and colleagues in 2011 where bacterial clones selected for phage resistance displayed acquisition of CRISPR spacers as compared to uninfected ones.[125] Indeed, the number of spacers present significantly correlated to the degree of resistance displayed by a strain. Furthermore, they analyzed the function of the two Cas proteins, Cas7 and Cas9. They discovered that Cas7 was involved in acquisition of resistance because strains carrying virus derived spacers were resistant to viral infection even if Cas7 was deleted suggesting that it was not involved in the mechanism of resistance. On the other hand, the Cas9 sequence had two nuclease domains (HNH and RuvC) was required for the mechanism of resistance. Another observation from the paper was that viruses with single base changes within their protospacer sequences had managed to overcome CRISPR bacterial resistance.

The first report of engineering the CRISPR system came when Jon van der Oost, Eugene Koonin and colleagues introduced the CRISPR system of one *E. coli* strain into another that lacked its own CRISPR system for biochemical characterization.[126] The CRISPR-Cas system they were analyzing performs the functions of Cas9 using Cas3 nuclease and several other proteins that are a part of the cascade complex. They knocked out each component and showed that the components of the cascade complex were involved in the processing of the CRISPR RNA. Additionally, they engineered spacer sequences from four essential genes of the lambda phage and demonstrated that resistance to lambda phage was conferred onto strains that were engineered to have these spacer sequences. Essentially, they created a bacterial flu shot. The authors made two types of CRISPR arrays: anti-sense (complementary to the coding DNA and the mRNA) and sense (only complementary to the other strand of DNA). The sense version of the CRISPR showed much stronger effects indicating that the CRISPR system targeted DNA, not RNA and they proposed this as a hypothesis in their paper. Scientists working in the phage community, however, were convinced that the CRISPR system targets DNA since RNAi would not be able to counter the explosive expansion of the viral particles post-infection. Luciano Marrafani, a Ph.D. student at the University of Chicago at the time, hypothesized that it must target DNA like restriction enzymes. Later in his post-doctoral fellowship under Erik Sontheimer at the Northwestern University, he observed that a strain of *Stapylococcus epidermis* contained a spacer with a sequence similar to a nickase enzyme found in a plasmid in *Stapylococcus aureus*. He demonstrated that the plasmids were not able to transfer to *S. epidermis* and disruption of either the spacer or the nickase sequence made the plasmid capable of transfer.[127] This generated indirect evidence that the CRISPR system targets DNA but Marrafani and Sontheimer wanted direct evidence. To that end they inserted a self-splicing intron in the nickase gene of the plasmid, if the CRISPR system targeted the mRNA, the interference would be unaffected by this change. However, if the system targeted DNA, the interference would be abolished. In this way, they showed clearly that indeed, DNA was the target. Since they had engineered this system successfully, they realized that it was essentially a restriction enzyme with the added bonus of being highly programmable.

The fact that the CRISPR system generates DSBs was not obvious since the beginning. Sylvian Moineau and his colleagues continued to probe for the mechanism of CRISPR; however, they were unsuccessful for a while because the CRISPR system proved to be so efficient and rapid that they could never see the destruction of foreign DNA in action. As luck would have it, they observed that some strains of *S. thermophilus* displayed partial resistance to plasmid transformation and one of these strains contained linearized plasmids.[128] For some reason, the process of interference through CRISPR had slowed down in this particular strain enough for them to see the intermediate products. Sequencing of these linearized plasmids showed that a blunt end was generated through cleavage of the strand precisely three nucleotides upstream of the Protospacer Adjacent Motif (PAM) sequence. They also displayed that invading viral DNA is also cleaved at this precise location relative to the PAM and the number of spacers matched the number of cuts that had been generated. These data definitively showed that the CRISPR system is guided by the crRNA to cleave DNA at precise locations. In spite of massive research efforts, one critical piece of the CRISPR system had yet to be uncovered. It was Emmanuelle Charpentier and Jörg Vogel who discovered this missing piece, even though they had no intention to do so.[129] Charpentier was interested in inter-genic regulatory RNA sequences in microbes and Vogel focused on creating comprehensive catalogues of microbial genome using Next Generation Sequencing. The two collaborated and combined their efforts to sequence the genome of *Streptococcus pyogenes*, an organism that Charpentier was already working on. Their efforts produced a remarkable finding, there was a novel small RNA transcribed from a region right next to the CRISPR locus, and it had near-perfect sequence similarity to the CRISPR repeats. Interestingly, these novel RNAs were the third most abundant class of RNAs, right after ribosomal and transfer RNA. They hypothesized that since this novel RNA showed sequence similarity to the CRISPR repeats, its product must hybridize to the precursor crRNA possibly for their processing into mature crRNA. They confirmed this hypothesis using gene deletion experiments that showed that deletion of this newly found tracrRNA was involved in crRNA processing and consequently was critical for CRISPR functionality. Further studies revealed that not only was this tracrRNA involved in crRNA processing, it was also a critical part of the Cas9 nuclease complex required to cleave DNA.[130]

Now that the essential components of the CRISPR system had been uncovered, Virginijus Siksnys at the Institute of Applied Enzymology in Vilnius had renewed interest in his two-decade long exploration of restriction enzymes. He and his collaborators wanted to test whether all the components of the CRISPR system from *S. thermophilus* could be functionally re-constituted in another distant microbe. Much to their surprise and delight, simply transferring that CRISPR locus to another microbe was enough to confer resistance to the respective invading phage or plasmid.[131] They also showed that only the Cas9 protein was needed for interference to take place and its two nuclease (HNH and RuvC) domains were essential for this function. Syksnys and his colleagues then took their experiments even further to see whether this system can be established *in vitro*.[132] They isolated the crRNA-Cas9 complex from *S. thermophilus* using streptavidin tags to study its action outside of an organism. They demonstrated that this complex could cleave DNA at the precise location it cleaved *in vivo* (3 nucleotides upstream of PAM) and that the crRNA could be trimmed down to 20 nucleotides and still retain its function. Furthermore, more remarkable was their finding that they were able to customize the spacers to target any DNA sequence of interest *in vitro*. Through introduction of mutations in each of Cas9 nuclease domains, they showed that HNH domain was responsible for the cleavage of the strand that is complementary to crRNA and RuvC cleaves the other strand. While Syksnys and his colleagues were conducting their study, Charpentier was also interested in the biochemical characterization of the CRISPR system. While at an American Society of Microbiology conference in Peurto Rico she met Jennifer Doudna, a leading structural biologist who had been solving crystal structure of the CRISPR Cascade complex. The two scientists decided to collaborate and like Syksnys, they showed that the CRISPR system could be re-constituted *in vitro* and the two nuclease domains cleave one strand of the target DNA each.[130] More importantly, they showed that crRNA and tracrRNA could be fused to generate a single guide RNA (sgRNA) that retains the functionality of the CRISPR system. The use of sgRNA would eventually become widely adapted in the scientific community to implement the CRISPR system in various models and studies. Both the groups independently conducted these studies and published their results very close to each other. Charpentier and Doudna's paper appeared in the journal *Science* in June 2012 whereas Syksnys's paper which was delayed due to rejection from *Cell* was published in the *Proceedings of the National Academy of Sciences* in September 2012.[132]

Tremendous progress had been made in the field of CRISPR and the next frontier that needed to be achieved was implementing this system in mammalian cells. The scientific community had realized that this system could be used as a powerful gene manipulating tool if could be executed in mammalian cells. However, this was a daunting challenge since mammalian organisms were far more complicated that microbes and their genomes are 1000-fold longer. Furthermore, the genome was tightly packed in complicated chromosomal structures and previous efforts to apply other microbial systems to them had been unsuccessful. Consequently, experts had been doubtful about this as late as 2012. Feng Zhang, who had previously adapted TALENs to mammalian cells[87] was still on the lookout for better methods to alter the mammalian genome. After listening to a talk on CRISPR, he decided he was going to attempt to adapt this system to human cells and by April 2011 he was able to demonstrate a modest reduction in luminescence in human embryonic kidney cells carrying plasmids with the luciferase gene. The next step was to optimize the system to increase the efficiency by increasing the amount of Cas9 that reached the nucleus since he observed that the *S. thermofilus* Cas9 clumped at the nucleolus. He found that Cas9 from *S. pyogenes* was evenly distributed across the cell and that even though human cells lacked the endogenous RNAseIII it was still able to process crRNA using alternative mechanisms. Furthermore, he examined several forms of the tracrRNA to establish which of them was durable in human cells and by the middle of 2012 he had developed a robust CRISPR system utilizing three components. He modified several loci in mice and human genome using both deletions through the NHEJ pathway and insertions using HR. Furthermore, he demonstrated that system could be used in a multiplexed manner by using multiple spacers at the same time. After Charpentier and Doudna showed that crRNA and tracrRNA could be fused to form a short sgRNA, he also used that but found that it had lower efficiency *in vivo*. This problem was corrected when full length fusions were employed instead of the short sgRNA, because it restored a critical a hairpin loop structure essential for their functioning. Zhang communicated these results on 5 October 2012 and they were published in *Science* on 3 January 2013.[133] This paper went on to become the most cited paper in the CRISPR field and his reagents were distributed among the scientific world through Addgene, a non-profit organization. Later, Zhang conducted more studies after that and showed that creation of elaborate mouse models for cancer and inherited diseases was possible within weeks through CRISPR. He also performed genome wide screens in order to look for genes essential for biological functions. Furthermore, he showed that the system could be modified to reduce off-target effects, one of the biggest drawbacks associated with it. He also collaborated with Koonin to find two new classes of CRISPR, which included a type that did not need the tracrRNA to function and could function with the crRNA alone.

Along with Feng Zhang, George Church—an expert in synthetic biology and genomics at Harvard—was one of the pioneers of mammalian genome editing using CRISPR. Church had been interested in the possibility of editing the human genome and also the ethical debate around it. Keenly aware of the progress in the field of CRISPR, he embarked upon a journey to test various combinations of the crRNA-tracrRNA fusion. He manipulated several genomic locations using the NHEJ and HR and demonstrated that full-length crRNA-tracrRNA fusions worked better than short fusions.[134]

As these studies published in early January, the scientific community awoke to the phenomenon of CRISPR and google searches for the term "CRISPR" accelerated at an unprecedented speed. This trend has continued till date as genetic editing has spread like wildfire in all fields of biological sciences. Within 1 year, several groups around the world had reported genomic editing in varied species including both single celled and complex multicellular organisms including mice and monkeys. Along with scientific interest, commercial interest in the technique skyrocketed as the possibility of application into human therapy and agriculture became apparent. The ability to potentially edit the human genome created the awareness of not just human therapy, but also human enhancement, stirring the debate over the possibility of designer babies and the social and ethical issues that would entail. The field progressed at such blinding speed that both the scientific and regulatory authorities were left grappling to catch up. So profound are the repercussions of this technology that it has led to raging ethical debates and intense long-drawn patent battles. So far-reaching is the interest created around the technology that it reached the average person and has found a niche in popular culture, with several documentaries, movies and TV series based on it.

Apart from human applications, CRISPR has percolated into many diverse fields of biological sciences from creating high-yield or disease resistant crops to combating malaria causing mosquitoes to paper-strip diagnostics. In 2020, several groups around the world have reported CRISPR based rapid diagnostic tests for malaria, HIV and the ongoing SARS-Cov2 pandemic. Furthermore, the technology was used for a genome wide study to identify genes involved in resistance to SARS-Cov2.[135] Additionally, the 2020 Nobel prize in Chemistry has recently been awarded to Emmanuelle Charpentier and Jennifer Doudna for "the development of a method for genome editing."

An effort to capture the exponential growth of the field beyond the initial years in the limited space of this chapter is most likely to fall grossly short. I urge the readers to explore recent reviews to capture the sense of expansion of the field. Not only have the pioneers of the field continued to expand the horizon of the technology, innumerable scientists around the world have made immense contributions in throwing light upon this system and utilizing it to answer a vast variety of scientific enquiries.

Delivery of CRISPR-Cas: Gene editing using the CRISPR-Cas system depends upon the successful delivery of the CRISPR-Cas components to the nuclei of target cells. While this system is widely used *in vitro,* its *in vivo* applicability is drastically restricted due to target region accessibility.[136] The components of this system are macromolecules, due to which they are unable to passively transport into the cell.[137] Consequently, it even harder for them to enter the nucleus, where the target genetic material resides. The delivery platforms used for gene editing technologies can be roughly divided into: viral and non-viral. The viral method involves using viruses to deliver the DNA plasmid into the target cells and the non-viral methods include delivery of mRNA or protein complexes. Apart from these, several synthetic delivery vehicles, like gold nanoparticles or lipid/polymer formations have also been used.

Viral delivery methods are most commonly used for the transfer of CRISPR-Cas components to the cells. They are highly effective due to their inherent ability to transfer exogenous genetic material to cells. Different viral vectors such as adenoviral, lentiviral, and adeno-associated viruses have been used for the application of CRISPR-Cas9 to various models.[138]

The Adenovirus vectors have been extensively because they usually do not integrate their genetic material into the host genome and their expression is extra-chromosomal.[139,140] These vectors stimulate a significant immune response[141] as a result of the extended presence of viral vector components in the organisms. A newer approach to bypass the extended presence of viral vectors in gene therapy models was attempted through the development of a new generation higher-capacity adenoviral vector that does not have any viral genes.[142] However, this method is yet to be used in an *in vivo* model.

Unlike the AV vectors, the lentiviral vectors integrate their genetic material into the host genome. Numerous studies have reported the use of lentiviral vectors for *in vivo* editing of genes involved in cancer, for example, for knockout of $Trp^{53fl/fl}$ and *Kras* in lung cancer mice models.[143] Although this method of gene transfer is efficient, it is often non-specific and integration of the genetic material at random genomic loci is known to occur. Integration of genetic material near a proto-oncogene could lead to severe consequences in an *in vivo* model.[144] Newer lentiviral vectors that do not integrate their genetic material into the host genome have been developed to overcome this hurdle; however, this comes at the cost of efficiency.[145] Owing to these restrictions, the lentiviral vectors are currently most suitable for *in vitro* studies.

Another virus used for gene delivery is the adeno-associated virus, which bypasses random insertion by integrating into a specific AAVS1 site (Adeno-Associated Virus integration site 1) in the mammalian genome.[146] As a result, the majority of toxicities associated with random integration are avoided, making this method ideal for *in vivo* models. Additionally, this virus transfects both dividing and non-dividing cells. However, the AAV vector is limited by its low loading capacity. The total space available in the AAV vector is about 4.7 kbp and the Cas9 protein alone is about 4.3 kbp.[147] Due to this size limitation, it is necessary to use two separate AAV vectors to deliver the entirety of the CRISPR-Cas9 system. To counter this, smaller saCas9 proteins from *Staphylococcus aureus* have been devised, allowing a single AAV vector to carry both the CRISPR and Cas9 genes. However, these smaller saCas9 proteins have been reported to have higher immunogenicity than conventional *Streptococcus pyogenes* Cas9 proteins, hampering their clinical potential.[148]

Among non-viral delivery methods, transporting naked DNA into mammalian cells to use the cells endogenous translation machinery for production of the CRISPR-Cas9 components is indeed an attractive proposition. Furthermore, DNA itself is a robust molecule that is stable at a range of temperatures. The CRISPR-Cas9 genes are delivered into the cells in the form of a plasmid and lack any sort of viral machinery.[149] In 2014, hydrodynamic injections were first used by several groups independently to introduce DNA into the mouse liver.[150,151] However, limitations like low efficiency render it impractical for clinical application. Furthermore, it is usually highly expressed only in the liver, further restricting its use in models of other disease organs.

Alternative methods for *in vivo* DNA delivery are being reported, for example, in 2017, Li et al. reported the use of an artificial virus to deliver the CRISPR-Cas9 components *in vivo*. They constructed a "multi-functional nucleus-targeting core-shell" synthetic virus, that enabled infiltration of the plasmid into the target cell nucleus without the requirement of a Nuclear Localization Sequence (NLS).[152] The advantages and drawbacks of the use of artificial viruses are still being uncovered.

An inherent advantage of using plasmids for genetic delivery is that tissue targeting signals can be incorporated into the plasmid itself. For example, the CD68 promoter was used for specific delivery of the CRISPR-Cas9 components to the macrophages.[153]

Apart from hydrodynamic injections and artificial viruses, nanoparticles can also be used to deliver plasmids to *in vivo* tissue. There are now several types of nanoparticles reported: gold, lipid, and polyethyleneimine based polymers.

Another method to deliver the CRISPR-Cas9 components to tissue is to transport the mRNA directly into the cells. In contrast to delivery of DNA, this method is temporary and lacks continual expression of the transferred genetic material.[154] This method is also rapid because the expression of the CRISPR components happens in mere minutes.[155] However, mRNA is not highly stable and is susceptible to rapid degradation by RNAse enzymes present ubiquitously. This is especially problematic because the sgRNA molecular has to wait around until the Cas9 protein is translated. Consequently, degradation of sgRNA greatly hinders the editing efficiency of the system. Moreover, each CRISPR component needs to be separately delivered for gene editing to take place. Studies that use a combination of mRNA delivery and AAV viral delivery have been reported, however, the efficiency reported was quite low (~6%).[156] To achieve better efficiency, modified versions of the sgRNA have been reported, that display ~80% editing efficiency in the murine liver.[157] Although newer delivery systems are being reported, the main challenge of mRNA delivery is the need to separately deliver both the CRISPR components. To counter this, a Zwitterion Amino Lipid (ZAL) vector was used for co-delivery of the CRISPR components.[158] Despite big strides in this particular method of delivery, the path to the clinic is still riddled with various challenges that need to be overcome.

Delivery of DNA or RNA to the target tissue represents an indirect way of gene manipulation. A direct and transient method for delivery of CRISPR-Cas9 components is transporting the Cas9 ribonucleoprotein (RNP) and sgRNA directly to the cells.

Several studies delivering direct RNP have taken place using different delivery agents, for example, Liu and colleagues used lipofectamine® to deliver the Cas9 RNP in a murine model.[159] Sun et al. used a PEI coated polymeric nanoparticle for delivery of the RNP complex.[160] Other interesting RNP delivery include the transfer of a supramolecule of a Cas9 (modified to have a partial negative charge) in complex with the sgRNA directly into the cytosol of the cells. This complex has a nuclear localization sequence (NLS) that leads the RNP complex to the nucleus and the presence of the complex in approximately 90% of the cells. This study was conducted *in vitro*,[161] however, Lee et al. used another supramolecular delivery platform using gold nanoparticles (CRISPR-Gold) to successfully edit the CXCR4 gene in the murine muscle tissue.[162] The transient nature of these delivery systems leads to a reduction in the off-target effects and relatively lower immunogenicity. Additionally, degradation of the sgRNA is not an issue in this system since it is already in complex with the Cas9 protein. However, the technique is not easily translatable because production of the Cas9 protein is a laborious process and once it is isolated and purified, its nuclease activity is lost within a span of a few days.

3. Applications of gene editing technologies

3.1 *In vitro* models and screens

Since research in human subjects can only be observational, one of the central ways of studying the pathogenesis and underlying mechanisms of human diseases is generation of disease models in other organisms. These disease models play a critical role in the understanding the development of the disease and generation and testing of new therapeutics.

The generation of these models is complicated because there are several confounding factors that obscure the genotype-phenotype linkage relationship even when only a single factor (e.g., point mutation in a gene) is being analyzed. This can be overcome by using pair-matched patient tissue samples from normal and disease subjects to delineate the genetic-phenotypic linkage. However, it is challenging to obtain a large number of patient samples and in several types of cancer, as they are not available. Previously, to address this problem, cellular models have been generated by using plasmid vectors to artificially over-express the target (often mutant) cDNA. However, such plasmids express the gene of interest at much higher levels than found natively. Additionally, the wild type protein continues to be expressed at the genomic level, making this a less than ideal system to link phenotypes to mutations. Apart from the over-expression of mutant genes *via* plasmids, mutant knockout clones have also been produced through classical homologous recombination leading to the generation of a limited number of cell lines with identical genetic backgrounds except for the induced mutation.[163,164] The above-mentioned techniques have been highly useful in elucidating phenotypes of several mutations. However, these techniques are also slow and arduous, hurdles that have deterred extensive adoption of these techniques for the process of target identification and drug development. The discovery of the CRISPR-Cas gene-editing technology has radically changed this landscape. The generation of mutant cell lines with identical genetic backgrounds has been rendered so straightforward that in a matter of a few years, the process has become widespread and commonplace and is being performed by scientists across the world. Additionally, genetic manipulation through CRISPR is efficient in most cell types including cells of immune origin and stem cells.[165,166] Additionally, altering the cells with the CRISPR-Cas system has opened up avenues to study genetic mutations in any specific disease cell line model. For example, if a mutation is hypothesized to make a cancer cell line more susceptible to a drug, the CRISPR system can directly target this precise genomic location to test this hypothesis.[167] Such studies allow for the rapid establishment of a causative role for particular genes in the context of a particular disease without any obscuring factors. Although CRISPR is revolutionizing drug discovery and therapeutics, and the system is effective in almost all cell types, the rate at which HDR happens can widely differ across cell lines. For example, high HDR rates are challenging to achieve in cells that do not perform active mitosis, for example, neurons. Newer methods using non-homologous or microhomology-mediated integration of cassettes to circumvent the HDR pathways have been used to counter this hurdle.[168] Additionally, the CRISPR system is now being continuously expanded and improved and now has ways to perform direct base editing without the use of any templates. The initial base editors used a fusion of deaminases to Cas9, resulting in hybrid enzymatic proteins capable of performing transition mutations. However, newer Cas9 systems are now capable of performing all 12 bp edits (including transversion mutations) drastically expanding the molecular toolkit at the disposal of oncology researchers.[169] One can foresee a dramatic increase in CRISPR-Cas variants and their use in several drug development strategies.

A survey conducted on AstraZeneca's drug pipeline indicated that failures in Phase 2 clinical trials could be traced back to an absence of linkage of the disease to a genetic target and an association was found between the success of a project with identification of a genetic link to a disease state.[170] Moreover, studies that reached the approval stage were four times more likely to have a genetic linkage to the disease state.[171] Companies are therefore prioritizing genetic screens that identify genetic abnormalities in diseases. *In vitro* CRISPR screens are playing a major role in identifying various mutations and genetic relationships leading to effective therapeutics and treatment strategies.[172] Previously, technologies like RNAi and shRNA have been used in screens that probe the whole genome, but these screens have limited applicability due to low specificity and wide variation in gene knockdown.[173] However, the information gleaned from the RNAi screens has enabled the rapid advance of CRISPR screens; therefore, several types of CRISPR screens now exist.[174,175] The basic principle of screens is to generate a library of sgRNAs that targets every gene in the genome and then express one of these sgRNAs per cell in a culture. This culture, which in theory has cells expressing all the sgRNAs present in the library, is maintained for several generations. Sequencing of the sgRNA library is performed at the beginning and end of the screen to examine the comparative presence of each sgRNA. If the knockout of a particular gene lends the cell with a growth disadvantage, the concentration of the sgRNA will be depleted over time. Since rapid progress has now been made in this field, apart from these general screens, specific targeted sgRNA screening libraries are also available. For example, it is now possible to obtain kinase sgRNA screens, or screens that a sgRNA library to target all the proteins related to epigenome regulation.[176] Additionally, apart from the conventional knock-out screens, modified Cas9 can be used for up or downregulation of various genes in the screen. Other screens developed include screens to identify pairs of genes that interact with each other or display synthetic lethality. Furthermore, CRISPR has been used to map and characterize non-coding areas of the genome. Thus, CRISPR can be used in a myriad of innovative ways to explore druggable targets in human disease.

3.2 *In vivo* models and screens

Through the 1980s, a deeper understanding of DNA repair mechanisms coincided with the emergence of mouse embryonic stem cell and gene targeting technologies. Subsequently, the combination of all three fields led to the generation of the first genetically modified mice using conventional gene targeting.[177] Since then, homologous recombination-based gene targeting has become the gold standard for *in vivo* analysis of gene function and has been pivotal in our current understanding of various pathological and normal biological processes.[178] However, traditional methods to produce genetically modified organisms need tedious clonal expansions and complex selection methods to identify cells with the desired genetic edits. A ground-breaking study in 1994 discovered that induction of DSB using endonucleases increased the efficiency of HR mediated DNA repair by several orders of magnitude.[179] This led to extensive efforts to find programmable endonucleases to make genome editing a reality. The laborious and lengthy process of earlier gene-editing technologies combined with the equally arduous method for the generation of genetically modified animals made *in vivo* genome editing highly restrictive. The advent of the CRISPR-Cas technology has simplified the process drastically and enabled the editing of genomes of even those animals that were not genetically modifiable previously.

Although CRISPR has been used to generate animal models for various animals, the majority of the models created are mouse models (~61%)[180] and these models have been used to dissect the mechanism of various disease states including cancer. Indeed, a plethora of imperative pathological information can be gleaned from cellular models; however, there still are a multitude of disease characteristics that can only be deciphered at the organismal level, especially in a disease like cancer where the microenvironment and a multi-organ metastatic cascade play such a pivotal role. Although, in theory, all Cas9 versions described previously can be implemented in mice, only some of them have been used *in vivo* so far.

The mouse models generated using CRISPR can be Transplantation based models, Germline engineering-based, or somatic delivery based. The transplantation-based models are widely used in the analysis of cancers of the hematopoietic origin.

Apart from using *ex vivo* edited cells, another way is germline editing of zygotes and embryonic stem cells to generate genetic mutations at the organism level.[181] As with other traditional gene-editing models, the generation of these models has been time-consuming and laborious and the use of CRISPR has greatly accelerated the procedure. Apart from producing single mutations in zygotes, the CRISPR technology can also be used to test synergy or antagonism between two mutations in an organism. This can be done by CRISPR editing of embryonic stem cells of organisms that already contain one desired mutation.[182] Zygotes can also be engineered to produce whole-body knockouts[181]/knock-ins,[183] conditional[183] or inducible knockout.[184] Additionally, CRISPR can also be used to introduce changes at the chromosomal level and inversions/duplications/deletions of up to a million base pairs have been reported.[185] The generation of these CRISPR engineered mice is still time-consuming because several edited zygotes have to be created to produce one with the desired edit. However, promising advances are rapidly being made in this dynamic field. Another method of using CRISPR in animal models is through the direct editing of somatic cells. This is method is the most relevant and applicable to human therapeutics since disease treatment usually takes place in fully formed humans. However, delivery of the CRISPR components to the organ/tissue of interest remains a challenge since some organs (intestines, pancreas, etc.) are more inaccessible than others (lungs, liver are more accessible). One delivery method used is the transfection of the CRISPR component DNA into the cells. For example, CRISPR component containing DNA plasmids can be introduced into the tail vein using a hydrodynamic injection.[186] Furthermore, electroporation[187] or polyethyleneimine mediated transfection[188] of the pancreas and brain has been reported for delivering the plasmids into the cells. This approach of transferring naked DNA to the target tissue is fairly straightforward but is limited only to certain organs.

Therefore, alternate approaches like viral vectors are also widely used. Both non-integrating (adenovirus and adeno-associated virus) and integrating (lentivirus) viruses have been used for this purpose. Integrating viruses are advantageous in CRISPR based screens since they can tag the genome.

However, viral vectors are limited in the size of the genetic material they can carry, thereby limiting the length of the CRISPR component DNA. To counter this, a smaller Cas9 from *Staphylococcus aureus* was packaged into the adeno-associated virus and used to target the liver where a high editing efficiency of 40% was reported.[148] Another approach is to have mice that have the Cas9 gene already knocked-in.[189] This reduces the cargo load that the virus vector needs to carry and frees up space to have a more sophisticated vector design. Moreover, this Cas9 knock-in can also be conditional, furthering the experimental toolkit. An advantage of knocking in the Cas9 in the animal is that it does not elicit an immune response to Cas9 expression.

Lastly, CRISPR can also be delivered to somatic cells using nanoparticles made of gold, lipids, or other PEGylated materials.[190] Emerging studies show promising results with high efficiency in some organs. Additionally, these methods do not trigger an immune response and can carry large cargo loads.

Although *in vitro* studies have been invaluable in deciphering internal properties of diseased cells, they fail to account for two very important aspects of the disease, disease development and interactions with the host microenvironment. Since cells from *in vitro* models are already diseased, they provide limited information on how a normal cell turns into a diseased one. Additionally, the affected tissue is a three-dimensional structure, surrounded by an elaborate microenvironment, with which it has a deep and intricate relationship. To better recapitulate the progression of disease in the body, *in vivo* screens are now being used. These screens are particularly beneficial in diseases like cancer. The first *in vivo* CRISPR screen was a genome wide loss of function screen to identify genes involved in tumor growth and metastasis. A non-metastatic cell line was transfected with a sgRNA library targeting the whole genome, and these cells were then transplanted subcutaneously into nude mice. These mice were then monitored for occurrence of lung metastasis which were sequenced to determine the proportion of sgRNAs present in the metastatic lung tumors. Thus, several genes involved in metastasis to the lungs were identified.[22] After that, several screens have been performed in mice to identify oncogenes, tumor suppressor genes, immunotherapy related genes and synthetic lethal genes.

3.3 Gene therapy

In the late 1960s, it was discovered that viruses are able to insert their genetic material into the genome of the cells they infect.[191,192] This garnered interest among geneticists, and it became apparent the viruses could be used as an effective tool for transferring genetic material to human cells.[193] Of course, the viral particles would have to be stripped off their pathogenic genes which would need to be replaced by genes with therapeutic potential. Motivated by these discoveries, Rogers et al. conducted a first proof of concept study to demonstrate genetic transfer using viruses.[194] Further encouraged by the results of that study, they went further and performed the first direct gene therapy trial in humans. They used the wild-type Shope Pappilomavirus for introduction of the arginase genes in two subjects with a urea cycle disorder.[195,196] They rationalized that because the Shope Pappilomavirus contained the Arginase gene, introduction of the virus into the patients could correct this deficiency. Unfortunately, they did not see the effect they had hoped to see and the arginase levels and the clinical course of the two patients remained unchanged. It was later uncovered after the viral genome was sequenced that it did not encode the gene arginase.

Although tremendous progress was made, it was not until the early 1990s that gene therapy started to gain immense interest. After successfully introducing foreign genes into murine bone marrow and transplanting these cells into mice,[197] Martin Cline set out to test a therapeutic approach for β-thalassemia in humans. Although he tested his therapy in two patients, he did so without obtaining the required permissions from his institutional review board, stirring a controversy.[198] The first approved protocol for introduction of foreign genetic material into humans was granted approval by the Recombinant DNA Advisory Committee in 1988.[199] In this trial, no actual therapy was administered but extracted Tumor Infiltrating Lymphocytes were modified to include a Neomycin resistance gene so as to enable tracking of these cells.[200] Based on this trial a subsequent trial was planned to administer TILs modified to express tumor necrosis factor to two patients with advanced melanoma. It was observed that tumor cells were unable to grow at the site of injection and no tumor cells were found when the injection site was excised 3 weeks later.[201] Another early trial that was happening around the same time was for Adenosine Deaminase Deficiency (ADA-SCID) a disease leading to severe immunodeficiency.[202] This trial used extraction of white blood cells from patients and their re-introduction back into the patient after modification to express the adenosine deaminase gene. One patient displayed a temporary response, but that could have been a result of the enzyme replacement therapy she was on. These early trials ushered in a new era of gene therapy trials and several studies were conducted for various diseases. Despite modest outcomes, gene therapy was being regarded as a solution to several diseases with a genetic basis. The excitement over gene therapy continued to escalate until the death of Jesse Gelsinger in 1999.[203] The 18-year-old took part in a gene therapy trial for partial deficiency of ornithine transcarbamylase, a liver enzyme needed to clear excess nitrogen and proteins. He died due to multiorgan failure, 4 days after administration of a very high dose of adenovirus. This was the first death that could directly be linked to the gene therapy and the worst-case scenario for both Jesse and gene therapy progress.

As of today, a search on https://www.clinicaltrials.gov using the term "gene therapy" yields 4576 studies of which 1877 are currently ongoing. Of these active trials, 1073 are for cancer, making up the majority of the trials due to the genetic basis of the disease. Furthermore, there are currently 32 active trials for the search term "CRISPR, 13 for Zinc fingers and three for TALENs. Gene editing technologies are poised have a profound impact on not just gene therapy in humans, but other areas of human therapy too. For example, CRISPR-Cas has been used to inactivate porcine viruses that can be dangerous for humans too, to pave way for safe and efficacious organ transplants from animals to humans. Furthermore, research is now being conducted in several rare diseases including but not limited to Angelman syndrome, inherited ocular diseases, transthyretin amyloidosis, and Leishmaniasis.

3.4 Gene editing in plants and agriculture

Though traditional methods of plant breeding are much faster than half a century ago, they will still probably fall short in meeting the agricultural demands of an ever-increasing population, more so in view of our current environmental challenges. Conventional methods of plant breeding depend upon locating populations possessing sufficient variation and usage of traditional crossing techniques in order to introduce desirable traits into those populations. As long as plant breeding is conducted in this way, critical time and resource limitation will continue. Gene editing technologies have the potential to overcome these limitations by accelerating the process of plant breeding much more than imagined previously. Out of all the gene editing technologies, CRISPR has had the most profound impact on agriculture and plant biotechnology. Although, a word of caution is also put forward by Gao in his 2018 short paper. He states that "although sometimes subjected to exaggerated headlines, the use of CRISPR in agriculture should best be considered as simply a new breeding method that can produce identical results to conventional methods in a much more predictable, faster and even cheaper manner."[204]

In plants, the CRISPR systems associated with the Cas9 and Cpf1 nucleases are the best studied and characterized. The editing of plant genomes has been broadly classified into three types: Site Directed Nuclease (SDN)1, 2 and 3. SDN1 is repair of CRISPR generated DSBs through the NHEJ pathway without any donor template. SDN2 is replacement of short sequences into the genome and SDN3 is insertion or replacement of longer sequences.[205]

CRISPR can be applied in several ways in agriculture, for example, it can be utilized as a tool for crop yield increase, improving quality, disease resistance and herbicide resistance. Manipulation of the cytokinin homeostasis is a practical way in which cereal yield can be increased. Editing of the cytokinin activation enzyme has been shown to enhance grain yield in rice under a range of environmental settings.[206] Furthermore, deletion of the cytokinin oxidase/dehydrogenase (CKX) enzyme responsible for degradation of cytokinin led to phenotypes with high yield in wheat.[207] Similarly, there are several other genes like amino acid permease 3, *O. sativa*, etc. that can be manipulated to increase yields in various crops.[208–211] Furthermore, genes like *ENO*,[212] *CLV*[213]etc. can be edited to increase yield in fruit crops. Along with yield, crop quality can also be enhanced and are critical in agriculture. Low amylose content in grains correlates with enhanced cooking and eating quality and deletion of granule-bound starch synthase 1 (GBSS1) using CRISPR-Cas led to generation of waxy maize varieties.[214] Furthermore, altering the GBSS1 amino acid sequence using CRISPR based cytidine base editors in lines of rice led to continuously low amylose crops.[215] Similar strategies can be easily extrapolated into other crops and amylose content can be manipulated to be higher or lower as per requirement because high amylose content is beneficial for human health. To that end, rice crops with higher amylose content have also been reported.[216] Another example of crop quality manipulation is gluten proteins which can lead to coeliac disease in some individuals. Conventional breeding methods are incapable of sufficiently reducing gluten content as there are more than 100 loci encoding gluten proteins in wheat. However, CRISPR-Cas has been used to target a conserved region with genes of the gluten family to create low gluten lines with up to 85% loss in immunoreactivity.[217] Along with these traits, several other traits, like phytic acid[218] and carotenoid[219] can now be effectively tweaked using CRISPR.

Not only can CRISPR-Cas be harnessed to improve crop quality and yield, it can also be adapted to make crops more resistant to infections that constantly threaten world food supply. There are two ways of preventing infections, first to introduce genes that directly prevent infections and second to manipulate genes that make the crop susceptible to that infection. The former method is useful; however, over time the pathogen evolves to overcome this disease resistance with reciprocal methods. Therefore, disruption of factors that make the crop susceptible to the infection might be more favorable approach against infectious stress. Bacterial blight, a disease caused by *Xanthomonas oryzae*, has time and again threatened global rice supply. The products of the *SWEET* genes are required for the crop to be susceptible to this infection, and mutating the promoter region of these genes has led to generation of rice lines with broad spectrum resistance to that pathogen.[220] Likewise, editing the promoter region of *LOB1* makes citrus plants resistant to *Xanthomonas Citri*.[221] Other diseases in wheat include fungal infection from *Blumeriagraminis* that causes powdery mildew. Simultaneous manipulation of three*Enhanced disease resistance 1* (EDR1) gene homologs that encode a MAP kinase responsible for inhibition against powdery mildew has been reported to give rise to crops with enhanced resistance to the pathogen.[222] Another set of three gene homologs called *mildew resistance locus O* (*MLO*) have been edited to create plants that show broad spectrum resistance to powdery mildew.[223] Similarly, manipulating this gene in tomatoes created resistance to its version of powdery mildew caused by *Oidiumneo lycopersici*.[224] In addition to bacterial resistance, the CRISPR-Cas system can be utilized to create resistance to plant viruses. Indeed, the system has been harnessed to confer resistance to caulimovirus[225] and geminivirus[226] and also RNA viruses. As mentioned earlier, host susceptibility manipulation is also an effective way to counter infections and has been used against viruses too. For example, the potyvirus uses the eukaryotic translation initiation factor 4E (eIF4E) and also its isoforms for its own translation.[227] Disruption of this gene (as it is not essential in plants) generated resistance to potyviruses in cucumber with compromising the fitness of the plants. Weeds are another challenge that affect crop productivity and soil quality. Traditional weed prevention with genetic methods include introduction of foreign genes like *bar* that encodes phophinothricin *N*-acetyltransferase and bestows herbicide resistance. However, using CRISPR-Cas to alter endogenous genes targeted by the herbicide is appealing due to its flexibility, speed and lack of foreign genes. Analysis of naturally occurring mutations in the Acetolactate synthase (ALS) has shown that some mutations in this key enzyme involved in branched-chain amino acid biosynthesis bring about herbicide tolerance.[228] For instance, change in particular bases through CBE bestowed rice crops with herbicide resistance without affecting ALS activity. Other precise mutations have been reported in both rice and wheat to generate crops with specific herbicide resistance.[229]

Apart from these applications, CRISPR can be used to improve existing breeding technologies. CRISPR has been applied to various breeding technologies like haploid induction, generating make sterile lines, fixation of hybrid vigor and manipulation of self-incompatibility. Due to limitation of space these topics will not be discussed in this chapter. The reader is directed towards a review by Zhu et al. for further reading.[230]

3.5 Cancer therapeutics

Targeting carcinogenic viral infections: Since the CRISPR-Cas system is an adaptive immune mechanism against viral infections in bacteria, they have an inherent potential as a tool against carcinogenic viral infections. Several viral infections, such as Human Papillomavirus (HPV), Hepatitis B Virus (HBV), Hepatitis C Virus (HCV), Epstein-Barr Virus (EBV) play a causative role in the pathogenesis of liver cancer, cervical cancer, nasopharyngeal carcinoma, Burkitt's and Hodgkins Lymphoma. Treating viral infections with gene editing technologies could be particularly beneficial since the viral genome does not share sequence similarity with the human genome, therefore, this intervention may generate fewer off-target events as compared to targeting endogenous locations on the human genome.

The hallmark of the Hepatitis B Virus (HBV) is the persistence of a circular viral mini-chromosome inside the host nucleus. This covalently closed circular DNA (cccDNA) acts as a template for generation of the virus and poses the main challenge in HBV management because current treatment strategies are not capable of eliminating it. Although vaccines against HBV are available, the diseases is yet to be controlled due to ineffective vaccine coverage. Furthermore, established HBV infections still need to be treated. In 2014, CRISPR-Cas9 was first used to target the HBV cccDNA both *in vitro* and *in vivo*, where reduction of core HBV proteins in liver cells and cccDNA in hepatocellular mice tumors was noted after systemic CRISPR hydrodynamic injection.[231] Numerous other studies on *in vivo* targeting of have now been reported editing of the viral genome and its consequent depletion in mice.[232]

Human Papillomavirus (HPV) related carcinogenesis is mainly related to the expression of E6 and E7 proteins. Like the HBV, there exist vaccines for HPV; however, inadequate coverage and treatment of established infections still pose significant challenges. Zhen et al. have reported inactivation of the HPV genes *in vitro* thereby restoring the function of the p53 and Rb proteins. Several other studies report *in vivo* intervention using CRISPR-Cas9, for example, systemic injection of CRISPR components (packaged in PEGylated liposomes) against HPV oncogenes E6 and E7 showed efficient deletion of the target genes, reduction in tumor size, and improvement in survival of treated animals.[233–235] Similarly, reduction in E6 expression and tumor burden was noted when the CRISPR guide RNA carrying Adeno Associated Viral (AAV) Vector was injected subcutaneously into Cas9 transfected cervical tumors.[236]

EBV belongs to the Gamma Herpesvirus family and infects primarily B cells and epithelial cells in over 90% of the population and establishes lifelong infections. EBV plays a causative role in a broad range of cancers like Burkitt's Lymphoma, Hodgkin's Lymphoma, and nasopharyngeal carcinoma.[237] Like HPV and HBV, there have been many studies that target the latent EBV genome using CRISPR-Cas9.[238,239] However, *in vivo* studies have yet to be conducted for EBV.

Cancer Immunotherapy: The growth of genome editing using CRISPR-Cas9 has arrived at an interesting intersection with the emerging field of cancer immunotherapy. This is particularly the case for Chimeric Antigen Receptor T-Cells (CAR-T Cells). This approach consists of extraction and *in vitro* expansion of tumor cell-specific T-cells that are re-introduced back into the patient. These T-cells, can be modified to have receptors that target the tumor, have displayed great clinical potential in trials for several lymphomas and leukemias and are now also being explored for solid tumors.[240] Currently, there are two CAR-T therapies approved by the FDA for the treatment of B-Cell Acute Lymphoblastic Leukemia and Diffuse Large B-Cell Lymphoma. CRISPR engineered CAR-T therapy is now being tested for several other antigens in solid tumors and holds great promise as a treatment strategy. Indeed, a basic search conducted using the keyword "CRISPR" on the registry of clinical trials in the United States (www.clinicaltrials.gov) yields 31 active/recruiting trials, of which 17 (55%) are related to cancer. Furthermore, of those 17 cancer trials, 15 aim to test treatment strategies using CRISPR engineered CAR-T cells as therapy for various malignancies. This demonstrates that CRISPR engineered T Cell therapy is at the most advanced stage in CRISPR based therapeutics.

3.6 Epigenome targeting

Epigenetic genome modifications are indeed a very important heritable level of gene regulation that cannot be ignored.[241,242] The epigenetic status can regulate whether or not it is expressed and to what extent. Promoter regions of silenced genes show distinctive methylation/acetylation signatures. For example, DNA methylation or methylation of Lysine residues 9 and 27 of the third histone are associated with a lack of expression. On the other hand, histone acetylation is associated with transcriptional activation. Several deregulated epigenetic modifications on genomic DNA or associated histones have been implicated in several human diseases including cancer.

Owing to such efficient genome editing capabilities, the CRISPR system has now been re-purposed by replacement of the nuclease with functional units of other enzymes, effectively using the conventional gRNA as a docking site for various genome engineering tools.[243,244] Introducing mutations in the two active domains of the Cas9 protein, namely, RuvC and HNH domains, creates an enzyme with deactivated nuclease activity. The resulting deactivated CRISPR-Cas9 system (dCas9) effectively binds to the target site without introducing double-stranded breaks. The addition of transcriptional activators or repressors to dCas9 can induce up or down-regulation of endogenous genes without knocking them out.[244] Furthermore, the dCas9 can be coupled with the catalytic domains of various enzymes capable of altering methylation or acetylation of histones present in the gene promoters. For example, the fusion protein generated using the catalytic domain of the histone demethylase LSD1 and dCas9 was used to repress genes associated with pluripotency maintenance.[245] Similarly, H3K9 trimethylation (H3K9me3) in the enhancer region of the human beta-globin locus using the dCas9-KRAB fusion protein repressed gene expression successfully.[246] Besides repression, transcriptional activation can also be done using this system, for example, dCas9 was fused with the catalytic domain of human acetyltransferase p300 to increase H3K27 acetylation of promoter and enhancer regions of target genes, thereby increasing their expression.[247] Similarly, dCas9 can be fused to a whole array of catalytic domains from various enzymes (e.g., histone methyltransferase—PRDM9, DNA Methylase—TET1, DNA Methyltransferase 3A—DNMT3A). There are several methods to apply this transcriptional regulation to cells. For example, re-activation of epigenetically silenced genes can be performed using these fusion proteins; however, the effectiveness, stability, and duration of the effect is highly dependent on the CpG island methylation status. Multiplexing of the dCas9 fusion proteins also makes it possible to simultaneously induce/repress the methylation or acetylation status of the same gene, creating a powerful tool for gene regulation. For instance, the demethylating dCas9 fusion proteins can individually produce a 30%–60% reduction in the methylation of the CpG islands.[248,249] However, the use of KRAB, DNMT3A, and DNMT3L together achieve up to 100% CpG island methylation resulting in a stable lack of expression of the target gene.[250] This repression achieved using this multiplexed approach is also resistant to external transcriptional activating stimuli (including dCas9 coupled promoter acetylation), however it can be reversed using dCas9 coupled demethylase proteins. In this way, the CRISPR system enables stable, heritable epigenomic changes.

4. Drawbacks of gene editing technologies

4.1 Off-target effects

As the CRISPR technology is based on the binding of complementary sequences, there is an inherent risk of non-specific binding of the sgRNA at locations other than the target gene although with lesser affinity. Various studies have demonstrated high levels of off-target effects using Cas9,[251–253] an observation that could have devastating effects on patients. For example, off-target editing by Cas9 could induce oncogenic mutations, chromosomal translocations, deletions, or insertions that could further complicate the treatment of a cancer patient. Detecting and minimizing these off-target effects is imperative to make this technology worthy of being in the clinic. Usually, off-target effects are detected by sensing DSBs or sequence analysis for Cas9 induced changes.[251]

These off-target effects can be broadly classified into two types: sgRNA dependent and sgRNA independent. The sgRNA dependent effects mainly occur due to the sequence similarity of some parts of the genomic DNA to the sequence of interest and can be detected easily using techniques mentioned below. However, sgRNA independent effects are usually base changes that are often harder to detect and define.

These off-target effects vary widely depending upon cell type and organism, and detecting their actual frequency is fraught with challenges due to limitations of current testing techniques. State of the art sequencing techniques are used to identify the induction of DSB. These detection techniques can be roughly classified into two types: unbiased and biased. As the names suggest, the unbiased method analyses the whole genome for possible changes whereas the biased approach only examines the predicted sites.

For example, in Digenome-Seq, an unbiased approach, the genomic DNA is incubated *in vitro* with sgRNA and Cas9 and the generated mismatches are compared to the target DNA sequence.[254] Now, improvements on the Digenome-seq, namely, SITE-Seq[255] and CIRCLE-seq[256] have also been reported. Notably, a newer method called GUIDE-seq[257] introduces a short sequence tag at all the DSB that are generated after the genomic DNA-sgRNA-Cas9 incubation, making detection easier. Thus, massive strides have been made in the detection ability of off-target effects. All of these methods positively contribute to the progress of CRISPR as a therapeutic.

Mitigation of these off-target effects has been done in various ways, for example, modification of sgRNAs. Conventional sgRNAs are 20 nucleotides in length and various truncations have been reported that have varying effects on the efficiency of the sgRNA. Even though conventional Cas9 is 20 bp in length, it is still able to bind to target sites at sequence mismatches of up to 10 bp. Truncation of the 5′ end increases the sensitivity of the mismatches and overcomes this problem.[258] This is due to an increase in the requirement of binding energy after 5′ truncation; therefore, the truncated sgRNA displays less affinity for off-target effects. The addition of 2 guanine nucleotides at the 5′ end has also been reported to display lower off-target effects, however, on-site efficiency could also lower due to this.[259] Therefore, despite current advances in mitigation of off-target effects, there are still several challenges that need to be overcome for this technology to be successfully applied to patients in the clinic.

4.2 Delivery of components

Several genetic diseases require diseased individuals to be treated at a stage where the pathology has partially or fully-developed and gene therapy would mean correcting genetic aberrations in the body of the patient. Delivery of the gene editing components to target cells of the diseased individual is the hurdle preventing widespread use of CRISPR as a therapeutic agent. Effective gene targeting consists of successful transfection into cells, minimal toxicity caused by the delivery method, and rapid clearance of the delivery components after treatment. It is challenging to achieve all these criteria simultaneously using current delivery methods available.[260] The currently used viral vectors have several advantages and disadvantages associated with them. Though these vectors perform satisfactorily in the laboratory setting and also have certain advantages in the clinical setting like low risk of carcinogenesis and low immunogenicity, they also have certain problems associated with them.[261] For example, critical genes might be disrupted by the integration of the transgene into the genome. Additionally, the loading capacity of most viral vectors is limited making simultaneous inclusion of the sgRNA, Cas9 and transgene sequences a very big challenge. Furthermore, the chronic presence of the vector components can lead to detrimental effects like an accumulation of off-target effects. The alternative to this delivery method is the delivery of the CRISPR machinery through ribonucleoprotein (RNP) complex.[262] These complexes degrade rapidly and demonstrate higher efficiency than other methods that deliver DNA or RNA. However, delivery of the complex to target tissue in individuals is difficult and is done using electroporation in *in vitro* models. Alternative methods of delivery include microinjections and transfection using lipids. The microinjection technique has been successfully utilized in delivering the RNP complex to animal embryonic cells and lipofectamine reagents have been extensively used *in vitro* with excellent results. However, there still is a lack of a technique that enables direct somatic editing, without which CRISPRs journey to the clinic cannot be completed.

5. Ethics of gene editing technologies

The invention of a scientific tool as effective as the CRISPR/Cas is seldom an isolated event. As has happened with most trailblazing scientific breakthroughs, ethicists and regulators have been left grappling, trying to cope with the very nature and far-reaching repercussions of this novel technology. On the face of it, it is a simple technique to correct faulty DNA, but so pervasive are the effects of this technology that ethicists, governments, theologians, feminist scholars and the general public are now in debate over this technology.[263] Since the technology can alter the human genome and anything that possesses DNA, there are multiple stakeholders who must be involved in the discussion of what this CRISPR storm means to the general public and to the future of humanity.

Furthermore, so widespread and accessible is the use of the technique that as of January 2020, a Do-It-Yourself CRISPR kit for microbial gene editing was available in the United States for about $180 and can be performed so long as someone has water and a microwave. One can argue that the ubiquity of the technique stands to democratize science; however, there also can be several uses of the technology that are shocking to say the least. For example, in 2018, a Chinese scientist Dr. He Jiankui announced that his lab had successfully engineered the CCR5 gene in embryos using CRISPR, leading to the birth of twin baby girls, given the pseudonyms Lulu and Nana. The mutation was intended to introduce resistance to the Human Immunodeficiency Virus (HIV) infection. The Southern University of Science & Technology (Shenzen, China), where he worked at the time responded that that He had "seriously violated academic ethics and codes of conduct". Statements were issued by every scientific authority in the world, widely condemning the study. Although the use of genome editing technology in humans is inevitable, and will happen sooner or later, there are several issues that arise.

First, we simply do not know enough about the cellular effects of CRISPR yet. There are a plethora of studies pointing to the off-target effects caused by this technology. Furthermore, we can only detect off-target effects of the known areas of the genome. There are still several regions of the genome that are yet to be studied and we are now discovering several non-coding regions of the genome with critical functions. It is impossible, at this stage, to predict and test comprehensively the changes that the CRISPR-Cas system causes at the molecular and cellular level. Effects of the genetic editing on the babies could still be waiting to manifest, a horrifying thought indeed. Since this was but an experiment, if unintended effects do happen, who would take the burden of the moral responsibility? The experiment was conducted without the consent of the subject, and they will have to suffer the consequences of this mis-calculated risk for the entirety of their lifetime. Furthermore, if some negative side effects were to happen, who would take the financial responsibility for medical management of the off-target effects? These are questions that should have been asked and thought about before the experiments were conducted.

Second, decisions about such an enormous discovery cannot be taken on an individual basis. As mentioned earlier, this particular technique has such far-reaching consequences that almost every segment of the public will get affected by it in some way or another. Can scientists assume the responsibility to make those decisions on behalf of humanity as a whole? What is worse, can one scientist alone take that decision on his/her own? The ease and widespread access of this technique is as much a boon as it is a bane. There are ethical committees and regulations in place to prevent such experiments from happening. Clearly, those guidelines and regulations were bypassed. Since one of the mainstays of this gene editing method is its simplicity, it is quite easy to carry out without detection, as seen in this particular case.

Third, we must take a page out of one failed gene therapy experiment and learn from it. In September 1999, a gene therapy dose escalation safety trial for a liver enzyme deficiency disease recruited 18 patients. 17 patients had already been given the dose and fared well. Unfortunately, Jesse Gelsinger, the 18th patient to receive the adenoviral vector containing correction genes passed away due to multi-organ failure as a result of the therapy.[203] This was the first death recorded as a direct effect of gene therapy. Prior to this, the field of gene therapy had garnered immense hope, optimism and speculation. Much like CRISPR, at the time gene therapy was touted as salvation for the millions of people suffering from genetic disorders. One failed human experiment with worst outcome stopped the field of gene therapy in its tracks, eroded public trust in the technology and set the field 10 years behind.

Lastly, as Ted Peters maintains in his 1995 paper,[264] in gene editing there are two main issues that need sifting. The first issue is somatic vs germline editing and the second is therapy vs enhancement. For the most part somatic gene editing, altering genetic sequences in differentiated cells of already developed humans, is therapeutic in nature. Somatic gene editing aims to correct any aberrations in an organism and these changes are not passed on to the next generation. Germline editing on the other hand takes place in zygotes or embryos, changes that will pass on to the next generation. This means that we are not changing just one individual, but all their generations to come. The experiment then crosses the line into being out of the control of the experimenter. Furthermore, germline editing tends to be enhancement centric. Today it is a matter of routine procedure to undergo *in vitro* fertilization for overcoming infertility. Screening of embryos for possible genetic abnormalities to select the healthiest embryo has already been reported.[265,266] It is also now possible to select the sex of the embryo, which has its own set of ethical challenges in countries like India where female feticide is so rampant that pre-natal sex determination is illegal. How long will it take for CRISPR to percolate into this realm and offer future parents the option to edit a particular gene for conferring disease resistance or even a preferred eye color? Not very long indeed.

Another application of CRISPR that can have profound ethical issues associated with it is "gene drive" experiments. It involves manipulating a target gene such that a higher proportion of the progeny will carry that gene. In effect, it allows the spread of that particular genetic manipulation throughout the population. This can be used for various applications like agricultural pest control, eradication of invasive species and rescuing endangered species. Although the concept has been known for more than a decade, the process has really been simplified by the advent of CRISPR-Cas. This has led to various scientists designing gene drives outside of the laboratory to possibly solve various challenges. For example, the most talked about gene drive is the eradication of malaria causing mosquitoes. According to the WHO, in 2018, there were 228 million malaria cases of which 93% were reported in the African region. There were approximately 405,000 deaths from malaria globally of which 67% were children under the age of 5. Malaria is indeed a significant public healthcare burden with a solution right in front of our eyes: there is no malaria if there are no mosquitoes. There is compelling rationale for attempting to eradicate malaria causing mosquitoes. About 20 years ago, a team lead by Professor Andrea Crisanti at the Imperial College of London modified the *Anopheles* mosquito genome[267] and made the information public 2 years later.[268] This development uncovered a new era in the fight against malaria. As several groups started working on this problem, new modifications of genetic traits essential for mosquito fertility were discovered. However, it was quickly realized that without the ability to spread these modifications throughout the wild population the developments were futile. Thus, innovative strategies for propagating these traits in the wild were formulated. The challenge, however, was that these traits were self-limiting. Self-limiting traits do not sustain in the population and are phased out in due time because of natural selection and the cost of fitness they incur. Furthermore, releasing sufficient numbers of mosquitoes into the wild to have actual effect on a population would not be feasible. These difficulties were solved by gene drives. Between 1940 and 1960, three groups demonstrated strategies for propagation of a gene through a population. Chris Curtis (University of Bristol, UK)[269] and Alexander Serebrovsky[270] (Moscow University, Russia) observed a phenomenon called genetic underdominance. They noted that particular chromosomal translocations could propagate and eliminate a population because they could cause heterozygous inferiority. Furthermore, Bill Hamilton (Imperial College London) postulated that elimination of a population was possible if a sex distorter was genetically encoded.[271] For example, if the Y chromosome was linked with an extreme bias for the male sex. These strategies are now collectively called gene drives where a particular genetic element biases its inheritance above other genes, thereby increasing its frequency in the current or future generations. The concept of gene drives is self-sustaining and does not need genetically modified organisms to be released multiple times. The advent of gene editing technologies has transformed the field of gene drives. Using CRISPR, it is now possible to modify mosquito genomes such that a higher frequency of the female progeny is sterile.[272] Over time, female sterility is expected to spread through the population, giving rise to enough sterile females to cause a population collapse. Therefore, CRISPR-Cas has now given a real possibility for eradication of malaria[273]; however, this comes with its own set of ethical dillemmas. Gene drives force inheritance through the population leading to the programming of not just an individual organism, but of entire populations or species if it is implemented outside the laboratory. However, global ecology is much like the game of Jenga, each ecological piece fitting together perfectly and resting stably on top of another. Eradication of a particular population or species *via* gene drives is much like pulling out one piece of Jenga. One never really knows if the interlinked system will be able to withstand this manipulation or collapse. Furthermore, if it does collapse, which way will the ecological Jenga tower fall and how will it affect the Jenga block that represents humans? The long-term effects of such population manipulation simply cannot be predicted. Furthermore, who has the authority to take decisions like that? Since malaria is challenge faced by so many countries together, and the gene drive once implemented will likely not be limited to one country, the decision for implementation must not solely be made by the government of the country where the drive will be initiated. However, co-operation of multiple countries poses another formidable challenge.

Considering that different countries will have different levels of regulation on this technology, it can be speculated that tight regulations in one country will not guaranty absence of misuse. International treaties to regulate CRISPR pose additional challenges due to problems in implementation in several countries simultaneously.

6. Conclusions

Massive strides have been made in the gene editing field in the past decade and the CRISPR-Cas system is indeed a breakthrough that has accelerated all aspects of biological research. The initial years of the CRISPR-Cas development were full of skepticism and resistance and it was not until the early 2010s that the scope of the technology was realized. The far-reaching consequences of such a simple gene editing technology have still not been fully realized and future holds great potential. However, there are hurdles like off target effects and delivery of components that need to be overcome before gene editing technologies can fully realize their potential. Several ethical issues also need to be paid attention to. Nevertheless, CRISPR stands elegantly poised to revolutionize therapeutics and it is certainly only a matter of time before the drawbacks of the technology are improved upon and applied in a better way.

References

1. Muller HJ. Artificial transmutation of the gene. *Science*. 1927;66(1699):84–87.
2. Auerbach C, Robson JM. Chemical production of mutations. *Nature*. 1946;157 (3984):302.
3. Rothstein RJ. One-step gene disruption in yeast. *Methods Enzymol*. 1983;101:202–211.
4. Scherer S, Davis RW. Replacement of chromosome segments with altered DNA sequences constructed in vitro. *Proc Natl Acad Sci U S A*. 1979;76(10):4951–4955.
5. Smithies O, Gregg RG, Boggs SS, Koralewski MA, Kucherlapati RS. Insertion of DNA sequences into the human chromosomal β-globin locus by homologous recombination. *Nature*. 1985;317(6034):230–234.
6. Thomas KR, Folger KR, Capecchi MR. High frequency targeting of genes to specific sites in the mammalian genome. *Cell*. 1986;44(3):419–428.
7. Mansour SL, Thomas KR, Capecchi MR. Disruption of the proto-oncogene int-2 in mouse embryo-derived stem cells: a general strategy for targeting mutations to non-selectable genes. *Nature*. 1988;336(6197):348–352.
8. Gaj T, Gersbach CA, Barbas III CF. ZFN, TALEN, and CRISPR/Cas-based methods for genome engineering. *Trends Biotechnol*. 2013;31(7):397–405.
9. Carroll D. Genome engineering with targetable nucleases. *Annu Rev Biochem*. 2014;83:409–439.
10. Youds JL, Boulton SJ. The choice in meiosis–defining the factors that influence crossover or non-crossover formation. *J Cell Sci*. 2011;124(4):501–513.
11. Latt SA. Sister chromatid exchange formation. *Annu Rev Genet*. 1981;15(1):11–53.
12. Rudin N, Sugarman E, Haber JE. Genetic and physical analysis of double-strand break repair and recombination in Saccharomyces cerevisiae. *Genetics*. 1989;122(3):519–534.
13. Choulika A, Perrin A, Dujon B, Nicolas J-F. Induction of homologous recombination in mammalian chromosomes by using the I-SceI system of Saccharomyces cerevisiae. *Mol Cell Biol*. 1995;15(4):1968–1973.
14. Rouet P, Smih F, Jasin M. Introduction of double-strand breaks into the genome of mouse cells by expression of a rare-cutting endonuclease. *Mol Cell Biol*. 1994;14(12): 8096–8106.
15. Plessis A, Perrin A, Haber JE, Dujon B. Site-specific recombination determined by I-SceI, a mitochondrial group I intron-encoded endonuclease expressed in the yeast nucleus. *Genetics*. 1992;130(3):451–460.
16. Chapman JR, Taylor MR, Boulton SJ. Playing the end game: DNA double-strand break repair pathway choice. *Mol Cell*. 2012;47(4):497–510.
17. Griffith F. The significance of pneumococcal types. *Epidemiol Infect*. 1928;27(2): 113–159.
18. Dawson MH, Sia RH. In vitro transformation of pneumococcal types: I A technique for inducing transformation of pneumococcal types in vitro. *J Exp Med*. 1931; 54(5):681.

19. Alloway JL. The transformation in vitro of R pneumococci into S forms of different specific types by the use of filtered pneumococcus extracts. *J Exp Med*. 1932; 55(1):91–99.
20. Alloway JL. Further observations on the use of pneumococcus extracts in effecting transformation of type in vitro. *J Exp Med*. 1933;57(2):265–278.
21. Avery OT, MacLeod CM, McCarty M. Studies on the chemical nature of the substance inducing transformation of pneumococcal types: induction of transformation by a desoxyribonucleic acid fraction isolated from pneumococcus type III. *J Exp Med*. 1944;79(2):137–158.
22. Smithies O, Koralewski MA, Song KY, Kucherlapati RS. Homologous recombination with DNA introduced into mammalian cells. *Cold Spring Harb Symp Quant Biol*. 1984;49:161–170. https://doi.org/10.1101/sqb.1984.049.01.019.
23. Mak TW. Gene targeting in embryonic stem cells scores a knockout in Stockholm. *Cell*. 2007;131(6):1027–1031.
24. Smithies O. Forty years with homologous recombination. *Nat Med*. 2001;7(10): 1083–1086.
25. Brinster RL, Braun RE, Lo D, Avarbock MR, Oram F, Palmiter RD. Targeted correction of a major histocompatibility class II E alpha gene by DNA microinjected into mouse eggs. *Proc Natl Acad Sci U S A*. 1989;86(18):7087–7091.
26. Kucherlapati RS, Eves EM, Song K-Y, Morse BS, Smithies O. Homologous recombination between plasmids in mammalian cells can be enhanced by treatment of input DNA. *Proc Natl Acad Sci U S A*. 1984;81(10):3153–3157.
27. Hasty P, Rivera-Perez J, Chang C, Bradley A. Target frequency and integration pattern for insertion and replacement vectors in embryonic stem cells. *Mol Cell Biol*. 1991; 11(9):4509–4517.
28. Kok FO, Shin M, Ni C-W, et al. Reverse genetic screening reveals poor correlation between morpholino-induced and mutant phenotypes in zebrafish. *Dev Cell*. 2015; 32(1):97–108.
29. Morgens DW, Deans RM, Li A, Bassik MC. Systematic comparison of CRISPR/Cas9 and RNAi screens for essential genes. *Nat Biotechnol*. 2016;34(6):634–636.
30. Lin A, Giuliano CJ, Sayles NM, Sheltzer JM. CRISPR/Cas9 mutagenesis invalidates a putative cancer dependency targeted in on-going clinical trials. *elife*. 2017;6:e24179.
31. Bak RO, Porteus MH. CRISPR-mediated integration of large gene cassettes using AAV donor vectors. *Cell Rep*. 2017;20(3):750–756.
32. Lombardo A, Cesana D, Genovese P, et al. Site-specific integration and tailoring of cassette design for sustainable gene transfer. *Nat Methods*. 2011;8(10):861–869.
33. Dever DP, Bak RO, Reinisch A, et al. CRISPR/Cas9 β-globin gene targeting in human haematopoietic stem cells. *Nature*. 2016;539(7629):384–389.
34. Voit RA, Hendel A, Pruett-Miller SM, Porteus MH. Nuclease-mediated gene editing by homologous recombination of the human globin locus. *Nucleic Acids Res*. 2014; 42(2):1365–1378.
35. Hubbard N, Hagin D, Sommer K, et al. Targeted gene editing restores regulated CD40L function in X-linked hyper-IgM syndrome. *Blood*. 2016;127(21):2513–2522.
36. Arias-Fuenzalida J, Jarazo J, Qing X, et al. FACS-assisted CRISPR-Cas9 genome editing facilitates Parkinson's disease modeling. *Stem Cell Rep*. 2017;9(5):1423–1431.
37. Bak RO, Dever DP, Reinisch A, Hernandez DC, Majeti R, Porteus MH. Multiplexed genetic engineering of human hematopoietic stem and progenitor cells using CRISPR/Cas9 and AAV6. *elife*. 2017;6:e27873.
38. DeWitt MA, Magis W, Bray NL, et al. Selection-free genome editing of the sickle mutation in human adult hematopoietic stem/progenitor cells. *Sci Transl Med*. 2016;8(360):360ra134–360ra134.

39. Jacquier A, Dujon B. An intron-encoded protein is active in a gene conversion process that spreads an intron into a mitochondrial gene. *Cell.* 1985;41(2):383–394.
40. Cohen-Tannoudji M, Robine S, Choulika A, et al. I-SceI-induced gene replacement at a natural locus in embryonic stem cells. *Mol Cell Biol.* 1998;18(3):1444–1448.
41. Fenina M, Simon-Chazottes D, Vandormael-Pournin S, et al. I-Sce I-mediated double-strand break does not increase the frequency of homologous recombination at the Dct locus in mouse embryonic stem cells. *PLoS One.* 2012;7(6):e39895.
42. Pan FC, Chen Y, Loeber J, Henningfeld K, Pieler T. I-SceI meganuclease-mediated transgenesis in Xenopus. *Dev Dyn Off Publ Am Assoc Anat.* 2006;235(1):247–252.
43. Thermes V, Grabher C, Ristoratore F, et al. I-SceI meganuclease mediates highly efficient transgenesis in fish. *Mech Dev.* 2002;118(1–2):91–98.
44. Wang Y, Zhou X-Y, Xiang P-Y, et al. The meganuclease I-SceI containing nuclear localization signal (NLS-I-SceI) efficiently mediated mammalian germline transgenesis via embryo cytoplasmic microinjection. *PLoS One.* 2014;9(9):e108347.
45. Grizot S, Smith J, Daboussi F, et al. Efficient targeting of a SCID gene by an engineered single-chain homing endonuclease. *Nucleic Acids Res.* 2009;37(16):5405–5419.
46. Ménoret S, Fontanière S, Jantz D, et al. Generation of Rag1-knockout immunodeficient rats and mice using engineered meganucleases. *FASEB J.* 2013;27(2):703–711.
47. Geurts AM, Cost GJ, Freyvert Y, et al. Knockout rats via embryo microinjection of zinc-finger nucleases. *Science.* 2009;325(5939):433.
48. Bibikova M, Carroll D, Segal DJ, et al. Stimulation of homologous recombination through targeted cleavage by chimeric nucleases. *Mol Cell Biol.* 2001;21(1):289–297.
49. Urnov FD, Miller JC, Lee Y-L, et al. Highly efficient endogenous human gene correction using designed zinc-finger nucleases. *Nature.* 2005;435(7042):646–651.
50. Mani M, Smith J, Kandavelou K, Berg JM, Chandrasegaran S. Binding of two zinc finger nuclease monomers to two specific sites is required for effective double-strand DNA cleavage. *Biochem Biophys Res Commun.* 2005;334(4):1191–1197.
51. Rémy S, Tesson L, Ménoret S, Usal C, Scharenberg AM, Anegon I. Zinc-finger nucleases: a powerful tool for genetic engineering of animals. *Transgenic Res.* 2010; 19(3):363–371.
52. Carbery ID, Ji D, Harrington A, et al. Targeted genome modification in mice using zinc-finger nucleases. *Genetics.* 2010;186(2):451–459. https://doi.org/10.1534/genetics.110.117002.
53. Meyer M, de Angelis MH, Wurst W, Kühn R. Gene targeting by homologous recombination in mouse zygotes mediated by zinc-finger nucleases. *Proc Natl Acad Sci U S A.* 2010;107(34):15022–15026. https://doi.org/10.1073/pnas.1009424107.
54. Lillico SG, Proudfoot C, King TJ, et al. Mammalian interspecies substitution of immune modulatory alleles by genome editing. *Sci Rep.* 2016;6. https://doi.org/10.1038/srep21645, 21645.
55. Liu X, Wang Y, Tian Y, et al. Generation of mastitis resistance in cows by targeting human lysozyme gene to β-casein locus using zinc-finger nucleases. *Proc Biol Sci.* 2014;281(1780): 20133368. https://doi.org/10.1098/rspb.2013.3368.
56. Radecke S, Radecke F, Cathomen T, Schwarz K. Zinc-finger nuclease-induced gene repair with oligodeoxynucleotides: wanted and unwanted target locus modifications. *Mol Ther.* 2010;18(4):743–753. https://doi.org/10.1038/mt.2009.304.
57. Maier DA, Brennan AL, Jiang S, et al. Efficient clinical scale gene modification via zinc finger nuclease-targeted disruption of the HIV co-receptor CCR5. *Hum Gene Ther.* 2013;24(3):245–258. https://doi.org/10.1089/hum.2012.172.
58. Tebas P, Stein D, Tang WW, et al. Gene editing of CCR5 in autologous CD4 T cells of persons infected with HIV. *N Engl J Med.* 2014;370(10):901–910. https://doi.org/10.1056/NEJMoa1300662.

59. Hermann M, Maeder ML, Rector K, et al. Evaluation of OPEN zinc finger nucleases for direct gene targeting of the ROSA26 locus in mouse embryos. *PLoS One*. 2012;7 (9):e41796. https://doi.org/10.1371/journal.pone.0041796.
60. Swarthout JT, Raisinghani M, Cui X. Zinc finger nucleases: a new era for transgenic animals. *Ann Neurosci*. 2011;18(1):25–28. https://doi.org/10.5214/ans.0972.7531.1118109.
61. Boch J, Bonas U. Xanthomonas AvrBs3 family-type III effectors: discovery and function. *Annu Rev Phytopathol*. 2010;48:419–436. https://doi.org/10.1146/annurev-phyto-080508-081936.
62. Bonas U, Stall RE, Staskawicz B. Genetic and structural characterization of the avirulence gene avrBs3 from Xanthomonas campestris pv. vesicatoria. *Mol Gen Genet MGG*. 1989;218(1):127–136. https://doi.org/10.1007/BF00330575.
63. Zhu W, Yang B, Chittoor JM, Johnson LB, White FF. AvrXa10 contains an acidic transcriptional activation domain in the functionally conserved C terminus. *Mol Plant Microbe Interact*. 1998;11(8):824–832. https://doi.org/10.1094/MPMI.1998.11.8.824.
64. Zhu W, Yang B, Wills N, Johnson LB, White FF. The C terminus of AvrXa10 can be replaced by the transcriptional activation domain of VP16 from the herpes simplex virus. *Plant Cell*. 1999;11(9):1665–1674. https://doi.org/10.1105/tpc.11.9.1665.
65. Yang B, Zhu W, Johnson LB, White FF. The virulence factor AvrXa7 of Xanthomonas oryzae pv. oryzae is a type III secretion pathway-dependent nuclear-localized double-stranded DNA-binding protein. *Proc Natl Acad Sci U S A*. 2000;97(17):9807–9812. https://doi.org/10.1073/pnas.170286897.
66. Sugio A, Yang B, Zhu T, White FF. Two type III effector genes of Xanthomonas oryzae pv. oryzae control the induction of the host genes OsTFIIAgamma1 and OsTFX1 during bacterial blight of rice. *Proc Natl Acad Sci U S A*. 2007;104 (25):10720–10725. https://doi.org/10.1073/pnas.0701742104.
67. Boch J, Scholze H, Schornack S, et al. Breaking the code of DNA binding specificity of TAL-type III effectors. *Science*. 2009;326(5959):1509–1512. https://doi.org/10.1126/science.1178811.
68. Moscou MJ, Bogdanove AJ. A simple cipher governs DNA recognition by TAL effectors. *Science*. 2009;326(5959):1501. https://doi.org/10.1126/science.1178817.
69. Morbitzer R, Römer P, Boch J, Lahaye T. Regulation of selected genome loci using de novo-engineered transcription activator-like effector (TALE)-type transcription factors. *Proc Natl Acad Sci U S A*. 2010;107(50):21617–21622. https://doi.org/10.1073/pnas.1013133107.
70. Christian M, Cermak T, Doyle EL, et al. Targeting DNA double-strand breaks with TAL effector nucleases. *Genetics*. 2010;186(2):757–761. https://doi.org/10.1534/genetics.110.120717.
71. Li T, Huang S, Jiang WZ, et al. TAL nucleases (TALNs): hybrid proteins composed of TAL effectors and FokI DNA-cleavage domain. *Nucleic Acids Res*. 2011; 39(1):359–372. https://doi.org/10.1093/nar/gkq704.
72. Kurita T, Moroi K, Iwai M, et al. Efficient and multiplexable genome editing using Platinum TALENs in oleaginous microalga, Nannochloropsis oceanica NIES-2145. *Genes Cells*. 2020;25(10):695–702. https://doi.org/10.1111/gtc.12805.
73. Liu J, Li C, Yu Z, et al. Efficient and specific modifications of the Drosophila genome by means of an easy TALEN strategy. *J Genet Genomics Yi Chuan Xue Bao*. 2012;39 (5):209–215. https://doi.org/10.1016/j.jgg.2012.04.003.
74. Wood AJ, Lo T-W, Zeitler B, et al. Targeted genome editing across species using ZFNs and TALENs. *Science*. 2011;333(6040):307. https://doi.org/10.1126/science.1207773.

75. Sander JD, Cade L, Khayter C, et al. Targeted gene disruption in somatic zebrafish cells using engineered TALENs. *Nat Biotechnol.* 2011;29(8):697–698. https://doi.org/10.1038/nbt.1934.
76. Tesson L, Usal C, Ménoret S, et al. Knockout rats generated by embryo microinjection of TALENs. *Nat Biotechnol.* 2011;29(8):695–696. https://doi.org/10.1038/nbt.1940.
77. Shen Y, Xu K, Yuan Z, et al. Efficient generation of P53 biallelic knockout Diannan miniature pigs via TALENs and somatic cell nuclear transfer. *J Transl Med.* 2017;15(1):224. https://doi.org/10.1186/s12967-017-1327-0.
78. Ma S, Zhang S, Wang F, et al. Highly efficient and specific genome editing in silkworm using custom TALENs. *PLoS One.* 2012;7(9):e45035. https://doi.org/10.1371/journal.pone.0045035.
79. Watanabe T, Ochiai H, Sakuma T, et al. Non-transgenic genome modifications in a hemimetabolous insect using zinc-finger and TAL effector nucleases. *Nat Commun.* 2012;3:1017. https://doi.org/10.1038/ncomms2020.
80. Higuchi K, Kazeto Y, Ozaki Y, et al. Targeted mutagenesis of the ryanodine receptor by Platinum TALENs causes slow swimming behaviour in Pacific bluefin tuna (Thunnus orientalis). *Sci Rep.* 2020;10(1):9351. https://doi.org/10.1038/s41598-020-65964-4.
81. Fu L, Wen L, Shi Y-B. Role of thyroid hormone receptor in amphibian development. *Methods Mol Biol.* 2018;1801:247–263. https://doi.org/10.1007/978-1-4939-7902-8_20.
82. Kwon H, Arends BR, Smith RC. Late-phase immune responses limiting oocyst survival are independent of TEP1 function yet display strain specific differences in Anopheles gambiae. *Parasit Vectors.* 2017;10(1):369. https://doi.org/10.1186/s13071-017-2308-0.
83. Yamamoto DS, Sumitani M, Hatakeyama M, Matsuoka H. Malaria infectivity of xanthurenic acid-deficient anopheline mosquitoes produced by TALEN-mediated targeted mutagenesis. *Transgenic Res.* 2018;27(1):51–60. https://doi.org/10.1007/s11248-018-0057-2.
84. Homma Y, Toga K, Daimon T, Shinoda T, Togawa T. A mitochondrial phosphatase PTPMT1 is essential for the early development of silkworm, Bombyx mori. *Biochem Biophys Res Commun.* 2020;530(4):713–718. https://doi.org/10.1016/j.bbrc.2020.07.124.
85. Wei J, Wagner S, Maclean P, et al. Cattle with a precise, zygote-mediated deletion safely eliminate the major milk allergen beta-lactoglobulin. *Sci Rep.* 2018;8(1):7661. https://doi.org/10.1038/s41598-018-25654-8.
86. Geissler R, Scholze H, Hahn S, et al. Transcriptional activators of human genes with programmable DNA-specificity. *PLoS One.* 2011;6(5):e19509. https://doi.org/10.1371/journal.pone.0019509.
87. Zhang F, Cong L, Lodato S, Kosuri S, Church GM, Arlotta P. Efficient construction of sequence-specific TAL effectors for modulating mammalian transcription. *Nat Biotechnol.* 2011;29(2):149–153. https://doi.org/10.1038/nbt.1775.
88. Mahfouz MM, Li L, Piatek M, et al. Targeted transcriptional repression using a chimeric TALE-SRDX repressor protein. *Plant Mol Biol.* 2012;78(3):311–321. https://doi.org/10.1007/s11103-011-9866-x.
89. Bultmann S, Morbitzer R, Schmidt CS, et al. Targeted transcriptional activation of silent oct4 pluripotency gene by combining designer TALEs and inhibition of epigenetic modifiers. *Nucleic Acids Res.* 2012;40(12):5368–5377. https://doi.org/10.1093/nar/gks199.
90. Valton J, Dupuy A, Daboussi F, et al. Overcoming transcription activator-like effector (TALE) DNA binding domain sensitivity to cytosine methylation. *J Biol Chem.* 2012;287(46):38427–38432. https://doi.org/10.1074/jbc.C112.408864.
91. Deng D, Yin P, Yan C, et al. Recognition of methylated DNA by TAL effectors. *Cell Res.* 2012;22(10):1502–1504. https://doi.org/10.1038/cr.2012.127.

92. Mussolino C, Morbitzer R, Lütge F, Dannemann N, Lahaye T, Cathomen T. A novel TALE nuclease scaffold enables high genome editing activity in combination with low toxicity. *Nucleic Acids Res.* 2011;39(21):9283–9293. https://doi.org/10.1093/nar/gkr597.
93. McGinn J, Marraffini LA. Molecular mechanisms of CRISPR-Cas spacer acquisition. *Nat Rev Microbiol.* 2019;17(1):7–12. https://doi.org/10.1038/s41579-018-0071-7.
94. Faure G, Makarova KS, Koonin EV. CRISPR-Cas: complex functional networks and multiple roles beyond adaptive immunity. *J Mol Biol.* 2019;431(1):3–20. https://doi.org/10.1016/j.jmb.2018.08.030.
95. Koonin EV, Makarova KS. Origins and evolution of CRISPR-Cas systems. *Philos Trans R Soc Lond Ser B Biol Sci.* 2019;374(1772): 20180087. https://doi.org/10.1098/rstb.2018.0087.
96. Jiang F, Doudna JA. CRISPR-Cas9 structures and mechanisms. *Annu Rev Biophys.* 2017;46:505–529. https://doi.org/10.1146/annurev-biophys-062215-010822.
97. Barrangou R, Horvath P. A decade of discovery: CRISPR functions and applications. *Nat Microbiol.* 2017;2:17092. https://doi.org/10.1038/nmicrobiol.2017.92.
98. Jackson SA, McKenzie RE, Fagerlund RD, Kieper SN, Fineran PC, Brouns SJJ. CRISPR-Cas: adapting to change. *Science.* 2017;356(6333):eaal5056. https://doi.org/10.1126/science.aal5056.
99. Mohanraju P, Makarova KS, Zetsche B, Zhang F, Koonin EV, van der Oost J. Diverse evolutionary roots and mechanistic variations of the CRISPR-Cas systems. *Science.* 2016;353(6299):aad5147. https://doi.org/10.1126/science.aad5147.
100. Klompe SE, Sternberg SH. Harnessing "a billion years of experimentation": the ongoing exploration and exploitation of CRISPR-Cas immune systems. *CRISPR J.* 2018;1(2):141–158. https://doi.org/10.1089/crispr.2018.0012.
101. Wright AV, Nuñez JK, Doudna JA. Biology and applications of CRISPR systems: harnessing nature's toolbox for genome engineering. *Cell.* 2016;164(1–2):29–44. https://doi.org/10.1016/j.cell.2015.12.035.
102. Hille F, Charpentier E. CRISPR-Cas: biology, mechanisms and relevance. *Philos Trans R Soc Lond Ser B Biol Sci.* 2016;371(1707). https://doi.org/10.1098/rstb.2015.0496.
103. Koonin EV, Makarova KS, Wolf YI. Evolutionary genomics of defense systems in archaea and bacteria. *Annu Rev Microbiol.* 2017;71:233–261. https://doi.org/10.1146/annurev-micro-090816-093830.
104. Makarova KS, Wolf YI, Koonin EV. Classification and nomenclature of CRISPR-Cas systems: where from here? *CRISPR J.* 2018;1(5):325–336. https://doi.org/10.1089/crispr.2018.0033.
105. Makarova KS, Haft DH, Barrangou R, et al. Evolution and classification of the CRISPR-Cas systems. *Nat Rev Microbiol.* 2011;9(6):467–477. https://doi.org/10.1038/nrmicro2577.
106. Koonin EV, Makarova KS. CRISPR-Cas: evolution of an RNA-based adaptive immunity system in prokaryotes. *RNA Biol.* 2013;10(5):679–686. https://doi.org/10.4161/rna.24022.
107. Makarova KS, Wolf YI, Koonin EV. Comparative genomics of defense systems in archaea and bacteria. *Nucleic Acids Res.* 2013;41(8):4360–4377. https://doi.org/10.1093/nar/gkt157.
108. Makarova KS, Wolf YI, Koonin EV. The basic building blocks and evolution of CRISPR-CAS systems. *Biochem Soc Trans.* 2013;41(6):1392–1400. https://doi.org/10.1042/BST20130038.
109. Makarova KS, Koonin EV. Annotation and classification of CRISPR-Cas systems. *Methods Mol Biol.* 2015;1311:47–75. https://doi.org/10.1007/978-1-4939-2687-9_4.
110. Makarova KS, Wolf YI, Alkhnbashi OS, et al. An updated evolutionary classification of CRISPR-Cas systems. *Nat Rev Microbiol.* 2015;13(11):722–736. https://doi.org/10.1038/nrmicro3569.

111. Koonin EV, Makarova KS, Zhang F. Diversity, classification and evolution of CRISPR-Cas systems. *Curr Opin Microbiol.* 2017;37:67–78. https://doi.org/10.1016/j.mib.2017.05.008.
112. Makarova KS, Wolf YI, Iranzo J, et al. Evolutionary classification of CRISPR-Cas systems: a burst of class 2 and derived variants. *Nat Rev Microbiol.* 2020;18(2):67–83. https://doi.org/10.1038/s41579-019-0299-x.
113. Kazlauskiene M, Kostiuk G, Venclovas Č, Tamulaitis G, Siksnys V. A cyclic oligonucleotide signaling pathway in type III CRISPR-Cas systems. *Science.* 2017;357(6351):605–609. https://doi.org/10.1126/science.aao0100.
114. Ishino Y, Shinagawa H, Makino K, Amemura M, Nakata A. Nucleotide sequence of the iap gene, responsible for alkaline phosphatase isozyme conversion in Escherichia coli, and identification of the gene product. *J Bacteriol.* 1987;169(12):5429–5433. https://doi.org/10.1128/jb.169.12.5429-5433.1987.
115. Stern MJ, Ames GF, Smith NH, Robinson EC, Higgins CF. Repetitive extragenic palindromic sequences: a major component of the bacterial genome. *Cell.* 1984;37(3):1015–1026. https://doi.org/10.1016/0092-8674(84)90436-7.
116. Newbury SF, Smith NH, Robinson EC, Hiles ID, Higgins CF. Stabilization of translationally active mRNA by prokaryotic REP sequences. *Cell.* 1987;48(2):297–310. https://doi.org/10.1016/0092-8674(87)90433-8.
117. Nakata A, Amemura M, Makino K. Unusual nucleotide arrangement with repeated sequences in the Escherichia coli K-12 chromosome. *J Bacteriol.* 1989;171(6):3553–3556. https://doi.org/10.1128/jb.171.6.3553-3556.1989.
118. Hermans PW, van Soolingen D, Bik EM, de Haas PE, Dale JW, van Embden JD. Insertion element IS987 from Mycobacterium bovis BCG is located in a hot-spot integration region for insertion elements in Mycobacterium tuberculosis complex strains. *Infect Immun.* 1991;59(8):2695–2705. https://doi.org/10.1128/IAI.59.8.2695-2705.1991.
119. Groenen PM, Bunschoten AE, van Soolingen D, van Embden JD. Nature of DNA polymorphism in the direct repeat cluster of Mycobacterium tuberculosis; application for strain differentiation by a novel typing method. *Mol Microbiol.* 1993;10(5):1057–1065. https://doi.org/10.1111/j.1365-2958.1993.tb00976.x.
120. Mojica FJ, Juez G, Rodríguez-Valera F. Transcription at different salinities of Haloferax mediterranei sequences adjacent to partially modified PstI sites. *Mol Microbiol.* 1993;9(3):613–621. https://doi.org/10.1111/j.1365-2958.1993.tb01721.x.
121. Jansen R, van Embden JDA, Gaastra W, Schouls LM. Identification of genes that are associated with DNA repeats in prokaryotes. *Mol Microbiol.* 2002;43(6):1565–1575. https://doi.org/10.1046/j.1365-2958.2002.02839.x.
122. Mojica FJM, Díez-Villaseñor C, García-Martínez J, Soria E. Intervening sequences of regularly spaced prokaryotic repeats derive from foreign genetic elements. *J Mol Evol.* 2005;60(2):174–182. https://doi.org/10.1007/s00239-004-0046-3.
123. Pourcel C, Salvignol G, Vergnaud G. CRISPR elements in Yersinia pestis acquire new repeats by preferential uptake of bacteriophage DNA, and provide additional tools for evolutionary studies. *Microbiology.* 2005;151(Pt. 3):653–663. https://doi.org/10.1099/mic.0.27437-0.
124. Bolotin A, Quinquis B, Sorokin A, Ehrlich SD. Clustered regularly interspaced short palindrome repeats (CRISPRs) have spacers of extrachromosomal origin. *Microbiology.* 2005;151(Pt. 8):2551–2561. https://doi.org/10.1099/mic.0.28048-0.
125. Barrangou R, Fremaux C, Deveau H, et al. CRISPR provides acquired resistance against viruses in prokaryotes. *Science.* 2007;315(5819):1709–1712. https://doi.org/10.1126/science.1138140.
126. Brouns SJJ, Jore MM, Lundgren M, et al. Small CRISPR RNAs guide antiviral defense in prokaryotes. *Science.* 2008;321(5891):960–964. https://doi.org/10.1126/science.1159689.

127. Marraffini LA, Sontheimer EJ. CRISPR interference limits horizontal gene transfer in staphylococci by targeting DNA. *Science*. 2008;322(5909):1843–1845. https://doi.org/10.1126/science.1165771.
128. Garneau JE, Dupuis M-È, Villion M, et al. The CRISPR/Cas bacterial immune system cleaves bacteriophage and plasmid DNA. *Nature*. 2010;468(7320):67–71. https://doi.org/10.1038/nature09523.
129. Deltcheva E, Chylinski K, Sharma CM, et al. CRISPR RNA maturation by trans-encoded small RNA and host factor RNase III. *Nature*. 2011;471(7340):602–607. https://doi.org/10.1038/nature09886.
130. Jinek M, Chylinski K, Fonfara I, Hauer M, Doudna JA, Charpentier E. A programmable dual-RNA-guided DNA endonuclease in adaptive bacterial immunity. *Science*. 2012;337(6096):816–821. https://doi.org/10.1126/science.1225829.
131. Sapranauskas R, Gasiunas G, Fremaux C, Barrangou R, Horvath P, Siksnys V. The Streptococcus thermophilus CRISPR/Cas system provides immunity in Escherichia coli. *Nucleic Acids Res*. 2011;39(21):9275–9282. https://doi.org/10.1093/nar/gkr606.
132. Gasiunas G, Barrangou R, Horvath P, Siksnys V. Cas9-crRNA ribonucleoprotein complex mediates specific DNA cleavage for adaptive immunity in bacteria. *Proc Natl Acad Sci U S A*. 2012;109(39):E2579–E2586. https://doi.org/10.1073/pnas.1208507109.
133. Cong L, Ran FA, Cox D, et al. Multiplex genome engineering using CRISPR/Cas systems. *Science*. 2013;339(6121):819–823.
134. Mali P, Yang L, Esvelt KM, et al. RNA-guided human genome engineering via Cas9. *Science*. 2013;339(6121):823–826.
135. Daniloski Z, Jordan TX, Wessels H-H, et al. Identification of required host factors for SARS-CoV-2 infection in human cells. *Cell*. 2021;184(1):92–105. https://doi.org/10.1016/j.cell.2020.10.030.
136. Stewart MP, Sharei A, Ding X, Sahay G, Langer R, Jensen KF. In vitro and ex vivo strategies for intracellular delivery. *Nature*. 2016;538(7624):183–192.
137. Belting M, Sandgren S, Wittrup A. Nuclear delivery of macromolecules: barriers and carriers. *Adv Drug Deliv Rev*. 2005;57(4):505–527.
138. Schmidt F, Grimm D. CRISPR genome engineering and viral gene delivery: a case of mutual attraction. *Biotechnol J*. 2015;10(2):258–272.
139. Jager L, Ehrhardt A. Persistence of high-capacity adenoviral vectors as replication-defective monomeric genomes in vitro and in murine liver. *Hum Gene Ther*. 2009;20(8):883–896.
140. Kozarsky KF, Wilson JM. Gene therapy: adenovirus vectors. *Curr Opin Genet Dev*. 1993;3(3):499–503.
141. Wang AY, Peng PD, Ehrhardt A, Storm TA, Kay MA. Comparison of adenoviral and adeno-associated viral vectors for pancreatic gene delivery in vivo. *Hum Gene Ther*. 2004;15(4):405–413.
142. Ehrke-Schulz E, Schiwon M, Leitner T, et al. CRISPR/Cas9 delivery with one single adenoviral vector devoid of all viral genes. *Sci Rep*. 2017;7(1):1–11.
143. Blasco RB, Karaca E, Ambrogio C, et al. Simple and rapid in vivo generation of chromosomal rearrangements using CRISPR/Cas9 technology. *Cell Rep*. 2014;9(4):1219–1227.
144. Popescu NC, Zimonjic D, DiPaolo JA. Viral integration, fragile sites, and proto-oncogenes in human neoplasia. *Hum Genet*. 1990;84(5):383–386.
145. Shaw A, Cornetta K. Design and potential of non-integrating lentiviral vectors. *Biomedicines*. 2014;2(1):14–35.
146. Linden RM, Ward P, Giraud C, Winocour E, Berns KI. Site-specific integration by adeno-associated virus. *Proc Natl Acad Sci U S A*. 1996;93(21):11288–11294.
147. Luther DC, Lee YW, Nagaraj H, Scaletti F, Rotello VM. Delivery approaches for CRISPR/Cas9 therapeutics in vivo: advances and challenges. *Expert Opin Drug Deliv*. 2018;15(9):905–913.

148. Ran FA, Cong L, Yan WX, et al. In vivo genome editing using Staphylococcus aureus Cas9. *Nature*. 2015;520(7546):186–191.
149. Rehman ZU, Hoekstra D, Zuhorn IS. Mechanism of polyplex-and lipoplex-mediated delivery of nucleic acids: real-time visualization of transient membrane destabilization without endosomal lysis. *ACS Nano*. 2013;7(5):3767–3777.
150. Xue W, Chen S, Yin H, et al. CRISPR-mediated direct mutation of cancer genes in the mouse liver. *Nature*. 2014;514(7522):380–384.
151. Yin H, Xue W, Chen S, et al. Genome editing with Cas9 in adult mice corrects a disease mutation and phenotype. *Nat Biotechnol*. 2014;32(6):551–553.
152. Li L, Song L, Liu X, et al. Artificial virus delivers CRISPR-Cas9 system for genome editing of cells in mice. *ACS Nano*. 2017;11(1):95–111.
153. Luo Y-L, Xu C-F, Li H-J, et al. Macrophage-specific in vivo gene editing using cationic lipid-assisted polymeric nanoparticles. *ACS Nano*. 2018;12(2):994–1005.
154. Nelles DA, Fang MY, O'Connell MR, et al. Programmable RNA tracking in live cells with CRISPR/Cas9. *Cell*. 2016;165(2):488–496.
155. Zetsche B, Volz SE, Zhang F. A split-Cas9 architecture for inducible genome editing and transcription modulation. *Nat Biotechnol*. 2015;33(2):139–142.
156. Yin H, Song C-Q, Dorkin JR, et al. Therapeutic genome editing by combined viral and non-viral delivery of CRISPR system components in vivo. *Nat Biotechnol*. 2016;34 (3):328–333.
157. Yin H, Song C-Q, Suresh S, et al. Structure-guided chemical modification of guide RNA enables potent non-viral in vivo genome editing. *Nat Biotechnol*. 2017;35 (12):1179–1187.
158. Miller JB, Zhang S, Kos P, et al. Non-viral CRISPR/Cas gene editing in vitro and in vivo enabled by synthetic nanoparticle co-delivery of Cas9 mRNA and sgRNA. *Angew Chem Int Ed*. 2017;56(4):1059–1063.
159. Zuris JA, Thompson DB, Shu Y, et al. Cationic lipid-mediated delivery of proteins enables efficient protein-based genome editing in vitro and in vivo. *Nat Biotechnol*. 2015;33(1):73–80.
160. Sun W, Ji W, Hall JM, et al. Efficient delivery of CRISPR-Cas9 for genome editing via self-assembled DNA nanoclews. *Angew Chem Int Ed Eng*. 2015;54(41):12029.
161. Mout R, Ray M, Yesilbag Tonga G, et al. Direct cytosolic delivery of CRISPR/Cas9-ribonucleoprotein for efficient gene editing. *ACS Nano*. 2017;11(3):2452–2458.
162. Lee K, Conboy M, Park HM, et al. Nanoparticle delivery of Cas9 ribonucleoprotein and donor DNA in vivo induces homology-directed DNA repair. *Nat Biomed Eng*. 2017;1(11):889–901.
163. Shirasawa S, Furuse M, Yokoyama N, Sasazuki T. Altered growth of human colon cancer cell lines disrupted at activated Ki-ras. *Science*. 1993;260(5104):85–88.
164. Sur S, Pagliarini R, Bunz F, et al. A panel of isogenic human cancer cells suggests a therapeutic approach for cancers with inactivated p53. *Proc Natl Acad Sci U S A*. 2009;106(10):3964–3969.
165. Schumann K, Lin S, Boyer E, et al. Generation of knock-in primary human T cells using Cas9 ribonucleoproteins. *Proc Natl Acad Sci*. 2015;112(33):10437–10442.
166. Grobarczyk B, Franco B, Hanon K, Malgrange B. Generation of isogenic human iPS cell line precisely corrected by genome editing using the CRISPR/Cas9 system. *Stem Cell Rev Rep*. 2015;11(5):774–787.
167. Kasap C, Elemento O, Kapoor TM. DrugTargetSeqR: a genomics-and CRISPR-Cas9–based method to analyze drug targets. *Nat Chem Biol*. 2014;10(8):626–628.
168. Paix A, Wang Y, Smith HE, et al. Scalable and versatile genome editing using linear DNAs with microhomology to Cas9 Sites in Caenorhabditis elegans. *Genetics*. 2014;198(4):1347–1356.

169. Anzalone AV, Randolph PB, Davis JR, et al. Search-and-replace genome editing without double-strand breaks or donor DNA. *Nature*. 2019;576(7785):149–157.
170. Cook D, Brown D, Alexander R, et al. Lessons learned from the fate of AstraZeneca's drug pipeline: a five-dimensional framework. *Nat Rev Drug Discov*. 2014;13(6): 419–431.
171. Nelson MR, Tipney H, Painter JL, et al. The support of human genetic evidence for approved drug indications. *Nat Genet*. 2015;47(8):856–860.
172. Miles LA, Garippa RJ, Poirier JT. Design, execution, and analysis of pooled *in vitro* CRISPR/Cas9 screens. *FEBS J*. 2016;283(17):3170–3180. https://doi.org/10.1111/febs.13770.
173. Echeverri CJ, Beachy PA, Baum B, et al. Minimizing the risk of reporting false positives in large-scale RNAi screens. *Nat Methods*. 2006;3(10):777–779.
174. Wang T, Wei JJ, Sabatini DM, Lander ES. Genetic screens in human cells using the CRISPR-Cas9 system. *Science*. 2014;343(6166):80–84.
175. Gilbert LA, Larson MH, Morsut L, et al. CRISPR-mediated modular RNA-guided regulation of transcription in eukaryotes. *Cell*. 2013;154(2):442–451.
176. Klann TS, Black JB, Chellappan M, et al. CRISPR–Cas9 epigenome editing enables high-throughput screening for functional regulatory elements in the human genome. *Nat Biotechnol*. 2017;35(6):561.
177. Koller BH, Hagemann LJ, Doetschman T, et al. Germ-line transmission of a planned alteration made in a hypoxanthine phosphoribosyltransferase gene by homologous recombination in embryonic stem cells. *Proc Natl Acad Sci U S A*. 1989;86(22): 8927–8931.
178. Capecchi MR. Gene targeting in mice: functional analysis of the mammalian genome for the twenty-first century. *Nat Rev Genet*. 2005;6(6):507–512.
179. Rouet P, Smih F, Jasin M. Expression of a site-specific endonuclease stimulates homologous recombination in mammalian cells. *Proc Natl Acad Sci U S A*. 1994;91 (13):6064–6068.
180. Ma X, Wong AS-Y, Tam H-Y, Tsui SY-K, Chung DL-S, Feng B. In vivo genome editing thrives with diversified CRISPR technologies. *Zool Res*. 2018;39(2):58.
181. Wang H, Yang H, Shivalila CS, et al. One-step generation of mice carrying mutations in multiple genes by CRISPR/Cas-mediated genome engineering. *Cell*. 2013;153 (4):910–918.
182. Huijbers IJ, Del Bravo J, Ali RB, et al. Using the GEMM-ESC strategy to study gene function in mouse models. *Nat Protoc*. 2015;10(11):1755.
183. Yang H, Wang H, Shivalila CS, Cheng AW, Shi L, Jaenisch R. One-step generation of mice carrying reporter and conditional alleles by CRISPR/Cas-mediated genome engineering. *Cell*. 2013,154(6):1370–1379.
184. Dow LE, Fisher J, O'rourke KP, et al. Inducible in vivo genome editing with CRISPR-Cas9. *Nat Biotechnol*. 2015;33(4):390–394.
185. Boroviak K, Doe B, Banerjee R, Yang F, Bradley A. Chromosome engineering in zygotes with CRISPR/C as9. *Genesis*. 2016;54(2):78–85.
186. Weber J, Öllinger R, Friedrich M, et al. CRISPR/Cas9 somatic multiplex-mutagenesis for high-throughput functional cancer genomics in mice. *Proc Natl Acad Sci U S A*. 2015;112(45):13982–13987.
187. Maresch R, Mueller S, Veltkamp C, et al. Multiplexed pancreatic genome engineering and cancer induction by transfection-based CRISPR/Cas9 delivery in mice. *Nat Commun*. 2016;7(1):1–13.
188. Zuckermann M, Hovestadt V, Knobbe-Thomsen CB, et al. Somatic CRISPR/Cas9-mediated tumour suppressor disruption enables versatile brain tumour modelling. *Nat Commun*. 2015;6(1):1–9.

189. Platt RJ, Chen S, Zhou Y, et al. CRISPR-Cas9 knockin mice for genome editing and cancer modeling. *Cell.* 2014;159(2):440–455.
190. Sun T, Patil R, Galstyan A, et al. Blockade of a laminin-411–notch axis with CRISPR/Cas9 or a nanobioconjugate inhibits glioblastoma growth through tumor-microenvironment cross-talk. *Cancer Res.* 2019;79(6):1239–1251.
191. Temin HM. Mixed infection with two types of Rous sarcoma virus. *Virology.* 1961;13:158–163. https://doi.org/10.1016/0042-6822(61)90049-6.
192. Sambrook J, Westphal H, Srinivasan PR, Dulbecco R. The integrated state of viral DNA in SV40-transformed cells. *Proc Natl Acad Sci U S A.* 1968;60(4):1288–1295. https://doi.org/10.1073/pnas.60.4.1288.
193. Tatum EL. Molecular biology, nucleic acids, and the future of medicine. *Perspect Biol Med.* 1966;10(1):19–32. https://doi.org/10.1353/pbm.1966.0027.
194. Rogers S, Pfuderer P. Use of viruses as carriers of added genetic information. *Nature.* 1968;219(5155):749–751. https://doi.org/10.1038/219749a0.
195. Rogers S, Lowenthal A, Terheggen HG, Columbo JP. Induction of arginase activity with the Shope papilloma virus in tissue culture cells from an argininemic patient. *J Exp Med.* 1973;137(4):1091–1096. https://doi.org/10.1084/jem.137.4.1091.
196. Terheggen HG, Lowenthal A, Lavinha F, Colombo JP, Rogers S. Unsuccessful trial of gene replacement in arginase deficiency. *Eur J Pediatr.* 1975;119(1):1–3. https://doi.org/10.1007/BF00464689.
197. Mercola KE, Stang HD, Browne J, Salser W, Cline MJ. Insertion of a new gene of viral origin into bone marrow cells of mice. *Science.* 1980;208(4447):1033–1035. https://doi.org/10.1126/science.6246577.
198. Beutler E. The cline affair. *Mol Ther.* 2001;4(5):396–397. https://doi.org/10.1006/mthe.2001.0486.
199. Rosenberg SA, Aebersold P, Cornetta K, et al. Gene transfer into humans—immunotherapy of patients with advanced melanoma, using tumor-infiltrating lymphocytes modified by retroviral gene transduction. *N Engl J Med.* 1990;323(9):570–578. https://doi.org/10.1056/NEJM199008303230904.
200. Rosenberg SA. Gene therapy for cancer. *JAMA, J Am Med Assoc.* 1992;268(17):2416. https://doi.org/10.1001/jama.1992.03490170088031.
201. Rosenberg SA, Anderson WF, Blaese M, et al. The development of gene therapy for the treatment of cancer. *Ann Surg.* 1993;218(4):455–463. discussion 463-464 https://doi.org/10.1097/00000658-199310000-00006.
202. Blaese RM, Culver KW, Miller AD, et al. T lymphocyte-directed gene therapy for ADA- SCID: initial trial results after 4 years. *Science.* 1995;270(5235):475–480. https://doi.org/10.1126/science.270.5235.475.
203. Stolberg SG. The biotech death of Jesse Gelsinger. *N Y Times Mag.* 1999;136–140:149–150. Published online November 28.
204. Gao C. The future of CRISPR technologies in agriculture. *Nat Rev Mol Cell Biol.* 2018;19(5):275–276. https://doi.org/10.1038/nrm.2018.2.
205. Yin K, Gao C, Qiu J-L. Progress and prospects in plant genome editing. *Nat Plants.* 2017;3:17107. https://doi.org/10.1038/nplants.2017.107.
206. Wang C, Wang G, Gao Y, et al. A cytokinin-activation enzyme-like gene improves grain yield under various field conditions in rice. *Plant Mol Biol.* 2020;102(4–5):373–388. https://doi.org/10.1007/s11103-019-00952-5.
207. Zhang Z, Hua L, Gupta A, et al. Development of an Agrobacterium-delivered CRISPR/Cas9 system for wheat genome editing. *Plant Biotechnol J.* 2019;17(8):1623–1635. https://doi.org/10.1111/pbi.13088.
208. Liu J, Chen J, Zheng X, et al. GW5 acts in the brassinosteroid signalling pathway to regulate grain width and weight in rice. *Nat Plants.* 2017;3:17043. https://doi.org/10.1038/nplants.2017.43.

209. Zeng Y, Wen J, Zhao W, Wang Q, Huang W. Rational improvement of rice yield and cold tolerance by editing the three genes OsPIN5b, GS3, and OsMYB30 with the CRISPR-Cas9 system. *Front Plant Sci.* 2019;10:1663. https://doi.org/10.3389/fpls.2019.01663.
210. Zhang Y, Li D, Zhang D, et al. Analysis of the functions of TaGW2 homoeologs in wheat grain weight and protein content traits. *Plant J Cell Mol Biol.* 2018;94(5):857–866. https://doi.org/10.1111/tpj.13903.
211. Zhou J, Xin X, He Y, et al. Multiplex QTL editing of grain-related genes improves yield in elite rice varieties. *Plant Cell Rep.* 2019;38(4):475–485. https://doi.org/10.1007/s00299-018-2340-3.
212. Yuste-Lisbona FJ, Fernández-Lozano A, Pineda B, et al. ENO regulates tomato fruit size through the floral meristem development network. *Proc Natl Acad Sci U S A.* 2020;117(14):8187–8195. https://doi.org/10.1073/pnas.1913688117.
213. Rodríguez-Leal D, Lemmon ZH, Man J, Bartlett ME, Lippman ZB. Engineering quantitative trait variation for crop improvement by genome editing. *Cell.* 2017;171(2):470–480.e8. https://doi.org/10.1016/j.cell.2017.08.030.
214. Gao H, Gadlage MJ, Lafitte HR, et al. Superior field performance of waxy corn engineered using CRISPR-Cas9. *Nat Biotechnol.* 2020;38(5):579–581. https://doi.org/10.1038/s41587-020-0444-0.
215. Xu Y, Lin Q, Li X, et al. Fine-tuning the amylose content of rice by precise base editing of the Wx gene. *Plant Biotechnol J.* 2021;19(1):11–13. https://doi.org/10.1111/pbi.13433.
216. Sun Y, Jiao G, Liu Z, et al. Generation of high-amylose rice through CRISPR/Cas9-mediated targeted mutagenesis of starch branching enzymes. *Front Plant Sci.* 2017;8:298. https://doi.org/10.3389/fpls.2017.00298.
217. Sánchez-León S, Gil-Humanes J, Ozuna CV, et al. Low-gluten, nontransgenic wheat engineered with CRISPR/Cas9. *Plant Biotechnol J.* 2018;16(4):902–910. https://doi.org/10.1111/pbi.12837.
218. Khan MSS, Basnet R, Islam SA, Shu Q. Mutational analysis of OsPLDα1 reveals its involvement in phytic acid biosynthesis in rice grains. *J Agric Food Chem.* 2019;67(41):11436–11443. https://doi.org/10.1021/acs.jafc.9b05052.
219. Dong OX, Yu S, Jain R, et al. Marker-free carotenoid-enriched rice generated through targeted gene insertion using CRISPR-Cas9. *Nat Commun.* 2020;11(1):1178. https://doi.org/10.1038/s41467-020-14981-y.
220. Oliva R, Ji C, Atienza-Grande G, et al. Broad-spectrum resistance to bacterial blight in rice using genome editing. *Nat Biotechnol.* 2019;37(11):1344–1350. https://doi.org/10.1038/s41587-019-0267-z.
221. Peng A, Chen S, Lei T, et al. Engineering canker-resistant plants through CRISPR/Cas9-targeted editing of the susceptibility gene CsLOB1 promoter in citrus. *Plant Biotechnol J.* 2017;15(12):1509–1519. https://doi.org/10.1111/pbi.12733.
222. Zhang Y, Bai Y, Wu G, et al. Simultaneous modification of three homoeologs of TaEDR1 by genome editing enhances powdery mildew resistance in wheat. *Plant J Cell Mol Biol.* 2017;91(4):714–724. https://doi.org/10.1111/tpj.13599.
223. Wang Y, Cheng X, Shan Q, et al. Simultaneous editing of three homoeoalleles in hexaploid bread wheat confers heritable resistance to powdery mildew. *Nat Biotechnol.* 2014;32(9):947–951. https://doi.org/10.1038/nbt.2969.
224. Nekrasov V, Wang C, Win J, Lanz C, Weigel D, Kamoun S. Rapid generation of a transgene-free powdery mildew resistant tomato by genome deletion. *Sci Rep.* 2017;7(1):482. https://doi.org/10.1038/s41598-017-00578-x.
225. Liu H, Soyars CL, Li J, et al. CRISPR/Cas9-mediated resistance to cauliflower mosaic virus. *Plant Direct.* 2018;2(3):e00047. https://doi.org/10.1002/pld3.47.
226. Ji X, Si X, Zhang Y, Zhang H, Zhang F, Gao C. Conferring DNA virus resistance with high specificity in plants using virus-inducible genome-editing system. *Genome Biol.* 2018;19(1):197. https://doi.org/10.1186/s13059-018-1580-4.

227. Chandrasekaran J, Brumin M, Wolf D, et al. Development of broad virus resistance in non-transgenic cucumber using CRISPR/Cas9 technology. *Mol Plant Pathol.* 2016;17(7):1140–1153. https://doi.org/10.1111/mpp.12375.
228. Powles SB, Yu Q. Evolution in action: plants resistant to herbicides. *Annu Rev Plant Biol.* 2010;61:317–347. https://doi.org/10.1146/annurev-arplant-042809-112119.
229. Li C, Zong Y, Wang Y, et al. Expanded base editing in rice and wheat using a Cas9-adenosine deaminase fusion. *Genome Biol.* 2018;19(1):59. https://doi.org/10.1186/s13059-018-1443-z.
230. Zhu H, Li C, Gao C. Applications of CRISPR-Cas in agriculture and plant biotechnology. *Nat Rev Mol Cell Biol.* 2020;21(11):661–677. https://doi.org/10.1038/s41580-020-00288-9.
231. Lin S-R, Yang H-C, Kuo Y-T, et al. The CRISPR/Cas9 system facilitates clearance of the intrahepatic HBV templates in vivo. *Mol Ther–Nucleic Acids.* 2014;3, e186.
232. Dong C, Qu L, Wang H, Wei L, Dong Y, Xiong S. Targeting hepatitis B virus cccDNA by CRISPR/Cas9 nuclease efficiently inhibits viral replication. *Antivir Res.* 2015;118:110–117.
233. Zhen S, Hua L, Takahashi Y, Narita S, Liu Y-H, Li Y. In vitro and in vivo growth suppression of human papillomavirus 16-positive cervical cancer cells by CRISPR/Cas9. *Biochem Biophys Res Commun.* 2014;450(4):1422–1426.
234. Kennedy EM, Kornepati AVR, Goldstein M, et al. Inactivation of the human papillomavirus E6 or E7 gene in cervical carcinoma cells by using a bacterial CRISPR/Cas RNA-guided endonuclease. *J Virol.* 2014;88(20):11965–11972. https://doi.org/10.1128/JVI.01879-14.
235. Jubair L, Fallaha S, McMillan NA. Systemic delivery of CRISPR/Cas9 targeting HPV oncogenes is effective at eliminating established tumors. *Mol Ther.* 2019;27(12):2091–2099.
236. Yoshiba T, Saga Y, Urabe M, et al. CRISPR/Cas9-mediated cervical cancer treatment targeting human papillomavirus E6. *Oncol Lett.* 2019;17(2):2197–2206. https://doi.org/10.3892/ol.2018.9815.
237. Farrell PJ. Epstein–Barr virus and cancer. *Annu Rev Pathol Mech Dis.* 2019;14:29–53. https://doi.org/10.1146/annurev-pathmechdis-012418-013023.
238. Yuen K-S, Chan C-P, Kok K-H, Jin D-Y. Mutagenesis and genome engineering of Epstein–Barr virus in cultured human cells by CRISPR/Cas9. In: Reeves A, ed. *In Vitro Mutagenesis.* Springer New York; 2017:23–31. Methods in Molecular Biology; vol. 1498. https://doi.org/10.1007/978-1-4939-6472-7_2.
239. Yuen K-S, Chan C-P, Wong N-HM, et al. CRISPR/Cas9-mediated genome editing of Epstein–Barr virus in human cells. *J Gen Virol.* 2015;96(3):626–636. https://doi.org/10.1099/jgv.0.000012.
240. Maus MV, Grupp SA, Porter DL, June CH. Antibody-modified T cells: CARs take the front seat for hematologic malignancies. *Blood.* 2014;123(17):2625–2635.
241. Bannister AJ, Kouzarides T. Regulation of chromatin by histone modifications. *Cell Res.* 2011;21(3):381–395.
242. Cedar H, Bergman Y. Linking DNA methylation and histone modification: patterns and paradigms. *Nat Rev Genet.* 2009;10(5):295–304.
243. Dominguez AA, Lim WA, Qi LS. Beyond editing: repurposing CRISPR–Cas9 for precision genome regulation and interrogation. *Nat Rev Mol Cell Biol.* 2016;17(1):5.
244. Qi LS, Larson MH, Gilbert LA, et al. Repurposing CRISPR as an RNA-guided platform for sequence-specific control of gene expression. *Cell.* 2013;152(5):1173–1183.
245. Kearns NA, Pham H, Tabak B, et al. Functional annotation of native enhancers with a Cas9–histone demethylase fusion. *Nat Methods.* 2015;12(5):401–403.
246. Thakore PI, D'ippolito AM, Song L, et al. Highly specific epigenome editing by CRISPR-Cas9 repressors for silencing of distal regulatory elements. *Nat Methods.* 2015;12(12):1143–1149.

247. Hilton IB, D'ippolito AM, Vockley CM, et al. Epigenome editing by a CRISPR-Cas9-based acetyltransferase activates genes from promoters and enhancers. *Nat Biotechnol.* 2015;33(5):510–517.
248. McDonald JI, Celik H, Rois LE, et al. Reprogrammable CRISPR/Cas9-based system for inducing site-specific DNA methylation. *Biol Open.* 2016;5(6):866–874.
249. Vojta A, Dobrinić P, Tadić V, et al. Repurposing the CRISPR-Cas9 system for targeted DNA methylation. *Nucleic Acids Res.* 2016;44(12):5615–5628.
250. Amabile A, Migliara A, Capasso P, et al. Inheritable silencing of endogenous genes by hit-and-run targeted epigenetic editing. *Cell.* 2016;167(1):219–232.
251. Frock RL, Hu J, Meyers RM, Ho Y-J, Kii E, Alt FW. Genome-wide detection of DNA double-stranded breaks induced by engineered nucleases. *Nat Biotechnol.* 2015;33(2):179–186.
252. Hsu PD, Scott DA, Weinstein JA, et al. DNA targeting specificity of RNA-guided Cas9 nucleases. *Nat Biotechnol.* 2013;31(9):827–832.
253. Lin Y, Cradick TJ, Brown MT, et al. CRISPR/Cas9 systems have off-target activity with insertions or deletions between target DNA and guide RNA sequences. *Nucleic Acids Res.* 2014;42(11):7473–7485.
254. Kim D, Bae S, Park J, et al. Digenome-seq: genome-wide profiling of CRISPR-Cas9 off-target effects in human cells. *Nat Methods.* 2015;12(3):237–243.
255. Cameron P, Fuller CK, Donohoue PD, et al. Mapping the genomic landscape of CRISPR–Cas9 cleavage. *Nat Methods.* 2017;14(6):600–606.
256. Lazzarotto CR, Nguyen NT, Tang X, et al. Defining CRISPR–Cas9 genome-wide nuclease activities with CIRCLE-seq. *Nat Protoc.* 2018;13(11):2615–2642.
257. Tsai SQ, Zheng Z, Nguyen NT, et al. GUIDE-seq enables genome-wide profiling of off-target cleavage by CRISPR-Cas nucleases. *Nat Biotechnol.* 2015;33(2):187–197.
258. Fu Y, Sander JD, Reyon D, Cascio VM, Joung JK. Improving CRISPR-Cas nuclease specificity using truncated guide RNAs. *Nat Biotechnol.* 2014;32(3):279–284.
259. Cho SW, Kim S, Kim Y, et al. Analysis of off-target effects of CRISPR/Cas-derived RNA-guided endonucleases and nickases. *Genome Res.* 2014;24(1):132–141.
260. Chen M, Mao A, Xu M, Weng Q, Mao J, Ji J. CRISPR-Cas9 for cancer therapy: opportunities and challenges. *Cancer Lett.* 2019;447:48–55.
261. Soppe JA, Lebbink RJ. Antiviral goes viral: harnessing CRISPR/Cas9 to combat viruses in humans. *Trends Microbiol.* 2017;25(10):833–850. https://doi.org/10.1016/j.tim.2017.04.005.
262. Kim S, Kim D, Cho SW, Kim J, Kim J-S. Highly efficient RNA-guided genome editing in human cells via delivery of purified Cas9 ribonucleoproteins. *Genome Res.* 2014,24(6):1012–1019.
263. Gouw AM. The CRISPR apple on the tree of knowledge conference highlights: CRISPR in science, ethics, and religion. *Zygon.* 2020;55(2):444–458.
264. Peters T. "Playing god" and germline intervention. *J Med Philos.* 1995;20(4):365–386. https://doi.org/10.1093/jmp/20.4.365.
265. Genetic Screening Centers (PGD). *The Fertility Institutes Offers Preimplantation Genetic Diagnosis (PGD) of Embryos to Screen for Hereditary Diseases*; 2020. Accessed December 23, 2020 https://www.fertility-docs.com/programs-and-services/pgd-screening/genetic-diagnosis-of-embryos-for-diseases-using-pgd.php.
266. Vuković P, Peccatori FA, Massarotti C, Miralles MS, Beketić-Orešković L, Lambertini M. Preimplantation genetic testing for carriers ofBRCA1/2 pathogenic variants. *Crit Rev Oncol Hematol.* 2021;157(103201). https://doi.org/10.1016/j.critrevonc.2020.103201.
267. Catteruccia F, Nolan T, Loukeris TG, et al. Stable germline transformation of the malaria mosquito *Anopheles stephensi. Nature.* 2000;405(6789):959–962. https://doi.org/10.1038/35016096.
268. Holt RA, Subramanian GM, Halpern A, et al. The genome sequence of the malaria mosquito *Anopheles gambiae. Science.* 2002;298(5591):129–149.

269. Curtis CF. Possible use of translocations to fix desirable genes in insect pest populations. *Nature*. 1968;218(5139):368–369. https://doi.org/10.1038/218368a0.
270. Serebrovsky AS. On the possibility of a new method for the control of harmful insects. Sterile-male technique for eradication or control of harmful insects. In: *Proceedings of a Panel on Application of the Sterile-Male Technique for the Eradication or Control of Harmful Species of Insects*. Vienna: Joint FAO/IAEA Division of Atomic Energy in Food and Agriculture. 1968:123–237.
271. Hamilton WD. Extraordinary sex ratios. *Science*. 1967;156(3774):477–488. https://doi.org/10.1126/science.156.3774.477.
272. Hammond AM, Roberto G. Gene drives to fight malaria: current state and future directions. *Pathog Glob Health*. 2017;111(8):412–423. https://doi.org/10.1080/20477724.2018.1438880.
273. Nateghi Rostami M. CRISPR/Cas9 gene drive technology to control transmission of vector-borne parasitic infections. *Parasite Immunol*. 2020;42(9). https://doi.org/10.1111/pim.12762.

CHAPTER TWO

CRISPR-Cas9: A fascinating journey from bacterial immune system to human gene editing

Munshi Azad Hossain*

Department of Molecular, Cell, and Developmental Biology, University of California, Los Angeles, CA, United States

*Corresponding author: e-mail address: azadmh74@ucla.edu

Contents

Abstract

Clustered Regularly Interspaced Short Palindromic Repeat (CRISPR)-Cas system has been discovered as an adaptive-immune system in prokaryotes. Microbes like bacteria and archaea use CRISPR-Cas9 as a part of their defense mechanism to ward off the virus and cleave their DNA. Over the past decades, researchers have identified that this simple CRISPR-Cas9 system of bacteria can be utilized to cut any DNA. It is also possible to make precise editing in the genome of almost any organism. This discovery has revolutionized the CRISPR-Cas9 tools and made it one of the most precise gene editing technology known till date. The simple, versatile and programmable nature of CRISPR-Cas9 system that contains a single guide RNA and Cas9 enzyme, made it an attractive choice for genome editing application. Scientists in the field of molecular biology, genetics and medicine extensively use this transformative technology to study gene regulation and also for treatment of several incurable genetic diseases. Today, CRISPR-Cas9 is the most powerful breakthrough of the century for its immense potential to modulate gene expression in living cells and its application to medicine and human health. Recently, ethical challenges associated with the application of this technology to human health become a hot debate in the scientific community. In this chapter the brief history of development of CRISPR-Cas9 tools and its immense application potential have been discussed.

Progress in Molecular Biology and Translational Science, Volume 178
ISSN 1877-1173
https://doi.org/10.1016/bs.pmbts.2021.01.001

1. Introduction

Viruses and bacteria are the most abundant organisms on Earth.[1,2] Virus infects bacteria, hijack bacterial cellular machineries to assemble new viruses and often kill the bacteria. In response, bacteria have developed defensive mechanisms that target virus. One such already known evolved bacterial defense strategy is restriction endonuclease, which cleaves viral DNA. In counter defense, Viruses also undergo modification of their genome, which is resistant to restriction endonuclease. This constant warfare between virus and bacterial has important impact on the evolution of both virus and bacterial genome.[2–4]

Interestingly, CRISPR-Cas9 genome editing tool is also a bacterial defense strategy against viruses. It is an adaptive and inheritable immune system of bacteria. Using CRISPR-Cas system bacterial cells cleaves viral genomic DNA when attacked and incorporate snippets of viral DNA into its own genome at the CRISPR loci. When re-exposed to the same virus, the CRISPR loci gets transcribed into RNA that binds to the Cas9 nuclease enzyme, which, then cleaves the viral DNA thus acts as a memory of past invasion (Fig. 1). After decades of diligent research, scientists have been able to engineer this adaptive mechanism of bacteria to edit the genomes of higher eukaryotes including human beings, giving rise to an improved era of genome editing.[5–7]

Bacterial CRISPR-Cas system has become a versatile two components tool to edit genome of any higher eukaryotes.[5–8] CRISPR tool contains a programmable single guide RNA (sgRNA) and Cas9 enzyme. The sgRNA is a chimeric version of natural bacterial CrRNA and tracrRNA (Figs. 1 and 2). The sgRNA has two functional parts. The 5′ region (~20 bp) of the sgRNA does base pairing with the target DNA. The 3′ part forms a stem loop-structure which binds with the Cas9 enzyme.[9] The Cas9 enzyme is unique nuclease that contains two nuclease domains: a RuvC-like nuclease domain located at N-terminus and an HNH-like nuclease domain in the middle. The enzyme gets activated upon binding to sgRNA. Then sgRNA–Cas9 complex binds to its target DNA and Cas9 cleaves both strands of the DNA. The HNH domain cleaves the complementary strand where sgRNA makes base pairing, while the RuvC domain cuts the opposite non-complimentary strand.[9–11] The CRISPR mediated DNA cleavage is extremely precise, efficient and shows rare off-target. Several studies elucidating the mechanism of action of CRISPR-Cas9 system showed that the precision comes from the concerted action of the sgRNA guided

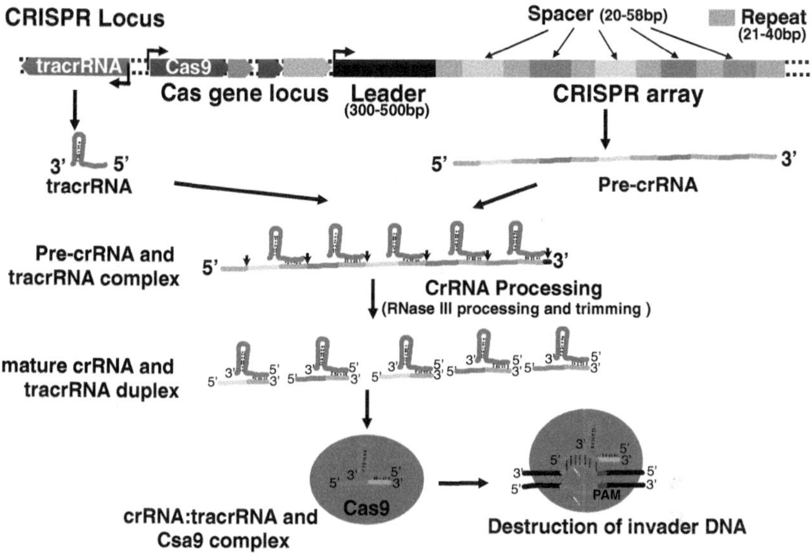

Fig. 1 Shows the DNA cleavage activity of the bacterial type II CRISPR-Cas system. CRISPR-Locus shows trans activating CRISPR RNA (tracrRNA) gene, CRISPR associated (Cas) genes locus, leader sequences and the CRISPR array comprises of repeat sequences (green) and spacers (variable colors). During CRISPR action the unique tracrRNA (orange) anneals to the repeat sequences of the long pre-crRNA and then processed by RNase III to form mature tracrRNA-crRNA duplex. It's then form complex with Cas9 enzyme and guide Cas9 enzyme to cleave invading viral DNA based on the complimentary between crRNA spacer and invader DNA preceding the protospacer adjacent motif (PAM) sequences.

activity of Cas9, Watson-crick base pairing of the sgRNA with target sequences, the presence of protospacer adjacent motif (PAM) sequences in the target DNA and above all precise structural modulation of Cas9 enzyme.[11–13]

It has been shown that DNA break in higher eukaryotes activates two endogenous cellular repairs pathways: non-homologous end-joining (NHEJ) and homology-directed repair (HDR). NHEJ repair pathway is error prone and introduces DNA bases of various lengths at the breakage site changing the DNA sequences. The addition or deletion ofDNA bases (INDENLS) leads to the permanent changes in DNA sequence that disrupts the translational reading frame of a coding sequence and subsequently impairs the expression of the target gene.[14,15] As CRISPR tool enables the scientist to precisely change the DNA sequences it can also be used to modify or mutate the binding sites of *trans*-acting factors in promoters or enhancers of a target DNA sequences. The HDR-mediated repair is more precise and can use exogenously supplied

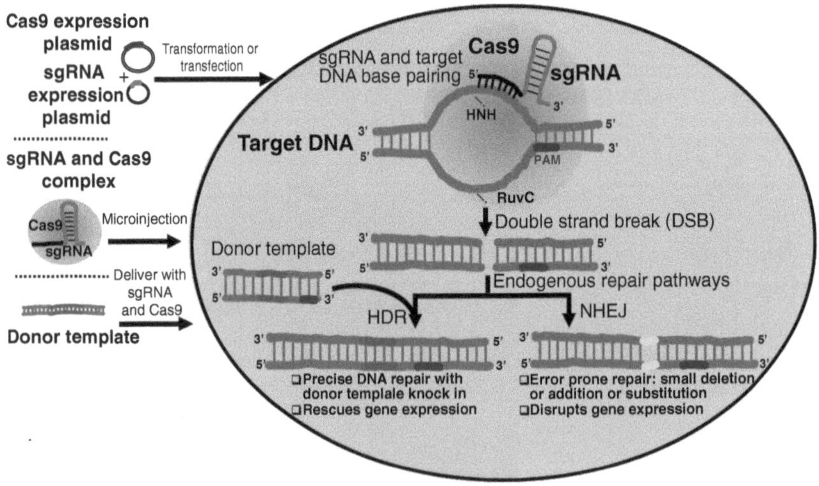

Fig. 2 Depicts the mechanism of CRISPR-Cas9 tool in higher eukaryotes genome editing. Two components: gRNA and Cas9 can be delivered into the cells by transformation or transfection of expression plasmids or by direct microinjection of sgRNA-Cas9 ribonucleoprotein complex. For Homology Directed Repair (HDR) Donor template flanking the region of the target DNA is required and can be delivered with sgRNA and Cas9. The sgRNA and Cas9 form complex inside the cells and interrogate the target DNA. The 5′ part of sgRNA (Black) base pairs with target DNA near PAM and Csa9 cleaves both strand of the DNA. First, HNH domain cuts the target strand (red scissor) and then RuvC domain cuts the non-target strand. DSB activates endogenous repair pathways. HDR incorporates desired donor template into the target strands and generate precise knock-in (green DNA sequences). The error-prone nonhomologous end joining (NHEJ) pathway leads to deletion or addition or substitution (INDELS) (yellow) at the breakage site and these changes in target DNA causes disruption of gene expression.

DNA "donor templates" to replace the DNA by homologous recombination at the breakage site.[16] So, for a given cell, using directed sgRNA, Cas9 enzyme and desired donor template, it is possible to introduce or repair a specific point mutation and to insert desired sequences at the target locus (Fig. 2).

Studies have shown that CRISPR-Cas9 system could be further improved by mutating the nuclease domain of Cas9 enzyme. A single mutation at D10A in RuvC domain impairs it nuclease activity.[17] This Cas9D10A enzyme can only cleave one strand of the DNA which specifically triggers more precise HDR repair pathway, not NHEJ. So, using mutant Cas9D10A and sgRNA, DNA repair can be performed through high-fidelity HDR pathway when provided with homologous repair template flanking the break site.[10,17]

A key property of Cas9 enzyme is its ability to bind DNA at sites defined by sgRNA base pairing and PAM sequences.[8] It has been shown that gRNA-Cas9 complex cleaves the DNA and attaches to the target site until replaced by cellular DNA repair machineries. Keeping these properties in mind, catalytically inactive version of Cas9 (dCas9) has been generated for regulating gene expression *in vivo*. The catalytically inactive Cas9 contains mutation at both the HNH and RuvC nuclease domain. These mutations abolish the nuclease activity of Cas9; however, it can bind to the target DNA guided by sgRNA.[18] Several studies have shown to use dCas9 to target transcription factors in perturbing *in vivo* gene expression, visualize genomic loci and many more applications.[19]

Within a short period of time from its discovery, the scientific community around the world has quickly adopted the CRISPR technology to delete, edit, or modify the genomic region of a vast array of cells and organisms.[8]CRISPR technology is easy to use and amenable than the existing genome editing tools: zinc finger nucleases (ZFN) based on eukaryotic transcription factors[20,21] or transcription activators-like effectors (TALEs) from *Xanthomonas* bacteria.[22–24] Both ZFNs and TALENs can be used to generate permanent mutation in the genome. These genome editing tools also specifically break DNA at the target site, activate cellular repair pathways and lead to alteration of DNA sequences at the target site.[25] However, ZFNs and TALENs approaches require substantial protein engineering for each DNA target site to be modified that is time consuming and costly. These methods are also not scalable for high throughput studies. In contrast, CRISPR-Cas9 system is easy to design and it requires only the construction of gRNA. Only by changing the target sequences in the sgRNA, CRISPR tool can be used to modify any DNA sequences in the genome.[8–10] The tools can be scalable to do multiplexing and genome wide screening by co-expressing multiple sgRNAs and Cas9 enzyme.[17] Today, CRISPR-Cas9 is the simplest, cheapest and most precise genome-editing tool. However, it took three decades of basic research in this area to unravel the mechanism and develop CRSIPR-Cas9 tools.

2. CRISPR-Cas9: An exciting journey from bacteria to human

Clustered Regularly Interspaced Short Palindromic Repeat (CRISPR) was first identified in *E. coli* in the late 1980s.[26] At that time, the interesting feature of the repetitive sequences was noticed, however, its function was not known. Soon after, CRISPR sequences were detected in several other bacterial species.[6] In the year 1993, CRIPSR was discovered in archaea, specifically, *Haloferax maditarranei*.[27] In that study, authors suggested that the repeat sequence could have a role in the regulation of gene expression. With the invention of automated DNA sequencing in the 1990s, CRISPR sequences were identified in many bacterial and archaeal genome. In the meantime, Mojica et al. suggest a common function of the repeat sequences in bacteria and archaea.[28] By comparative genomic analysis among different species Jensen et al. in 2002 named the repeat sequences as CRISPR and also defined the common characteristics of CRISPR: (i) its location in intergenic region, (ii) contain multiple short direct repeats (21–37 bp) with little sequence variation, (iii) repeats are interspaced with similar sized non-repetitive non-conserved sequences, (iv) a leader sequences of 300–500 bp were located one side of the repeat clusters. The repeat sequences and leader sequences are conserved among the species. Interestingly, genomic analysis also led to the discovery of CRISPR-associated genes (Cas), which were always found in the CRISPR loci of different bacteria (Fig. 1). The association of Cas gene always with CRISPR loci suggested a functional relationship between Cas gene and the genesis of CRISPR loci.[6,29] However, the function of the CRISPR sequences remained unknown until 2005.

The remarkable breakthrough in the field of CRISPR came from the three different studies in the year 2005. These studies collectively demonstrated that spacer sequences in the CRISPR loci derived from the extra-chromosomal or foreign DNA and bacteria acquire new spacer from bacteriophage DNA.[30–32] Interestingly, Bolotin and colleagues also found a correlation between number of spacer sequences in bacteria and its resistance to invading bacteria phages.[31] Further, these studies suggested two important features of CRISPR: (i) It could activate the capture of foreign DNA to create a past genetic invasion and (ii) CRISPR could be used to generate antisense RNA.[30,31]

The real breakthrough came when the function of the CRISPR-cas9 system was shown experimentally as acquired resistance to viruses in prokaryotes.[33] This remarkable discovery started with a simple observation during comparative sequence analysis of various *Streptococcus thermophilus* strains related to normal and phage resistant variants. They found that resistant *S. thermophilus* strain contains additional spacer sequences in the CRISPR locus than the sensitive strain. Using phage-host model system consisting of *S. thermophilus* and virulent bacteriophage, this group provided direct evidence that infection of bacteria with phage leads to the integration of new spacers derived from phage genomic sequences within a CRISPR region of bacteria. The addition of new spacer made the bacteria resistant to further infection by phage strain from where bacteria acquired the spacer. This study further demonstrated that phage-resistance ability of the bacteria could be modulated by addition and deletion of spacer sequences to CRSIPR locus and Cas-gene is necessary for the phage resistance.[33] Following this discovery, a series of insightful studies describing the molecular mechanism of CRISPR-Cas9 system helped to clarify the RNA guided mechanism of Cas9 gene DNA cleavage activity. By studying CRISPR system in *E. coli*, Brouns et al. showed that after transcription of CRISPR locus, Cas protein cleaved long CRISPR RNA into several small CRISPR RNAs (crRNA) retaining the virus-derived spacer. Then, these crRNAs serve as guide RNA to direct endonuclease activity of Cas protein and interfere with virus proliferation.[34] In the same year, Marraffini et al. demonstrated that *Staphylococcus epidermidis* CRISPR system can block horizontal plasmid conjugation and concluded that Cas enzyme targets DNA rather RNA.[35] However, investigation using purified *Pyrococcus furiosus* extract that contains crRNA and Cas protein, showed that CRISPR-Cas9 system can also cleave RNA *in vitro*.[36]The intricate mechanism of CRISPR-Cas9 system was further unraveled with identification of protospacer adjacent motif (PAM). Analysis of spacer sequences in *S. thermophilus* and homologous sequences in phages revealed the presence of PAM sequences in the phage DNA and mutation in PAM sequences in phage genome overcome the bacterial resistance.[37]

At this time, CRISPR researchers were wondering how CRISPR-Cas9 system discriminates between its own *vs* invader DNA as spacer sequences are present in both bacteria and virus genome. Identification of PAM sequences provided the clue. Absence of PAM sequences in CRISPR array prevents CRISPR system to cut its own DNA. Interestingly, in *Staphylococcus epidermidis* that contains different CRISPR system than *S. thermophilus*, the mismatch between crRNA and target sequences enabled CRISPR interference. In *S. epidermidis* crRNA contains ~8 bases at its 5′ end which base paired with CRISPR array but not with the *bona fide* target DNA. The mismatch in base pairing of crRNA with target DNA triggered plasmid interference.[38]

After the first experimental evidence for CRISPR immunity in *S. thermophilus*, the basic mechanism of CRISPR function and mechanism was established by 2010. In this year, Moineau and colleagues showed that CRISPR-Cas9 system in *S. thermophilus* specifically cleaves invading virus DNA and also plasmid DNA within proto-spacer motif, at a specific site. This study also revealed that endonuclease activity requires only Cas9 enzyme that is proto-spacer specific and orientation dependent.[39] Shortly after this study, Deltcheva et al. used differential RNA-sequencing technique to study crRNA biogenesis mechanism in another *Streptococcus* sp. and found a non-coding trans-activating CRISPR RNA called tracrRNA. The study further demonstrated that tracrRNA hybridizes with pre crRNA at its repeat sequences. The RNA duplex is then recognized by endoribonuclease RNase III and Cas9 enzyme to further process the trac-RNA and crRNA complex to maturation which facilitates Cas9 to find target DNA cleavage.[40] These two studies together established that CRISPR-Cas9 system cut the target DNA at specific site and it needs at least three components: tracrRNA, crRNA and Cas9 (Fig. 1).

After these findings, the genome editing potential of CRISPR-cas9 system becomes extremely attractive among scientific community. In 2011, Sapranauskas et al. showed that *S. thermophilus* CRISPR-cas9 can be transferred to *E. coli* and provide heterologous protection to plasmid transformation and phage infection. The study further demonstrated that mutation in Cas9 enzyme abolishes the plasmid interference in *E. coli* which implies that Cas9 is the sole enzyme necessary for CRISPR mediated immunity.[41] Soon after, the remarkable biochemical characterization in the history of CRISPR biology took place. In the year 2012, two groups independently showed that purified Cas9 and crRNA complex is capable of cleaving target DNA *in vitro*.[10,42] Jinek et al. further demonstrated that a programmable single guide RNA (sgRNA), chimeric version of trac-RNA and crRNA, can mimic the natural duplex structure of trac-RNA and crRNA and activates DNA cleavage by Cas9 enzyme. The sgRNA contains 20 nucleotides at its 5′ end that is variable and the 3′ tracrRNA portion is constant. The 20 nucleotides portion of the sgRNA mimic the natural crRNA and base pairs with target DNA (Fig. 2). Therefore, simply changing the 5′ end of sgRNA sequences, it could be programmed to target any sequences in the genome provided the target sequences lie immediate upstream of a PAM sequence.[10] These two biochemical studies are breakthrough in the field of CRISPR biology elucidating the mechanism of CRISPR-Cas9 activities and its potential as genome editing tool. Above all, the editing tools become simple and versatile two components system by the discovery of programmable sgRNA which along with Cas9 enzyme can cut at a specific site of the target DNA. Soon after these findings, scientific community start harnessing the tools for genome editing application in higher eukaryotes including mammalian cells (Table 1).

However, it was not confirmed whether such a bacterial DNA cleavage system would function in higher eukaryotes where DNA exists as highly ordered structure into chromatin in the nucleus. Also, in bacteria, there is no nucleus and cytoplasmic compartmentalization and CRISPR-Cas9

Table 1 Shows the major known discoveries in CRISPR biology.

	Discovery/event	Main features	Lead author and year of publication
Discovery of CRISPR and Cas genes in prokaryotes	First, identification CRISPR in *E. coli*.	• Repeat sequences in *E. Coli* • Function of these sequences were not understood.	Ishino et al. (1987)
	Identification of CRISPR in archaea	• Repeat sequences in *Haloferaxmeditarraei* • Authors suggested the function of repeat sequences in gene regulation	Mojica et al. (1993)
	Presences of CRISPR in prokaryotes	• Repeat elements are present 40% bacteria and 90% archaea • Suggested that repeats sequences might have similar function	Mojica et al. (2002)
	Characteristics of CRISPR	• Named repeat element as CRISPR-Clustered Regularly Interspaced Short Palindromic Repeat • Described the characteristics of CRISPR. • Discover the CRISPR associated genes (Cas) with CRISPR • Hypothesized that CRISPR and Cas genes could be functionally related	Jansen et al. (2002)
Function and mechanism of CRISPR: RNA guided Bacterial adaptive immune system	Prokaryotes defense system against virus	• CRISPR sequences originated from virus and extrachromosomal • Suggested a protective function for bacteria against foreign DNA through anti-sense RNA • Speculated Photospacer Adjacent Motif (PAM) might direct Cas9 activity	Mojica et al. (2005) Pourcel et al. (2005) Bolotin et al. (2005)
	CRISPR provides adaptive immunity to prokaryotes against virus	• Showed CRISPR-Cas9 system as bacterial acquired defense system against viruses • Spacer in CRISPR target foreign DNA • Cas enzymes regulates spacer acquisition and phages defense	Barrangou et al. (2007)

Continued

Table 1 Shows the major known discoveries in CRISPR biology.—cont'd

	Discovery/event	Main features	Lead author and year of publication
	CRISPR targets DNA and CRISPR RNA guides prokaryotic viral defense	• Transcription and processing of CRISPR region produced small RNAs which control Cas9 nuclease activity • Demonstrated that CRISPR-Cas9 target foreign DNA	Brouns et al. (2008) Marraffini et al. (2008)
	CRISPR immunity: Self *vs* invader DNA discrimination	• Mutation in PAM sequence overcome bacterial resistance • PAM sequences help bacterial to recognize own vs phage DNA	Deveau et al. (2008) Marraffini and Sontheimer (2010)
In vitro Characterization of Cas9 and gRNA	Activity of Cas9 enzyme *in vitro* and development of programmable sgRNA	• Purified Cas9 enzyme and CRISPR RNA can cleave DNA • CRISPR RNA can be programmed to make single gRNA to target specific region of DNA	Gasiunas et al. (2012) Jinek et al. (2012)
CRISPR-Cas9 in eukaryotic cell	CRISPR-Cas9 genome engineering eukaryotic cells	• Demonstration of genome editing in mammalian cells by heterologous expression gRNA and Cas9 enzyme • Multiplexing gRNA to target several genes at once	Jinek et al. (2013) Mali et al. (2013) Cong et al. (2013)
Structural elucidation of CRISPR-Cas9	Structure of Cas9 -gRNA and target DNA as complex	• Revealed the molecular mechanism of RNA-guided DNA targeting by Cas9 • Cas9-sgRNA complex undergo inactive to active conformational change upon recognition of PAM sequences	Nishimasu et al. (2014)
	Crystal structure of Cas9	• Guide RNA mediated activation of Cas9 enzyme	Jinek et al. (2014)

CRISPR-Cas9 to edit in animal and human genome	Generate mice with multiple mutations, reporter and conditional alleles	• Developed mice with multiple mutations with microinjection of CRISPR components with multiple gRNA in mouse zygotes • One step knock-in using CRISPR-component injection into mouse zygote • Showed off-target mutations are rare in mouse and embryonic stem cells	Wang et al. (2013) Yang et al. (2013)	
	Editing of genome in human embryo	• CRISPR-Cas9 effectively cleaved β-globin gene but HDR repair process was inefficient and off-target activity • Successfully repair mutation in *MYBPC3* by using CRISPR in heterozygous human embryo. No off-target mutation • Successfully corrected Marfan syndrome pathogenic mutation in human embryo with 89% efficiency, no off target and indels	Liang et al. (2015) Ma et al. (2017) Zeng et al. (2018)	
CRISPR-Baby	Birth of twin girls with CRISP edited genome	• Scientist edited *CCR5* gene in the embryo to prevent HIV infection • This is highly controversial and unethical	2019	
Nobel prize for CRISPR Discovery	Given Nobel prize for discovering CRISPR-Cas9 tool	• Jennifer A Doudna, Professor at UC Berkley and Emmanuelle M Charpentier, director at the Max Planck Institute for Infection Biology in Berlin, were awarded the 2020 Nobel Prize in Chemistry for their work on CRISPR-Cas9	2020	

cleaves invader DNA which are not structured. In the beginning of 2013, research from three different laboratories showed that CRISPR-Cas9 system could be successfully engineered to carry out genome editing in mammalian cells. Interestingly, in all three studies, bacterial Cas9 gene was codon optimized for suitable expression in mammalian cell and also nuclear localization signal was added to ensure localization of Cas9 protein into nucleus. In a simple and straight forward study, Jennifer Doudna and colleagues demonstrated that co-expression of sgRNA and Cas9 in HEK293T cells can target human clathrin light chain (CLTA) gene where gRNA was designed to interrogate. The double stranded DNA cleavage by Cas9 activates endogenous non-homologous end joining (NHEJ) repair in genomic DNA and leads to genome editing.[9] Around the same time two studies from Feng Zhang and colleagues and George M Church group came out in the same issue of Science journal.[17,43] Mali et al. showed that heterologous expression of Cas9 and programmable gRNA can edit genome in HEK293T, K562 cells as well as induced pluripotent stem cells.[43] Feng Zhang and colleagues showed that genome editing could also be accomplished by co-expression of mature crRNA, trac-RNA and Cas9 enzyme. This study further demonstrated that precise site-specific homology-directed repair (HDR) can be achieved by using gRNA and mutated Cas9 that can only cut one strand of the target DNA in mammalian genome. Interestingly, authors also showed programmability and wide applicability of the RNA guided nuclease activity can be achieved by expressing multiple gRNAs using single CRISPR array to target several genes at once.[17] Together, these studies established that the simple two components (gRNA and Cas9 enzyme) CRISPR-Cas9 system is enough to edit any gene in the genome of higher eukaryotes. The simple, programmable and versatile nature of CRISPR-Cas9 technology provides unprecedented flexibility and several advantages over the already exist genome editing tools like zinc finger (ZF) and transcription activator-like effectors (TALEs) which involved tedious protein engineering.[8] Soon after these discoveries, CRIPSR-Cas9 had triggered an evolution in the field of genome editing technology and several laboratories around the world to start using this RNA guided nuclease system to edit genome in different organisms.[44]

Despite the significant understanding of the mechanism of CRISPR-cas9 technology, it is still unknown how sgRNA and Cas9 complex find out the 20 base pairs target sequences within complex mammalian genome. Cas9 is a large (160 kDa) multi-domain and multi-functional enzyme containing nuclease activity. So, it is of paramount importance to make sure that CRISPR-Cas9 system doesn't perform any off-target mutation in the genome. Avoiding off-target mutation of Cas9 is extremely important particularly for clinical application where any mutation could lead to unpredictable results.[12] Therefore, understanding the underlying mechanism of CRISPR-Cas9 system at the atomic level becomes crucial for CRISPR biology and also for the development of more efficient and precise CRISPR based tools.

Cas9 contains two main nuclease domains: RuvC and HNH, each named after the known nuclease domain structure. Mutational analysis of each domain showed that the HNH domain cleaves target DNA strand complementary to gRNA and RuvC domain cleaves the non-target strand, opposite of the complementary strand.[12,45] The molecular mechanism of target DNA cleavage by CRISPR-Cas9 has been elucidated at the atomic level by the crystal structure of Cas9 enzyme with sgRNA and DNA as a complex.[11,13] Structure analysis showed that Cas9 enzyme remains as inactive state and upon binding to sgRNA Cas9 undergoes large conformational rearrangement, which converts Cas9 to active form. Then the active Cas9-sgRNA complex becomes competent to interrogate target sequences.[13,46] The DNA recognition by the sgRNA-Cas9 occurs in a step wise manner where 10 bp seed sequence of gRNA is ready for base pairing with target DNA and Cas9 enzyme interrogate the PAM sequences in the DNA. Initial binding of Cas9 to PAM leads to unwinding of DNA template by Cas9 enzyme and then complete base pairing with target sequence and sgRNA induces concerted conformational changes to Cas9.[12] The conformation change leads to the activation of HNH domain nuclease activity and cut the target strand. After the target strand cleavage, further confirmation changes activate RuvC nuclease domain, which cuts the non-target strand. After the cleavage of both DNA strands, cellular machineries replace Cas9-gRNA complex from the DNA.[12,47] More recently, in a kinetic study it has been shown that DNA unwinding at the target site (R-loop formation) is the rate-limiting step that leads to the quick cleavage of the both strands of DNA.[48] The better understanding of the intricate mechanism of gene editing by CRISP-Cas9 technology opens up a new avenue for its application to edit genome in animals and human.

3. Therapeutic applications of the CRISPR-Cas9: Technology for precise gene therapy

CRISPR-Cas9 gene-editing technology has dramatically changed the possibility of gene therapy technology and opens up remarkable therapeutic benefits for several human diseases. More specifically, CRISPR-Cas9 has enable scientists to envision the precision-based gene therapy treatment of genetic diseases that involve the repair of DAN or RNA to cure the disease (e.g. sickle cell anemia, cystic fibrosis, Duchenne muscular dystrophy *etc.*) at the roots.[49,50] Several model animals have been developed in order to test the potential applications of CRISPR-Cas9 technology to modify disease-causing gene in the genome.

First report of using CRISPR technology for gene targeting in mouse came out in the year 2013. Wang et al. demonstrated that by directly injecting CRISPR components (Cas9 mRNA or protein, multiple sgRNAs) into mouse zygote can generate mice carrying mutation in multiple target genes.[51] In a separate study, this group also showed that CRISPR technology could also be used to generate knock-in mouse in one-step injection of CRISPR components in the zygote. The study further showed that the knock-in is very precise and off-target mutations are rare in the targeted mouse.[52] Since then, CRISPR technology have been used to generate several other animal models and also many mouse model that are useful to study human diseases such as cancer.[7] However, it was not confirmed whether CRISPR technology would work to repair genetic mutation in the genome. In 2013, Wu et al. successfully showed that by co-injecting Cas9 mRNA, sgRNA and WT donor fragment into the mouse zygote, cataract causing mutation could be repaired. The resulting mice were able to transmit the repair into offspring and showed rare off-target event.[53] In another study, Long et al. used the similar approach to cure Duchenne muscular dystrophy disease in a model mouse. However, genetic correction was varied in the resultant mice that showed 2%–100% gene correction.[54] Further, an exciting research on post-natal mice showed that injection of Cas9 protein, specific sgRNA and single stranded DNA oligo, as a donor template through tail vein can correct the desired target mutation.[55] All these studies showed the future potential of CRISPR technology to pave the way to revolutionize medicine for new treatment for incurable diseases in human.

In the year 2015, CRISPR has been used to edit human genome in discarded embryo. The resultant embryo was very poorly edited and showed genetic mosaicism.[56] More recently, highly efficient and more precise editing in human diploid zygote was performed utilizing CRISPR technology. In this study authors corrected a causative mutation associated with congenital heart disease. More importantly, the study showed that repaired human embryos has negligible mosaicism and no off-target effect.[57] The year 2018 Zeng et al. successfully achieved highly efficient and precise correction of a mutation in *FBN1* gene in heterozygous human embryos using CRISPR tools. The authors claimed that about 89% of embryos were perfectly edited and genomic analysis of tested embryos by deep sequencing and whole genome sequencing showed no off-target and indels.[7,58] Collectively, all the these studies provide a fundamental platform to understand the CRISPR technology and brings it a step closer to clinical application for correcting inherited genetic mutations associated with incurable diseases.[49]

Remarkable advances have also been made in the field of cancer therapy using CRISPR-Cas9 mediated engineered T-cell and several clinical trials have been launched related to engineered T cells on lung cancer patients and esophageal cancer patients.[49,59] CRISPR technology has also been used in screening inhibitors of Receptor-interacting serine/threonine-protein kinase 1 (RIPK1) which could be used as a potential target for amplifying anti-tumor immune reactivity. Recently, one such inhibitor has been identified and has been approved for clinical trial to examine its effect on pancreatic tumor.[49] CRISPR technology has also been used for treating other diseases such as thalassemia, sickle cell disease and retinal dystrophy disorder.[49] Interestingly, CRISPR-Cas9 has been used to repair splicing defects of the *CEP290* mRNA that causes retinal dystrophy disorder.[60,61] The developed CRISPR-Cas9 components (EDIT-101) administered subretinally, which specifically restores the correct splicing of CEP290 RNA and cures the disorder. The developed drug EDIT-101 has been approved for clinical trial.[49] This is one of the successful examples of translation of CRISPR-Cas9 technology.

4. Therapeutic applications of the CRISPR-Cas9: Problems and challenges

Despite significant advances have been made in the field of therapeutic application of CRISPR-Cas9 technology; there are stills concerns about the safety and efficiency of this versatile gene editing methods. One of the big challenges is its off-target mutagenesis through gRNA directed binding at closely matched sequences in the genome.[8] Several advances have been made to detect off target activity of CRISPR such as using computer algorithms to design shorter and better gRNA,[62] and also designing of high fidelity Cas9 enzyme.[63] In addition, variety of genome wide sequencing methodologies such as GUIDE-seq, BLESS, HTGTS, Digenome-seq have been developed to readily detect the potential off-target mutation (s) in the edited genome.[64,65] Chimeric gRNA has also designed with partial DNA and RNA sequences and it has been demonstrated that Cas9 enzyme guided by DNA-RNA chimera show reduced off-target in edited genome in human cells.[66] Recently, CRISPR Guide RNA Assisted Reduction of Damage (GUARD) was developed to protect off target activity of Cas9 by blocking the off-target site through competition with an inactive short gRNA and Cas9 complex. Based on the fact that gRNA shorter than 16 nucleotides in length can bind to the target site but don't activate Cas9 nuclease, authors demonstrate that on target activity of the Cas9 was maintained while CRISPR GUARD RNAs were multiplexed to block off target.[63] Also, significant progresses have been made to engineer Cas9 enzyme, which has high DNA specificity with reduced off-target activity and broad PAM compatibility. Recently, Jennifer Doudna and colleagues designed a new hyper-accurate Cas9 variant (HypaCas9) that shows high specificity toward on-target activity in human cells.[67] One of the key limitations to most commonly used spCas9 enzyme is its specificity to act near PAM sequences NGG only. A variant of Cas9 enzyme Cas9 enzyme, xCas9 has been discovered recently which has specificity to broader PAM sequences with substantially lower off-target activity.[68] All these tools to reduced off-target activity have been developed and tested based on the standard human genomes as a reference. However, there are significant natural variations in genome sequences between different individuals. So for therapeutic applications of CRISPR-Cas9, sgRNA must be designed considering inter-individual genetic variations to maximize the accuracy and efficiency of precise gene editing.[69] Another major challenge for translational application of this technology is the natural immunogenicity against spCas9 enzyme in human T-cell. So, translational application should be coupled with immunogenicity test of the receiving patients.[70]

The application of CRISPR-Cas9 technology also raises ethical concerns for its ability to make permanent and inheritable changes in human genome.[7] In the history of CRISPR biology, the most unethical event, which provoked international outcry among the scientists, is the news of CRISPR baby. In the year, 2019 in a bombshell revelation, He Jiankui, a Chinese scientist, announced the birth of a twin girls (popularly known as CRISPR-Baby) with edited genome using CRISPR tool. He claimed that he edited a gene CCR5, which encodes a protein related to HIV infections (https://www.nature.com/articles/d41586-019-00673-1). Immediately, He's revelations raised deep concern among scientific community all over the world. China's National Health Commission ordered an investigation and following investigation, He was fired from the university and sentenced to jail (https://doi.org/10.1038/d41586-019-00673-1). In fact, gene editing technology in human is highly controversial for ethical reasons and currently not approved by the US National Institutes of Health (NIH).[49] In the wake of this controversy, world health organization (WHO) established a global multi-disciplinary expert panel to oversight and governance the scientific, ethical, social and legal challenges associated with human genome editing (https://www.who.int/ethics/topics/human-genome-editing/en/). The 2020 noble prize-winning scientist for CRISPR, Jennifer Doudna says "we need to be thinking about these broader implications of a powerful technology and how to develop them responsibly" (https://www.ft.com/content/6d063e48-4359-11ea-abea-0c7a29cd66fe). The future of this wonderful discovery will be dependent on how scientific community focus on the good side of the research, and we all need to be responsible for its good use.

5. Future research

CRISPR-Cas9 system is the easiest, simplest, highly efficient and versatile genome editing tools. Any scientist with a fundamental knowledge of biochemistry and molecular biology can design and perform the genome editing in any cells at ease. Editing can be performed using simple biochemical techniques and don't need any special instruments. For this reason, this simple but remarkably powerful technology becomes widely accepted among the scientific community and its potentials as a technology were already explored in several organisms to human cell and embryos. However, CRISPR is an evolving technology and despite its immense potentials, there are some challenges for its application.

Cas9 is a big protein and because of its size there are some difficulties in delivery. Therefore, significant protein engineering of Cas9 protein has been performing to improve the Cas9 enzyme. Recently, Mammoth Biosciences reported that they have developed a "hyper-compact" Cas enzyme, Casφ, which is half the size of the traditional Cas9. The smaller size of Cas enzyme would be a huge advantage in terms of delivery of ribonucleotide particles (https://endpts.com/eyeing-in-vivo-editing-mammoth-licenses-jennifer-doudnas-new-crispr-enzyme/).

CRISPR becomes the discovery of the century after three decades of basic research on bacterial immune system. Understanding the intricate mechanism underlying the bacterial system, lead to the discovery of the remarkable genome editing system, which revolutionizes the application of gene editing technology for the betterment for human life. Scientific communities have seen before how basic research on the fascinating natural biological system led to the invention of several modern biological tools, e.g., restriction enzymes. Interestingly, in nature, the small and simple prokaryotic organisms harbor smart technology and invention of CRISPR-Cas9 system is the latest example of such a smart technology.

References

1. Hendrix RW. Bacteriophage genomics. *Curr Opin Microbiol.* 2003;6(5):506–511.
2. Tao P, Wu X, Rao V. Unexpected evolutionary benefit to phages imparted by bacterial CRISPR-Cas9. *Sci Adv.* 2018;4(2):eaar4134.
3. Carte J, Wang R, Li H, Terns RM, Terns MP. Cas6 is an endoribonuclease that generates guide RNAs for invader defense in prokaryotes. *Genes Dev.* 2008;22(24): 3489–3496.
4. Chibani-Chennoufi S, Bruttin A, Dillmann ML, Brüssow H. Phage-host interaction: an ecological perspective. *J Bacteriol.* 2004;186(12):3677–3686.
5. Hsu Patrick D, Lander Eric S, Zhang F. Development and applications of CRISPR-Cas9 for genome engineering. *Cell.* 2014;157(6):1262–1278.
6. Ishino Y, Krupovic M, Forterre P, Margolin W. History of CRISPR-Cas from encounter with a mysterious repeated sequence to genome editing technology. *J Bacteriol.* 2018; 200(7):1–17.
7. Jacinto FV, Link W, Ferreira BI. CRISPR/Cas9-mediated genome editing: from basic research to translational medicine. *J Cell Mol Med.* 2020;24(7):3766–3778.
8. Doudna JA, Charpentier E. Genome editing. The new frontier of genome engineering with CRISPR-Cas9. *Science.* 2014;346(6213):1258096.
9. Jinek M, East A, Cheng A, Lin S, Ma E, Doudna J. RNA-programmed genome editing in human cells. *eLife.* 2013;2:1–9.
10. Jinek M, Chylinski K, Fonfara I, Hauer M, Doudna JA, Charpentier E. A programmable dual-RNA–guided DNA endonuclease in adaptive bacterial immunity. *Science.* 2012;337(6096):816.
11. Nishimasu H, Ran FA, Hsu Patrick D, et al. Crystal structure of Cas9 in complex with guide RNA and target DNA. *Cell.* 2014;156(5):935–949.
12. Jiang F, Doudna JA. CRISPR-Cas9 structures and mechanisms. *Annu Rev Biophys.* 2017;46:505–529.
13. Jinek M, Jiang F, Taylor DW, et al. Structures of Cas9 endonucleases reveal RNA-mediated conformational activation. *Science.* 2014;343(6176):1247997.
14. Bibikova M, Golic M, Golic KG, Carroll D. Targeted chromosomal cleavage and mutagenesis in Drosophila using zinc-finger nucleases. *Genetics.* 2002;161(3):1169–1175.
15. Gong C, Bongiorno P, Martins A, et al. Mechanism of nonhomologous end-joining in mycobacteria: a low-fidelity repair system driven by Ku, ligase D and ligase C. *Nat Struct Mol Biol.* 2005;12(4):304–312.
16. Bibikova M, Carroll D, Segal DJ, et al. Stimulation of homologous recombination through targeted cleavage by chimeric nucleases. *Mol Cell Biol.* 2001;21(1):289–297.
17. Cong L, Ran FA, Cox D, et al. Multiplex genome engineering using CRISPR/Cas systems. *Science.* 2013;339(6121):819–823.
18. Qi Lei S, Larson Matthew H, Gilbert Luke A, et al. Repurposing CRISPR as an RNA-guided platform for sequence-specific control of gene expression. *Cell.* 2013; 152(5):1173–1183.

19. Pandelakis M, Delgado E, Ebrahimkhani MR. CRISPR-based synthetic transcription factors in vivo: the future of therapeutic cellular programming. *Cell Syst.* 2020; 10(1):1–14.
20. Urnov FD, Miller JC, Lee Y-L, et al. Highly efficient endogenous human gene correction using designed zinc-finger nucleases. *Nature.* 2005;435(7042):646–651.
21. Miller JC, Holmes MC, Wang J, et al. An improved zinc-finger nuclease architecture for highly specific genome editing. *Nat Biotechnol.* 2007;25(7):778–785.
22. Boch J, Scholze H, Schornack S, et al. Breaking the code of DNA binding specificity of TAL-type III effectors. *Science.* 2009;326(5959):1509–1512.
23. Christian M, Cermak T, Doyle EL, et al. Targeting DNA double-strand breaks with TAL effector nucleases. *Genetics.* 2010;186(2):757–761.
24. Miller JC, Tan S, Qiao G, et al. A TALE nuclease architecture for efficient genome editing. *Nat Biotechnol.* 2010;29(2):143–148.
25. Mussolino C, Morbitzer R, Lütge F, Dannemann N, Lahaye T, Cathomen T. A novel TALE nuclease scaffold enables high genome editing activity in combination with low toxicity. *Nucleic Acids Res.* 2011;39(21):9283–9293.
26. Ishino Y, Shinagawa H, Makino K, Amemura M, Nakata A. Nucleotide sequence of the iap gene, responsible for alkaline phosphatase isozyme conversion in Escherichia coli, and identification of the gene product. *J Bacteriol.* 1987;169(12):5429–5433.
27. Mojica FJ, Juez G, Rodríguez-Valera F. Transcription at different salinities of Haloferax mediterranei sequences adjacent to partially modified PstI sites. *Mol Microbiol.* 1993; 9(3):613–621.
28. Mojica FJ, Díez-Villaseñor C, Soria E, Juez G. Biological significance of a family of regularly spaced repeats in the genomes of Archaea, Bacteria and mitochondria. *Mol Microbiol.* 2000;36(1):244–246.
29. Jansen R, Embden JD, Gaastra W, Schouls LM. Identification of genes that are associated with DNA repeats in prokaryotes. *Mol Microbiol.* 2002;43(6):1565–1575.
30. Mojica FJ, Díez-Villaseñor C, García-Martínez J, Soria E. Intervening sequences of regularly spaced prokaryotic repeats derive from foreign genetic elements. *J Mol Evol.* 2005;60(2):174–182.
31. Bolotin A, Quinquis B, Sorokin A, Ehrlich SD. Clustered regularly interspaced short palindrome repeats (CRISPRs) have spacers of extrachromosomal origin. *Microbiology.* 2005;151(pt 8):2551–2561.
32. Pourcel C, Salvignol G, Vergnaud G. CRISPR elements in Yersinia pestis acquire new repeats by preferential uptake of bacteriophage DNA, and provide additional tools for evolutionary studies. *Microbiology.* 2005;151(pt 3):653–663.
33. Barrangou R, Fremaux C, Deveau H, et al. CRISPR provides acquired resistance against viruses in prokaryotes. *Science.* 2007;315(5819):1709–1712.
34. Brouns SJJ, Jore MM, Lundgren M, et al. Small CRISPR RNAs guide antiviral defense in prokaryotes. *Science.* 2008;321(5891):960–964.
35. Marraffini LA, Sontheimer EJ. CRISPR interference limits horizontal gene transfer in staphylococci by targeting DNA. *Science.* 2008;322(5909):1843–1845.
36. Hale CR, Zhao P, Olson S, et al. RNA-guided RNA cleavage by a CRISPR RNA-Cas protein complex. *Cell.* 2009;139(5):945–956.
37. Deveau H, Barrangou R, Garneau JE, et al. Phage response to CRISPR-encoded resistance in Streptococcus thermophilus. *J Bacteriol.* 2008;190(4):1390–1400.
38. Marraffini LA, Sontheimer EJ. CRISPR interference: RNA-directed adaptive immunity in bacteria and archaea. *Nat Rev Genet.* 2010;11(3):181–190.
39. Garneau JE, Dupuis M-È, Villion M, et al. The CRISPR/Cas bacterial immune system cleaves bacteriophage and plasmid DNA. *Nature.* 2010;468(7320):67–71.
40. Deltcheva E, Chylinski K, Sharma CM, et al. CRISPR RNA maturation by trans-encoded small RNA and host factor RNase III. *Nature.* 2011;471(7340):602–607.

41. Sapranauskas R, Gasiunas G, Fremaux C, Barrangou R, Horvath P, Siksnys V. The Streptococcus thermophilus CRISPR/Cas system provides immunity in *Escherichia coli*. *Nucleic Acids Res*. 2011;39(21):9275–9282.
42. Gasiunas G, Barrangou R, Horvath P, Siksnys V. Cas9-crRNA ribonucleoprotein complex mediates specific DNA cleavage for adaptive immunity in bacteria. *Proc Natl Acad Sci U S A*. 2012;109(39):E2579–E2586.
43. Mali P, Yang L, Esvelt KM, et al. RNA-guided human genome engineering via Cas9. *Science*. 2013;339(6121):823–826.
44. Sander JD, Joung JK. CRISPR-Cas systems for editing, regulating and targeting genomes. *Nat Biotechnol*. 2014;32(4):347–355.
45. Chen H, Choi J, Bailey S. Cut site selection by the two nuclease domains of the Cas9 RNA-guided endonuclease. *J Biol Chem*. 2014;289(19):13284–13294.
46. Jiang F, Zhou K, Ma L, Gressel S, Doudna JA, Structural Biology. A. Cas9-guide RNA complex preorganized for target DNA recognition. *Science*. 2015;348(6242):1477–1481.
47. Sternberg SH, Redding S, Jinek M, Greene EC, Doudna JA. DNA interrogation by the CRISPR RNA-guided endonuclease Cas9. *Nature*. 2014;507(7490):62–67.
48. Gong S, Yu HH, Johnson KA, Taylor DW. DNA unwinding is the primary determinant of CRISPR-Cas9 activity. *Cell Rep*. 2018;22(2):359–371.
49. You L, Tong R, Li M, Liu Y, Xue J, Lu Y. Advancements and obstacles of CRISPR-Cas9 technology in translational research. *Mol Ther Methods Clin Dev*. 2019;13:359–370.
50. Naldini L. Gene therapy returns to centre stage. *Nature*. 2015;526(7573):351–360.
51. Wang H, Yang H, Shivalila CS, et al. One-step generation of mice carrying mutations in multiple genes by CRISPR/Cas-mediated genome engineering. *Cell*. 2013;153(4):910–918.
52. Yang H, Wang H, Shivalila Chikdu S, Cheng Albert W, Shi L, Jaenisch R. One-step generation of mice carrying reporter and conditional alleles by CRISPR/Cas-mediated genome engineering. *Cell*. 2013;154(6):1370–1379.
53. Wu Y, Liang D, Wang Y, et al. Correction of a genetic disease in mouse via use of CRISPR-Cas9. *Cell Stem Cell*. 2013;13(6):659–662.
54. Long C, McAnally JR, Shelton JM, Mireault AA, Bassel-Duby R, Olson EN. Prevention of muscular dystrophy in mice by CRISPR/Cas9-mediated editing of germline DNA. *Science*. 2014;345(6201):1184–1188.
55. Yin H, Xue W, Chen S, et al. Genome editing with Cas9 in adult mice corrects a disease mutation and phenotype. *Nat Biotechnol*. 2014;32(6):551–553.
56. Liang P, Xu Y, Zhang X, et al. CRISPR/Cas9-mediated gene editing in human tripronuclear zygotes. *Protein Cell*. 2015;6(5):363–372.
57. Ma H, Marti-Gutierrez N, Park S-W, et al. Correction of a pathogenic gene mutation in human embryos. *Nature*. 2017;548(7668):413–419.
58. Zeng Y, Li J, Li G, et al. Correction of the Marfan syndrome pathogenic FBN1 mutation by base editing in human cells and heterozygous embryos. *Mol Ther*. 2018;26(11):2631–2637.
59. June CH, O'Connor RS, Kawalekar OU, Ghassemi S, Milone MC. CAR T cell immunotherapy for human cancer. *Science*. 2018;359(6382):1361.
60. Dulla K, Aguila M, Lane A, et al. Splice-modulating oligonucleotide QR-110 restores CEP290 mRNA and function in human c.2991 + 1655A > G LCA10 models. *Mol Ther Nucleic Acids*. 2018;12:730–740.
61. Maeder ML, Stefanidakis M, Wilson CJ, et al. Development of a gene-editing approach to restore vision loss in Leber congenital amaurosis type 10. *Nat Med*. 2019;25(2):229–233.
62. Fu Y, Sander JD, Reyon D, Cascio VM, Joung JK. Improving CRISPR-Cas nuclease specificity using truncated guide RNAs. *Nat Biotechnol*. 2014;32(3):279–284.

63. Coelho MA, De Braekeleer E, Firth M, et al. CRISPR GUARD protects off-target sites from Cas9 nuclease activity using short guide RNAs. *Nat Commun*. 2020; 11(1):1–12. 4132.
64. Hsu PD, Scott DA, Weinstein JA, et al. DNA targeting specificity of RNA-guided Cas9 nucleases. *Nat Biotechnol*. 2013;31(9):827–832.
65. Zhang X-H, Tee LY, Wang X-G, Huang Q-S, Yang S-H. Off-target effects in CRISPR/Cas9-mediated genome engineering. *Mol Ther Nucleic Acids*. 2015;4:e264.
66. Yin H, Song C-Q, Suresh S, et al. Partial DNA-guided Cas9 enables genome editing with reduced off-target activity. *Nat Chem Biol*. 2018;14(3):311–316.
67. Chen JS, Dagdas YS, Kleinstiver BP, et al. Enhanced proofreading governs CRISPR–Cas9 targeting accuracy. *Nature*. 2017;550(7676):407–410.
68. Hu JH, Miller SM, Geurts MH, et al. Evolved Cas9 variants with broad PAM compatibility and high DNA specificity. *Nature*. 2018;556(7699):57–63.
69. Scott DA, Zhang F. Implications of human genetic variation in CRISPR-based therapeutic genome editing. *Nat Med*. 2017;23(9):1095–1101.
70. Wagner DL, Amini L, Wendering DJ, et al. High prevalence of *Streptococcus pyogenes* Cas9-reactive T cells within the adult human population. *Nat Med*. 2019;25(2):242–248.

CHAPTER THREE

Modification of Cas9, gRNA and PAM: Key to further regulate genome editing and its applications

Rahul Gupta[a], Darshana Gupta[b], Kazi Tawsif Ahmed[b], Dhritiman Dey[c], Rajveer Singh[c], Snehasikta Swarnakar[a], V. Ravichandiran[c], Syamal Roy[c], and Dipanjan Ghosh[c,*]

[a]Infectious Diseases and Immunology Division, CSIR-Indian Institute of Chemical Biology, Kolkatta, India
[b]Department of Botany, Visva-Bharati, West Bengal, India
[c]Department of Natural Products, National Institute of Pharmaceutical Research and Education (NIPER), Kolkata, India
*Corresponding author: e-mail address: dipanjan4u@gmail.com

Contents

Abstract

The discovery of CRISPR-Cas9 system has revolutionized the genome engineering research and has been established as a gold standard genome editing platform. This system has found its application in biochemical researches as well as in medical fields including disease diagnosis, development of therapeutics, etc. The enormous versatility of the CRISPR-Cas9 as a high throughput genome engineering platform, is derailed by its off-target activity. To overcome this, researchers from all over the globe have explored the system structurally and functionally and postulated several strategies to upgrade the system components including redesigning of Cas9 Nuclease and modification of guide RNA(gRNA) structure and customization of the protospacer adjacent motif. Here in this review, we portray the comprehensive overview of the strategies that has been adopted for redesigning the CRISPR-Cas9 system to enhance the efficiency and fidelity of the technology.

Progress in Molecular Biology and Translational Science, Volume 178
ISSN 1877-1173
https://doi.org/10.1016/bs.pmbts.2020.12.001

Abbreviations

bp	base pair
Cas/cas	CRISPR associated protein/gene
CRISPRs	clustered regularly interspaced short palindromic repeats
crRNAs	CRISPR RNAs
DNA	deoxyribonucleic acid
DRs	direct repeats
DSB	double-strand breaks
GC	guanine cytosine
gRNA	guide RNA
PAM	protospacer adjacent motif
pre-crRNA	precursor CRISPR RNA
RNA	ribonucleic acid
STs	sequence types
tracrRNA	trans-activating CRISPR RNA
WT	wild-type

1. Introduction

The targeted gene modification has been the prevailing approach for understanding the gene function and its genetic contributions in triggering a disease during the past decades. There has been tremendous progress in defined and targeted genome modification due to the finding and implementation of CRISPR-Cas9 system, derived from the prokaryotes.[1,2] CRISPR-Cas9 system consists of a CRISPR array, which is transcribed and modified to give rise to a crRNA and a tracrRNA, and a RNA-guided endonuclease (Cas9). For Genome editing, recently a single guide RNA (sgRNA), a synthetic fusion of tracrRNA and crRNA, are used to target the locus adjacent to protospacer-adjacent motif (PAM) sequence at the 3′ end of the targeted DNA sequence. While sgRNA guided Cas9 generates sequence specific double stranded DNA breaks (DSBs) resulting in induction of an endogenous cellular repair pathways namely non-homologous end-joining (NHEJ) and homology-directed repair (HDR).[3–5] The advent of CRISPR-Cas9 system, as a gold standard in genome editing platform (Fig. 1), has paved the way for the researchers to alter genes with unprecedented ease. It is used to study the gene function, imaging topology, and mutational effects.[6,7] This modification system provides an efficient modulation of genome editing in a broad spectrum of organisms. However, despite its potentiality, application of CRISPR-Cas9 system is often troublesome due to its action on unwanted sites that are similar to the target sequence, leading to off target activity.[8] To reduce the off-target effects of the CRISPR-Cas9 system, researchers

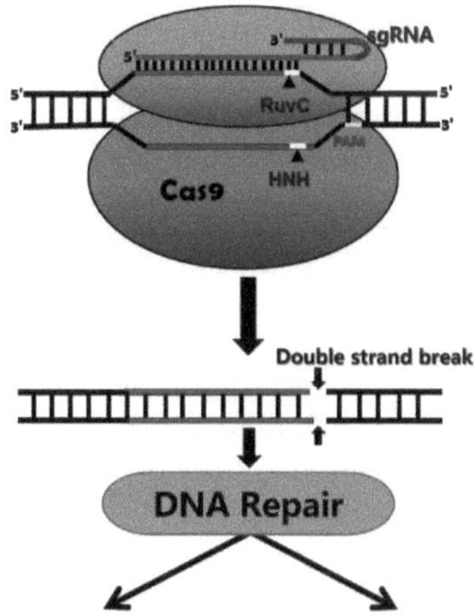

Fig. 1 Endonuclease Cas9 need a sgRNA [which is a chimeric, 20 bp long complementary to the target DNA, is having two RNA components—CRISPR RNA (crRNA; blue) and transactivating RNA (tracrRNA; pink), fused with a linker moiety], PAM sequence (yellow) (pivotal for DNA binding and cleavage). once the PAM is recognised by the Cas9, the two domain HNH (cuts the target strand) and RuvC (that cuts the non-target strand) starts DNA cleavage producing sequence specific double stranded DNA breaks (DSBs) resulting in induction of an endogenous cellular repair pathways: non-homologous end-joining (NHEJ) and homology-directed repair (HDR).

have adopted the following strategies: (i) Engineering of Cas9 protein/endonuclease, (ii) Modification of guides, and (iii) Customization of PAM sequences.[9–11] Hence, we emphasize the current engineering approaches of the CRISPR-Cas9 components which will facilitate precise genome editing and efficacy by reducing the off-target mutagenesis.

2. Cas9 engineering

Cas9 is a DNA endonuclease belonging to the type II CRISPR-Cas system which requires sgRNA guidance to generate DSBs in the target DNA.[1,12] However, researchers are now developing strategies for efficient and high-fidelity genome modification using this gold standard technology. Several variants of Cas9 have been discovered from different bacterial species but Cas9 from *Streptococcus pyogenes* (SpCas9) is the most popular endonuclease and has been highly explored both structurally and functionally. The generation of DSBs by the CRISPR-Cas9 system was found to be fatal in bacterial system when targeted towards the genome due to lack of effective repair machinery as compared to mammalian system and the result of genome editing found to be more better than bacterial system. The instance of modification by SpCas9 was observed when scientists used human codon optimized Cas9 protein and sgRNA components in a single plasmid along with SV40 nuclear localization signal (NLS). The transfection of that system resulted in an efficient genome editing by forming RNP complex (cas9 protein and sgRNA) inside the nucleus of mammalian cells while severe off-target mutagenesis occurred.[5] This lack of precision posed as a major obstacle when gene therapy-based applications were under consideration for this technology.[5] Therefore, there has been a tremendous effort to re-engineer SpCas9 nucleases to reduce off-target cleavage by reducing its size and increasing its fidelity and efficacy.[3,10,13] This further opened the path for the researchers to modify the SpCas9 nuclease domains genetically. Initial modifications included two nuclease activity sites, the HNH-like nuclease domain (cleaves target strand) and the RuvC-like nuclease domain (cleaves non-target strand). The substitution of Aspartic acid with Alanine in the RuvC domain (D10A) produces a nick in the DNA which is complementary to the targeted DNA strand while substitution of Histidine with Alanine in the HNH domain (H480A) produces a nick in the DNA which is non-complementary to the targeted DNA strand[2,5,14] (Fig. 2). This single amino acid change produce a protein called "*Nickase (nCas9)*" that could introduce single stranded breaks within the targeted DNA sequence. Moreover, several studies have revealed that this nickase enzyme, when used in a paired format, and also the use of NHEJ inhibitor generates a staggered

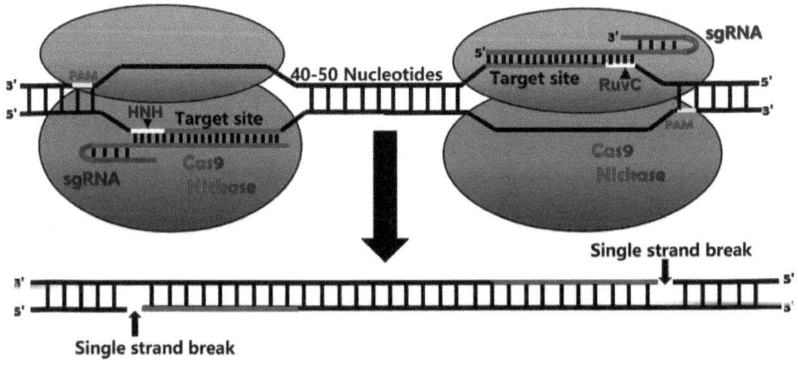

Fig. 2 Schematic representation of DNA double-stranded breaks induced by a pair of sgRNAs guiding Cas9 nickases. The nickase enzyme (nCas9) has a point mutation in one of the two functional domains (D10A in RuvC domain and H840A in HNH domain), thus generating a single stranded break at individual targets and a staggered double stranded break in a paired format at the same time.

double stranded break (DSBs) and favor's Homologous Recombination Repair pathway over Non-Homologous End Joining Repair Pathway which is preferable for precise editing.[1,5] Further, individual nicks experience high fidelity Base Excision Repair pathway which is safe and error free and thereby reduces the off-target effects by 100 to 1500 fold as compared to the wild type SpCas9.[13,15]

Recently, researchers are utilizing nCas9 as a "base editor" by fusing nCas9 complex with base conversion enzyme such as cytidine deaminase, adenine deaminase, uracil glycosylase inhibitor (UGI) which commendably converts C to T, A to G, and vice-versa without producing DSBs at the target site. Such approaches can be used to treat a single nucleotide mutation responsible for certain diseases[6,14–16] (Fig. 4). Not only that nCas9 are also being used for altering genetic code and incorporating early STOP codons in genes and this approach is called "CRISPR-STOP." CRISPR-STOP strategy found to be more efficient, safe and less deleterious as compared to knockout (KO) of a gene using WT-Cas9.[15] However, there is one drawback of such a strategy that it relies on the enzymes that convert one nucleotide to another.[16] Beyond nickases, researchers are utilizing these tools for additional purposes by inactivating both the domains of SpCas9 i.e. RuvC (D10A0 and HNH (H840A) domains, leads to a complete loss of the nuclease activity but retain its sequence specific DNA binding protein property and thus generating a deactivated or catalytically inactivated *dead Cas9 (dCas9)*[5,17,18](Fig. 3).

The fusion of the dCas9 protein with other nuclease domains such as *Fok*I was performed to establish dimerization within the molecule, thereby increasing the length of the recognition sequence. The strategy suffered the limitation of requiring the targeted sites to be within a distance of 15–25 bp.[5] Similarly, fusion proteins were created with Zinc Finger nucleases and

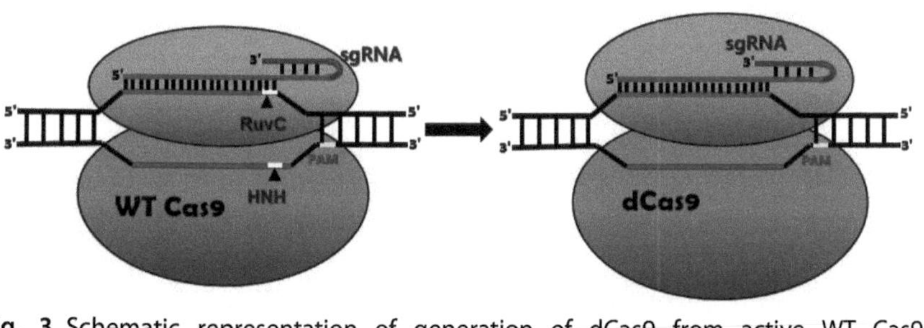

Fig. 3 Schematic representation of generation of dCas9 from active WT Cas9. Catalytically dead version of cas9, dCas9, lacks the endonuclease activity as both the nuclease domains (HNH and RuvC) are inactivated but it retains specific target DNA binding activity.

Transcription activator-like effector nucleases thereby reducing the size of the engineered protein, thus making it preferable for use with viral delivery systems.[17,18] Another viable strategy involved the use of structurally unstable protein domains (Dihydrofolate reductase and Estrogen Receptor 50) infusion with Cas9. The unstable protein domains would mark the fusion protein for proteasomal degradation in the absence of stabilizing molecules (Trimethoprim and 4-hydroxytamoxifen, respectively). This reduced the exposure of the Cas9 protein to the genome thereby minimizing off-target effects.[5] Beyond this dCas9 protein has been fused with multiple activator and repressor domains to analyze their effect on targeted DNA sequences and therefore, significantly being utilized for various regulatory purposes such as regulation of genes, editing of nucleic acid, epigenetic regulation, imaging of genome, chromatin topology, and mutation at gene[1,19] (Fig. 4). The protein itself has been used in gene repression leading to the creation of silencing libraries for multiple organisms including bacteria. Fusions of dCas9 with activator domains like VP64 and p65AD (activation

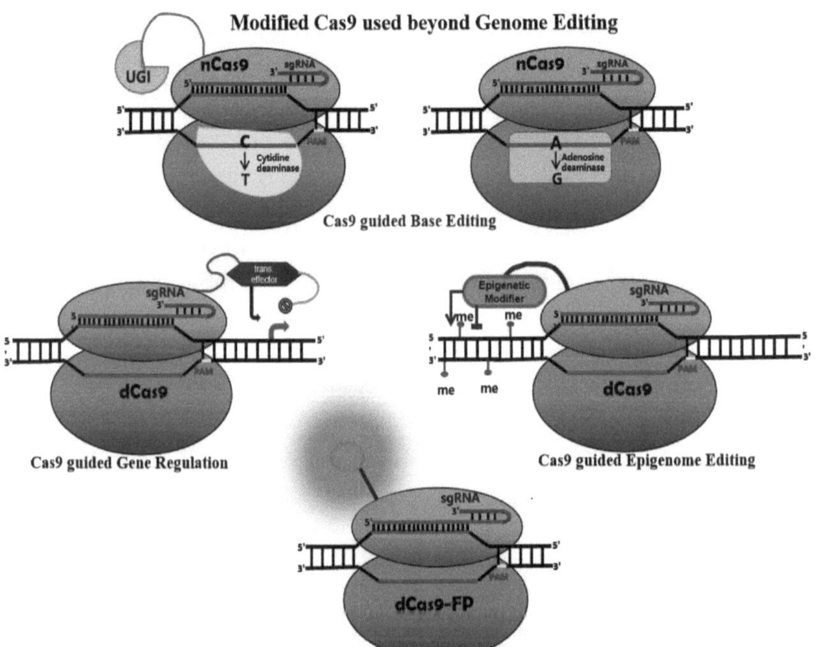

Fig. 4 Schematic representation of broad scope of CRISPR Cas9 based engineering beyond genome modification. WT Cas9 usually used to edit genomes because of active DNA cleavage activity but catalytically impaired Cas9 enzymes (nCas9 and dCas9) are being used to execute Cas9 guided targeted base editing, gene regulation, chromatin imaging, epigenome editing, and chromatin topology manipulations without double strand breaks.

domain) and repressor domains like KRAB (Kruppel associated box) and SID4X (mSin3 interaction domains) have been created for gene regulation in mammalian cells.[7,20] In order to control the induction of such transcriptional regulators, light-inducible domains were employed. The design involved two components, (i) dCas9 fused with Calcium Integrin Binding protein (CIB1) and, (ii) a Cryptochrome light-inducible heterodimerizing protein (Cry2) fused with an activator. Upon irradiation with blue light, CIB1 and Cry2 domains interact thereby recruiting the activator to the sequence where dCas9-gRNA complex is targeted. Similar to light-inducible systems, chemical inducible systems were also generated. The fusion of Cas 9 with Human Estrogen Receptor 2 domain (HRT2) at the N or C terminus, created a nuclease, responsive to 4-hydroxy tamoxifen (4-HT). The studies revealed a significant upregulation in the presence of the inducer. Multiple combinations of such inducers with their responsive domains have been under trials for applications in modulation of Cas9 activity.[14,21] The modifications of histones through methylation, phosphorylation, and acetylation-based events are essential for driving the regulation of a gene.[22] Such epigenetic regulation involves enzymes that can add or remove these functional groups from the histone structure. The fusion of functional domains of such enzymes with Cas9 has also generated a class of regulators guided to the targeted gene through sgRNAs.[1,20,23]

Further, the fusion of dCas9 with deaminases such as cytidine deaminase and adenine deaminase led to the development of base editors, with the aim to correct point mutations.[18] Multiple generations of fusion proteins led to the development of eA3A-BE3, a chimera of human APOBEC3a (apolipoprotein B mRNA editing enzyme, catalytic polypeptide like) cytidine deaminase linked with SaCas9n (*Staphylococcus aureus* Cas9 nickase enzyme) and a Uracil Glycosylase Inhibitor. The combination works by converting Cytidine to Uracil within the targeted sequence, uracil glycosylase inhibitor prevents removal of Uracil from the strand, Cas9 nickase cleaves the complementary strand and the endogenous base excision repair pathway corrects the nicked strand using the uracil containing strand as the template, thus converting a GC base pair into an AT base pair. The deactivated version of Cas9 (dCas9) has been extensively used for applications involving the regulation of gene expression.[3,7,18,24]

3. Structural based re-designing of Cas9

A recent investigation into the structure of the Cas9 protein-sgRNA-target DNA complex revealed the presence of four contact points, hydrogen bonds formed through N497, R661, Q695 and Q926 of Cas9 with the phosphate backbone of the DNA. Accordingly, scientists hypothesized that the energies conserved by such sequence-independent interactions were responsible for tolerance of variations in the targeted sequence thereby leading to off-target recognition. Experiments were conducted to analyze the effect of alanine substitutions of each residue mentioned above, individually and in parallel. The investigations revealed no variation in activity with respect to the targeted strand however, off-target events were significantly reduced in SpCas9 variants harboring alanine substitutions of all four residues. This version of Cas9 was named as SpCas9-HF1 and has created the possibility of engineering Cas9 and other orthogonal proteins at sites exhibiting non-sequence specific interactions with the target molecule. Further, the on-target efficiency of SpCas9-HF1 was found to be comparable with WT Cas9.[25]

Crystal structure-based information has been used to engineer residues present in the active site of protein molecules, often supported through molecular modeling approaches to identify critical interactions. Such inspections identified a positively charged groove within the complex, involved in stabilization of the non-template strand of the targeted DNA sequence. The hypothesis governing the neutralization of 31 positively charged residues present in this groove was driven by the concept of facilitating re-hybridization of the template and non-template strands of the targeted DNA sequence. This would create a requirement for higher energies to disrupt the Watson-Crick base pairing in the targeted dsDNA, hence reducing off-target RNA-DNA hybridization. Amongst the 31 single amino acid substitutions, five mutants displayed ten times lower off-target mutagenesis. However, significant non-specific events were yet detectable and thus, the five selected mutations were engineered in parallel. Through such combinatorial mutagenesis, 34 variants of Cas9 were analyzed, of which 8 presented undetectable off-target mutagenesis. The off-target activity and on-target activity of all combinatorial mutants were compared, thereby screening out mutants that had compromised on target-specific activity. This study generated three variants of SpCas9, namely, SpCas9 (K855A), SpCas9 (K810A/K1003A/R1060A), referred to as eSpCas9 (1.0) and SpCas9 (K848A/K1003A/R1060A), referred to as eSpCas9 (1.1).[2,5,25]

Following rational designing approaches to wangle the SpCas9 protein, the molecular dynamics of the DNA-RNA-protein complex formation were investigated. The information obtained from single molecule Fourier Resonance Energy Transfer (FRET) experiments revealed the conformational state of the HNH domain to be directly attributable to off-target cleavages. The functional residue H840 resides at a point distant from the catalytic site of the protein in its deactivated state, a conformational change in which relocates to the residue to the active site.[2,22] The mechanism by which mismatches at these locations can inactivate the HNH domain was yet unknown. To understand this aspect, Gaussian accelerated Molecular Dynamics (GaMD) simulation studies were performed, which revealed insights correlating with the practical observations.[23] The presence of 4 mismatches at PAM distal end causes the RNA-DNA hybrid to remain in an open state, thus favoring interactions with amino acid residues in the L2 loop of the HNH domain. These interactions eventually result in restricting the positioning of the H840 residue to the active site, thereby locking the HNH domain in an inactive state. However, reducing the number of mismatches to three fails to stably lock the H840 residue, thereby allowing the cleavage to take place, in the presence of mismatches. The work presented these interactions within the L2 loop as a new segment available for engineering, where the locking interactions can be amplified for the 3 mismatch-tolerant positions.[18,21,23]

Further, Babu et al. shown that an arginine-rich bridge helix (BH) existing in Cas9 endonuclease play a crucial role in providing higher selectivity in DNA targeting. They substituted two proline residues present inside the loop region of the BH of spCas9 and found that it helps to impart impaired DNA cleavage activity and resulting in accumulation of nicked products and thereby reducing target DNA linearization. Recently, Bratovič et al. demonstrated Cas9 tolerance to mismatches along the target DNA relies on arginine-rich bridge helix. They have shown that R63 and R66 significantly decreases the Cas9 specificity in presence of mismatches by stabilizing the R-loop of BH while Q768 found to decreases the sensitivity of Cas9 to PAM-distal mismatches. Further, the arginine substituted Cas9 variant (Cas9_R63A/Q768A) when used in human cells it increases the specificity of the Cas9 and thus providing an evidence that functional and structural based re-designing of Cas9 can play a crucial role in increasing Cas9 specificity and sensitivity for genome editing. However, all these strategies ominously help to improve on-target cleavage activity and thus reducing the off-target mutagenesis.[26,27]

4. Customizing PAM specificity by leveraging directed evolution

Due to the specific PAM recognition stringency of SpCas9, researchers are utilizing a site directed evolution approach to engineer PAM specificity of SpCas9 by subjecting to random mutation.[1,20] Three different sets of SpCas9 were obtained and named as VQR (D1135V/R1335Q/T1337R), VRER (D1135V/G1218R/R1335E/T1337R), and EQR (D1135E/R1335Q/T1337R) whose PAM specificity found to be 5′-NGA-3′, 5′-NGCG-3′, and 5′-NGAG-3′, respectively instead of 5′-NGG-3′ detected in the WT-SpCas9. This single amino acid random mutagenesis occurs in the PAM-interacting domain (PI) of SpCas9. This strategy significantly reduces the off-target cleavage as the PAM sequences for EQR (5′-NGCG-3′) and VRER (5′-NGAG-3′) SpCas9 found to be 23 times less ample in the human genome as compared to the standard 5′-NGG-3′ PAM sequences of SpCas9. Similarly, researchers have changed PAM specificity of SaCas9 from 5′-NNGRRT-3′ to 5′-NNNRRT-3′ by random mutagenesis and this approach resulted in a two to fourfold increase in the targeting range. A current approach to modify PAM specificity to develop a SpCas9 variant by using phage-assisted continuous evolution (PACE) technique. In this ω subunit of bacterial RNA polymerase was bonded with dCas9 which causes phage to propagate by binding to PAM sequence and Crystal structure reveals that it has mutation near the DNA-sgRNA interface i.e. R324L, S409I, and M694I which found to cleave the target DNA by identifying five PAM (5′-CAA-3′, 5′-GAA-3′, 5′-GAT-3′, 5′-NNG-3′, and 5′-NG-3′) sequences. Further, a single point mutation of SpCas9 (D1135E) instigated by directed evolution was found to recognize more 5′-NGG-3′ standard PAMs than WT-SpCas9. These techniques exhibited a substantial reduction in off-target effects as compared to the WT-SpCas9. All the above-directed evolution strategies are unbiased and do not requires structural data but it has a limitation of its own that is each variant of SpCas9 and its orthologs, which needs to develop distinctly with an effective PAM sequence.[14,28,29]

5. Guide RNA (gRNA) engineering

Although the 20-nt guide sequence of the sgRNA and the PAM adjacent to the target sequence in the genome tightly controls the targeting specificity of Cas9, potential off-target cleavage could still occur with mismatches of at least three to five base pair in the PAM-distal part of the guide sequence.[5,12] Previous studies have demonstrated that various *guide RNA* make-up can impact the cleavage of on-target and off-target sites. A typical sgRNA comprises of the seed sequence and non-seed sequence. Several studies have shown that the 10–12 base pairs adjacent to the PAM

(3′ end of the guide RNA), named as the "seed sequence" and is considered more important than the rest of the *guide RNA* sequence in terms of Cas9 specificity. However, the true "seed region" of one to five base pairs from the 3′ side of guide region proximal to the PAM was identified by chromatin immunoprecipitation followed by sequencing (ChIP-seq) of DNA bound to dead Cas9 (dCas9) in *murine embryonic stem cells*.[24,30] This was further confirmed by Yu Zhang and his colleagues, by targeting the human genome with *CRISPR/Cas9* followed by ChIP-seq analysis of the genome. The results obtained were consistent with previous observations that base-pairing next to the PAM is censorious for DNA targeting. Pelletier and his colleagues reported that the sequence distal to PAM (considered as a non-seed sequence) is important for the induction of a conformational change in Cas9 upon sgRNA binding.[8] Thus, using sgRNAs that emphasizes on the seed sequence and the PAM-distal target sequences equally, will reduce off-target editing. Sharp and colleagues have reported a well-defined seed region for target binding and abundant off-target binding sites, in their genome-wide binding maps of dCas9 study, conducted through ChIP-seq in mouse embryonic stem cells.[31] Further, mismatches of one to five base pairs at the 5′ end of sgRNA i.e. non-seed region are better endured for Cas9 mediated cleavage than those at the 3′ end. However, the degree of mismatch tolerance of single and double base pairs varies, depending on their position along with the guide *RNA-DNA* interface. Zhang et al. proposed that designing a short guide RNA complementary to the desired target site is highly preferable for high throughput applications. They demonstrated that ethe specificity of SpCas9-mediated DNA cleavage is a sequence and locus-dependent and governed by the position, quantity, and identity of mismatches within the duplex (DNA-gRNA hybrid).[8] The PAM-proximal 8–12 bp of the guide sequence is critical for specificity whereas the PAM-distal sequence also contributes to the overall specificity of SpCas9-mediated DNA cleavage. An interesting discovery reported that less active sgRNA is related to exceptionally low or high GC content. In a study of CRISPR/Cas9-mediated *mutagenesis* in *Drosophila*, Ren et al. observed that sgRNAs with minimum four GCs in the sequence of the six base pairs, proximal to the PAM sequence is highly effective, suggesting that efficient sgRNA can be selected on the basis of the GC content of the seed region. Furthermore, guanine is strongly preferred over cytosine as the first base adjacent to PAM, conversely, cytosine is favored over guanine, as the fifth base proximal to PAM, whereas *adenine* is favored in the middle of the sgRNA, and cytosine is disfavored at position 18. These design principles are based upon the in vivo formation of stable non-canonical structures called G-quadruplexes of the guanine-rich sequences, which contributes to sgRNA stability. However, Zhang et al. reported that sgRNAs with +67 or +85 nucleotide (nt) tracrRNA tails mediated DNA cleavage at all target sites tested, with up to fivefold higher levels of indels than the corresponding crRNA-tracrRNA duplexes. Northern blot analyses for the guide RNA truncations showed increased levels of expression for the longer tracrRNA sequences and suggested that higher sgRNA expression or stability is partially related to improved target cleavage and indicated that the tracrRNA tail is important for optimal SpCas9 expression and activity in vivo.[11,30–32]

6. Conclusion

The discovery of CRISPR as a genome editor has revolutionized the scientific world by showing its impact in a wider application ranging from drug development, cancer therapy, agriculture, HIV treatment, and many more. Since then scientist are putting their tremendous efforts to use this technology in diagnosis of various diseases and even in treatment. Hence, broadening its potential in medicinal therapeutics. Over the years, crystal structures of the CRISPR-associated (Cas) proteins have helped in building models to explain structural and functional linkages. A significant number of investigations were performed with Type II CRISPR, which gained popularity due to the ease with which it performs and permits modifications thus serving as a platform for engineering proteins toward desirable characteristics. Furthermore, the development of nCas9, and dCas9 has broadened its application beyond genome editing based on DNA methylation, base editor, chromatin imaging, optogenetics, and gene regulation at the genome level. Although the finding of different Cas endonucleases such as Cpf1 and Cas13 has also brought new role to CRISPR to edit and target even RNA with great specificity and also broadened the utilization of CRISPR in the area of diagnostics and therapeutic treatments against viral infection still await more exploration. Moreover, the allergenic potency of Cas9 in gene therapies in humans remain unknown therefore it gives a great opportunity to the researcher for perusing potential solution against such issues.

Acknowledgment

Mr. Kazi gratefully acknowledges the financial assistance in the form of fellowship (File No.:09/202(0070)/2017-EMR-I) from Council of Scientific & Industrial Research (CSIR), India. The authors would like to give they're thanks to the Department of

Pharmaceuticals, Ministry of Fertilizer, Government of India for providing the fellowship to Mr. Rajveer Singh. We also thank the DBT, Govt. of India [grant number BT/PR26301/GET/119/258/2017] and the Govt. of West Bengal Department of Biotechnology WBDBT [grant numbers BT/P/Budget/RD-74/2017] for their support.

Conflict of interest

The authors declare that there was no conflict of interest.

References

1. Adli M. The CRISPR tool kit for genome editing and beyond. *Nat Commun*. 2018;9(1). https://doi.org/10.1038/s41467-018-04252-2.
2. Jiang F, Doudna JA. CRISPR–Cas9 structures and mechanisms. *Annu Rev Biophys*. 2017;46(1):505–529. https://doi.org/10.1146/annurev-biophys-062215-010822.
3. Chen S. Minimizing off-target effects in CRISPR-Cas9 genome editing. *Cell Biol Toxicol*. 2019;35(5):399–401. https://doi.org/10.1007/s10565-019-09486-4.
4. Chen JS, Dagdas YS, Kleinstiver BP, et al. Enhanced proofreading governs CRISPR–Cas9 targeting accuracy. *Nature*. 2017;550(7676):407–410. https://doi.org/10.1038/nature24268.
5. Gupta D, Bhattacharjee O, Mandal D, et al. CRISPR-Cas9 system: a new-fangled dawn in gene editing. *Life Sci*. 2019;232:116636. https://doi.org/10.1016/j.lfs.2019.116636.
6. Kleinstiver BP, Pattanayak V, Prew MS, Tsai SQ, Nguyen NT, Joung JK. 731. High-fidelity CRISPR-Cas9 nucleases with no detectable genome-wide off-target effects. *Mol Ther*. 2016;24. https://doi.org/10.1016/s1525-0016(16)33539-0.
7. Shen C, Hsu M, Chang C, et al. Synthetic switch to minimize CRISPR off-target effects by self-restricting Cas9 transcription and translation. *Nucleic Acids Res*. 2018;47(3). https://doi.org/10.1093/nar/gky1165.
8. Zhang X, Tee LY, Wang X, Huang Q, Yang S. Off-target effects in CRISPR/Cas9-mediated genome engineering. *Mol Ther Nucl Acids*. 2015;4. https://doi.org/10.1038/mtna.2015.
9. Deng Z, Sun Y. Faculty opinions recommendation of improving CRISPR-Cas nuclease specificity using truncated guide RNAs. *Fac Opin Post Publ Peer Rev Biomed Lit*. 2014. https://doi.org/10.3410/f.718252428.793492591.
10. Hajiahmadi Z, Movahedi A, Wei H, et al. Strategies to increase on-target and reduce off-target effects of the CRISPR/Cas9 system in plants. *Int J Mol Sci*. 2019;20(15):3719. https://doi.org/10.3390/ijms20153719.
11. Ryan DE, Taussig D, Steinfeld I, et al. Improving CRISPR–Cas specificity with chemical modifications in single-guide RNAs. *Nucleic Acids Res*. 2017;46(2):792–803. https://doi.org/10.1093/nar/gkx1199.
12. Fu Y, Sander JD, Reyon D, Cascio VM, Joung JK. Improving CRISPR-Cas nuclease specificity using truncated guide RNAs. *Nat Biotechnol*. 2014;32(3):279–284. https://doi.org/10.1038/nbt.2808.
13. Li D, Zhou H, Zeng X. Battling CRISPR-Cas9 off-target genome editing. *Cell Biol Toxicol*. 2019;35(5):403–406. https://doi.org/10.1007/s10565-019-09485-5.
14. Ribeiro LF, Ribeiro LF, Barreto MQ, Ward RJ. Protein engineering strategies to expand CRISPR-Cas9 applications. *Int J Genom*. 2018;2018:1–12. https://doi.org/10.1155/2018/1652567.
15. Ran FA, Hsu PD, Lin CY, et al. Double nicking by RNA-guided CRISPR Cas9 for enhanced genome editing specificity. *Cell*. 2013;154(6):1380–1389.
16. Shen B, Zhang W, Zhang J, et al. Efficient genome modification by CRISPR-Cas9 nickase with minimal off-target effects. *Nat Methods*. 2014;11(4):399–402. https://doi.org/10.1038/nmeth.2857.

17. Rouet R, Oñate LD, Li J, Murthy N, Wilson RC. Engineering CRISPR-Cas9 RNA–protein complexes for improved function and delivery. *CRISPR J.* 2018;1(6): 367–378. https://doi.org/10.1089/crispr.2018.0037.
18. Brocken DJ, Tark-Dame M, Dame RT. dCas9: a versatile tool for epigenome editing. *Curr Issues Mol Biol.* 2017;26:15–32.
19. Richardson CD, Ray GJ, Dewitt MA, Curie GL, Corn JE. Enhancing homology-directed genome editing by catalytically active and inactive CRISPR-Cas9 using asymmetric donor DNA. *Nat Biotechnol.* 2016;34(3):339–344. https://doi.org/10.1038/nbt.3481.
20. Moon SB, Kim DY, Ko J, Kim Y. Recent advances in the CRISPR genome editing tool set. *Exp Mol Med.* 2019;51(11):1–11. https://doi.org/10.1038/s12276-019-0339-7.
21. Slaymaker IM, Gao L, Zetsche B, Scott DA, Yan WX, Zhang F. Rationally engineered Cas9 nucleases with improved specificity. *Science.* 2015;351(6268):84–88. https://doi.org/10.1126/science.aad5227.
22. Zhou L, Peng R, Zhang R, Li J. The applications of CRISPR/Cas system in molecular detection. *J Cell Mol Med.* 2018;22(12):5807–5815. https://doi.org/10.1111/jcmm.13925.
23. Ricci CG, Chen JS, Miao Y, et al. Deciphering off-target effects in CRISPR-Cas9 through accelerated molecular dynamics. *ACS Central Sci.* 2019;5(4):651–662. https://doi.org/10.1021/acscentsci.9b00020.
24. Nowak CM, Lawson S, Zerez M, Bleris L. Guide RNA engineering for versatile Cas9 functionality. *Nucleic Acids Res.* 2016;44(20):9555–9564. https://doi.org/10.1093/nar/gkw908.
25. Van Der Oost J, Westra ER, Jackson RN, Wiedenheft B. Unravelling the structural and mechanistic basis of CRISPR–Cas systems. *Nat Rev Microbiol.* 2014;12(7):479.
26. Bratovič M, Fonfara I, Chylinski K, et al. Bridge helix arginines play a critical role in Cas9 sensitivity to mismatches. *Nat Chem Biol.* 2020;16(5):587–595. https://doi.org/10.1038/s41589-020-0490-4.
27. Babu K, Amrani N, Jiang W, et al. Bridge helix of Cas9 modulates target DNA cleavage and mismatch tolerance. *Biochemistry.* 2019;58(14):1905–1917. https://doi.org/10.1021/acs.biochem.8b01241.
28. Kleinstiver BP, Prew MS, Tsai SQ, et al. Engineered CRISPR-Cas9 nucleases with altered PAM specificities. *Nature.* 2015;523(7561):481–485. https://doi.org/10.1038/nature14592.
29. Lee JK, Jeong E, Lee J, et al. Directed evolution of CRISPR-Cas9 to increase its specificity. *Nat Commun.* 2018;9(1). https://doi.org/10.1038/s41467-018-05477-x.
30. Yin H, Song C, Suresh S, et al. Structure-guided chemical modification of guide RNA enables potent non-viral in vivo genome editing. *Nat Biotechnol.* 2017;35(12): 1179–1187. https://doi.org/10.1038/nbt.4005.
31. Tycko J, Myer V, Hsu P. Methods for optimizing CRISPR-Cas9 genome editing specificity. *Mol Cell.* 2016;63(3):355–370. https://doi.org/10.1016/j.molcel.2016.07.004.
32. Ren X, Yang Z, Xu J, et al. Enhanced specificity and efficiency of the CRISPR/Cas9 system with optimized sgRNA parameters in Drosophila. *Cell Rep.* 2014;9:1151–1162.

CHAPTER FOUR

CRISPR/dCas system as the modulator of gene expression

Tawsif Ahmed Kazi and Swadesh Ranjan Biswas*
Department of Botany, Visva-Bharati, Santiniketan, West Bengal, India
*Corresponding author: e-mail address: swadesh_s2000@yahoo.co.in

Contents

Abstract

CRISPR/Cas has been a very exciting field of research because of its multifaceted applications in biological science for editing genome. This tool can be programmed to target any region of DNA of choice by designing gRNA. The potential of gRNA to recruit a CRISPR-associated protein at a specific genomic site allowed scientists to engineer genome of diverse species for research and development. The application of Cas9 has been further expanded with a recently developed catalytically inactive protein (dead Cas9). CRISPR/dCas system is widely used as a programmable vector to deliver functional cargo (transcriptional effectors) to the desired sites at the genome for targeted transcriptional repression (CRISPR interference, CRISPRi) or activation (CRISPR activation, CRISPRa). It is now possible to regulate gene expression in cells without altering the DNA sequence. These CRISPRi/a toolboxes have explored many unsolved biological issues. Further research on CRISPR system could help diagnose and treat various human diseases.

Progress in Molecular Biology and Translational Science, Volume 178
ISSN 1877-1173
https://doi.org/10.1016/bs.pmbts.2020.12.002

1. Introduction

CRISPR/Cas system is a powerful technology, used as gene editing tool to selectively modify DNA sequences at any specific location in the genome with far more precision than conventional genome modification technologies. The discovery of this system has revolutionized the development of molecular biology, with lucrative applications in medicine and biotechnology. In recent years, CRISPR/Cas has been a fascinating area of research, because of its potential application in targeted genome and epigenome engineering, to unravel the hidden science and facts behind the complex life processes. With the identification of CRISPR/Cas in bacteria, rigorous exploration has been done to identify newer and efficient CRISPR/Cas systems, to broaden the range of applications in genetic engineering and molecular biology. The CRISPR systems, found in 40% of bacteria and almost omnipresent in archaea,[1,2] are a part of adaptive immunity systems against the invading foreign genetic material. Till date, the CRISPR-Cas systems have been categorized into two classes and six major types,[1] among them, CRISPR/Cas9 system has received immense attention for genome engineering. CRISPR-Cas9 system is based on nucleolytic activity of the endonuclease protein, Cas9, which is guided to the desired site in the genome by a specificity determinant RNA, termed as guide RNA (gRNA). Apart from these, another sequence known as protospacer adjacent motif (PAM), present adjacent to the target site, is recognized by the CRISPR/Cas9 system and is crucial for the functionality of Cas9. The Cas9 protein binds to the target location in the presence of gRNA, with high precision and performs a double strand break followed by the introduction of indel mutations at the cleavage site, by non-homologous end joining repair mechanism of the cell. Using predesigned repair template, specific bases in the DNA can also be altered by homologous recombination mediated repair. The CRISPR/Cas systems and their mode of action has been reviewed in detail by Gupta et. al.[1]

The Cas9 protein is a multi-domain protein,[3] consisting of two nucleolytic domains (RuvC and HNH), that is responsible for the double strand break of the target DNA. The key residues responsible for the catalytic activity of Cas9 are D10 and H840 of RuvC and HNH domain respectively. Introduction of point mutation, to replace both the residues with Alanine, generated a nuclease deficient Cas9 protein termed as dead Cas9 or dCas9. Although the nuclease activity of dCas9 was lost, the protein was still capable of binding to DNA at a specified location in the genome with same precision, when guided by gRNA.[4] The development of dCas system has broadened the application of CRISPR-Cas9 technology beyond genome editing. Unlike permanent knock-out by genome engineering, CRISPR/dCas system triggers reversible modulation of gene expression, and to achieve this, two separate technologies have been developed, where the process of transcriptional repression is called as CRISPR interference (CRISPRi) and transcriptional activation is called CRISPR activation (CRISPRa). To date, these systems have been extensively employed, outside the gene editing context, for modulating the expression of several genes and to identify their cellular functions. In this chapter, we overview these gene expression regulation technologies with a clear insight of their applications in the field of biotechnology and medical research as well as future developments.

2. CRISPRi

The transcriptional landscape of an organism is regulated by a complex network of epigenetic modifications, transcription factors and chromatin organization. To understand the role of these elements, it is highly desired to have efficient tools to interrogate the mechanistic relationship between the regulatory components and gene expression. Manipulation of gene activity is currently carried out using siRNA/shRNA approaches, where it leads to the degradation of transcribed mRNAs. The drawbacks of these technologies, like significant off-target activity,[5] limited packaging capacity of the exogenous vector and the ability to target a single isoform of a gene,[6] limits the application of these toolboxes, as it may result in qualitative and quantitative differences between the properties of a single gene isoform and many isoforms expressed from the genome. The advent of CRISPR/dCas system has solved the aforementioned problems by virtue of its specificity and broader applicability. dCas9 residing on the DNA, acts as a road block and interferes with transcription elongation by physically blocking the RNA polymerase progression as well as it also hinders the binding of essential transcription factors. It is known that RNA polymerase activity on the promoter or RNA polymerase processivity along the coding sequence is sterically hindered by the dCas9, loaded on the DNA, resulting in transcriptional repression known as CRISPRi.[7]

In the era of CRISPR-Cas system, considerable interest has been centered on understanding genomic functions by interrogating multiple genes in a high throughput manner using CRISPRi.[8,9] This targeted gene silencing strategy was successful in prokaryotes, as evidenced from experiments in *E. coli* showing up to ~300-fold down-regulation of mRNA expression, when a single sgRNA was used to direct dCas9 and even ~1000-fold reduction was observed when two sgRNAs were combined to block transcription elongation.[7] CRISPRi has also shown its ability to partially or quantitatively repress the target genes, which paved the way for the development of new technologies to evaluate the functionality of essential genes for cell survival, morphology, and antibiotic resistance in bacteria.[10] Unlike genomic knockout, dCas9 mediated gene silencing strategies has a higher degree of freedom to modulate the gene expression and study the relation between quantitative gene expression and cellular function.[7] To enhance the effectivity of dCas9 for better transcriptional repression in mammalian cells, various transcriptional effector domains like KRAB (Krüppel associated box) domain of Kox1, the WRPW domain of Hes1, the CS (Chromo Shadow) domain of HP1α, enhancer of Zeste homolog 2 (EZH2)[11] and the SID4X domain were used in fusion with dCas9[8,9,12] and as a result, it was found that dCas9-KRAB fusion demonstrated relatively consistent and robust effects in targeted gene repression. The transcriptional repressor, KRAB, fused with dCas9 molecules, while targeting the gene regulatory regions like promoters or enhancers,[13] attracts histone deacetylases and methyltransferases, and blocks the mRNA synthesis, by adding epigenetic marks on inactive heterochromatin H3K9 and H3K27[11]

3. CRISPRa

Modulation of gene expression is comprised of both transcriptional repression and transcriptional activation. Upon successful establishment of dCas9 efficiency as CRISPRi, methods were developed to activate the gene expression using the similar phenomenon like CRISPRi, called as CRISPRa. Unlike CRISPRi, only dCas9-CRISPR system cannot be employed for the modulation of gene expression. Hence, effector domains such as transactivator domain of the Herpes simplex viral protein 16 (VP16), multiple copies of VP16 ($VP(16)_n$; n is the number of monomers),[8,14,15] p300,[16] or the trans-activator domain of nuclear factor kappa B (p65) or p65 with heat-shock factor 1 (HSF1),[8,9] were fused to dCas9,[8,14,17] and it was observed that transcriptional activation using VP16 monomer is ineffective, while decent but not great activation effects were observed with multimers of VP16 (VP48, VP64, VP160, VP192).[17] These CRISPRa systems mainly functions by acetylating histones in target regions (using catalytic subunit of p300 histone acetyltransferase) or by recruiting transcription factors for transcription activation. The efficacy of gene induction was increased with the development of a three component system (VPR), based on dCas9 protein linked with the protein complex, comprised of VP64, p65, and Rta, where Rta is an Epstein–Barr virus transcription factor.[18,19] It was also observed that thatching of multiple sgRNAs and localizing multiple copies of dCas9-transactivator complex, along the promoter region, enhanced the gene activation, because of the increased abundance of the activation signal at the site.[15,17]

With the preliminary success of CRISPRa, a plethora of strategies has been adopted by the scientists to improve the CRISPRa technology. Multiple copies of the trans-activator domains were fused with dCas9 to amplify the activation signal for improved gene activation. Later on Tanenbaum et al.[20] devised a dCas9-SunTag system in which the dCas9 is fused to a multimeric peptide array (SunTag), where each peptide binds to one single-chain variable fragment (scFv) fused with VP64. In contrast to the SunTag system, that enhances the activation signal by recruiting multiple copies of the same transactivator, an improved method, called the synergistic activation mediator (SAM) system was developed, where the dCas9-VP64 is implemented with a modified sgRNA containing two MS2 hairpins. This MS2 hairpin acts as an interaction site for the MS2 binding protein (MCP) fused with transactivator domain of p65 and the human heat shock factor 1 (HSF1).[9] With the accelerated development of science, day by day, more variants were designed by combining the SAM with SunTag, or using the SunTag system to recruit p65AD-HSF1 instead of VP64.[21] A direct comparative analysis revealed that SunTag, SAM or VPR can activate a range of genes in a variety of cellular contexts, including human, mouse and Drosophila cells.[22]

Upon setting the platform for dCas9 mediated gene regulation, strategies were aimed to control the spatiotemporal activity of the system, i.e., controlling the recruitment of the effector protein or DNA binding activity of the dCas9 in respect of space and/or time. This has been achieved by using split-(d)Cas9 or split-(d)Cas9-effector proteins, which are assembled into a functional DNA-binding complex in the presence of suitable sgRNA,[23] upon induction with chemical[24] or light.[25,26] In this strategy, engineered dCas9, splitted into individual domains are assembled in the presence of full length and mature sgRNA, as Cas9 domains are disassembled in the presence of truncated sgRNA.[23] To chemically induce the assembly of the N- and C-terminal domains of dCas9, both the domains (N- and C-terminal domains) were fused individually with FK506 binding protein 12 (FKBP) and FKBP rapamycin binding (FRB) domains, respectively, that dimerize upon rapamycin induction, and thus, assembly of the dCas domains occurs, and the functionality is regained. Likewise split-dCas9 has been reassembled by photo-inducible dimerization domains termed Magnets (pMag and nMag), in response to exposure to blue light.[25,26] In order to control effector recruitment, i.e., to bring together full-length dCas9 and effectors such as VP64 or p65 upon blue light irradiation, rather than split-(d)Cas9 reconstitution, the hetero-dimerization partners CRY2 and CIB1 have been used.[25,27] Though several technologies have been formulated for effective modulation of the gene expression, the functionality of these systems requires stringent evaluation for improvement.

4. Applications

dCas9 combined with several effector molecules and targeted transcription factors (TTFs) has enormous applications in the field of molecular biology and medical research for genome-wide modulation of transcription (Table 1). In a recent study, Gilbert et al.[38] screened Human K562 cells for genes that control cell growth by targeting nearly 1600 protein coding genes, using combinations of 10 sgRNAs per gene. In this study, the sgRNAs were designed to thatch around −50 to +300 bp for CRISPRi and −400 to −50 bp for CRISPRa, and the transcriptional repression was achieved using dCas9-KRAB and activation was done using dCas9-SunTag recruiting scFv–VP64 fusions.[38] A large number of new genes and regulatory mechanisms have been identified for cell growth and drug resistance, using CRISPRi/a. CRISPRi/a have been used to study the effect and sensitivity to cholera-diphtheria toxin on human myeloid leukemia K562 cells.[38] In cancer research, the target and the mode of action of an anticancer drug, rigosertib, was deciphered using CRISPRi/a.[39] Apart from identifying the function of individual genes, genetic interactions in a cell is also studied using the CRISPRi/a toolboxes, where more than one genes are perturbed simultaneously, to understand the combinatorial effect on the phenotype, which may be different from the case when genes are agitated individually, thus allowing us to understand the mechanism of gene interaction and explore the genetic network further.[40] To target more than one gene using CRISPRi, for high throughput study of genetic interactions, multiple sgRNAs, targeting the desired sites are delivered simultaneously

Table 1 Application of CRISPR/dCas9 toolboxes for epigenomic engineering.

Epigenetic modifications	Gene expression modulation	CRISPR tool	References
DNA methylation	Suppression	DNMT3A DNMT3A-3L SunTag-DNMT3A	28–31
DNA demethylation	Activation	dCas9-TET1 dCas9-SunTag-TET1 dCas9/MS2/MCP-TET1	32–34
Histone demethylation	Suppression	dCas9-LSD1	35
Histone acetylation	Activation	p300Core	16
Transcriptional factor recruitment	Activation	VP64 VP160/VP192 p65/p65-HSF1 SunTag-VP64 VPR SunTag-p65-HSF1 SAM TREE Casilio Scaffold	36
DNA Methylation	Suppression	dCas9-KRAB dCas9-KRAB-MeCP2 dCas9-EZH2	11,13,37

but the major drawback of using multiple sgRNAs is the unhealthy competition for dCas9 protein between the sgRNAs, hence, prior optimization is necessary.[41] In a recent study, to overcome the challenges of using multiple sgRNA for a single Cas9 protein, orthogonal Cas9 proteins from *Streptococcus pyogenes* and *Staphylococcus aureus* were used for dual screening of a gene pair[42]

In biological studies, for the development of potential medications, use of CRISPRi/a technology is increasing day by day and it has been found that these technologies has played a major role for the treatment of metabolic disorders and diseases related to gene malfunctioning such as diabetes mellitus, Duchenne muscular dystrophy (DMD), and haploinsufficiency-related disorders. CRISPRa approaches were also used to activate endogenous antiviral factors *APOBEC3B* and *APOBEC3G* to fight against human immunodeficiency virus (HIV) infection.[43] Furthermore, an altered strategy to combat HIV infection, based on CRISPRa system, was designed, known as "shock-and-kill," that works by reactivating latent HIV using CRISPRa and followed by death of infected cells by direct cytotoxic effects of the viral proteins or by immune clearance.[44–47] Recently, for the destruction of hepatitis-B virus (HBV) genomes, a CRISPRa approach was hired to activate *APOBECs*; and this method can also be used in combination with nuclease active Cas9 to treat HBV.[48,49] Obesity and Dravet syndrome has been treated *in vivo* using CRISPRa-toolbox.[50,51] In oncological studies, CRISPRi approaches have proved to be highly efficient and has been used to repress proto-oncogenes like *Granulin,*[52] and simultaneously activating tumor suppressors *PTEN, DKK3,* or *CHEK2*, resulting in the suppression of cancer cell proliferation.[53–55] To destroy the tumor cells by cytotoxic $CD8^+$ lymphocytes, tumor antigens were overexpressed using recently developed MAEGI approach, based on CRISPRa.[56]

5. Epigenome engineering and CRISPRi/a

The epigenetic modifications, DNA methylation or histone modifications, have profound impact on chromatin organization, gene expression and regulation, involved in numerous cellular processes such as differentiation, development and tumorigenesis.[57–60] Due to the lack of efficient tools, it has been a challenge to investigate the epigenetics behind the complex diseases. With the emergence of dCas9 toolbox, fusion of epigenetic enzymatic domains with the dCas9 has enabled the researchers to ensure desired locus-specific epigenetic modifications, which in turn has resulted in addressing several biological questions that remained unanswered (Table 2). Understanding the epigenetic modification that influences gene expression, has been crucial to development of novel technologies of treatment of human diseases.

In mammalian cells, methylation of unmethylated CpGs, is catalyzed by the methyl-transferases, DNMT3A and DNMT3B,[60] where DNMT3A is stimulated by the co-factor DNMT3L,[85] whereas TET1 protein initiates removal of methyl residues from DNA.[86] The enzyme complex comprised of a dCas9 fused to the catalytic domain of DNA methyltransferase (DNMT3A), has been used to induce site-specific methylation of cytosines in the CpG islands around the dCas9 target site (Transcription start site), in human and mice cells,[28,29,87] which in turn resulted in the repression of the tumor-suppressor gene CDKN2A, due to the increase in methylation of the CpG island spanning the Transcription start site (TSS).[28] Hypermethylation

Table 2 Applications of CRISPRi/a approaches in fundamental and therapeutic research.

Application domain	CRISPR/dCas system	Application	References
Fundamentals	dCas9-KRAB dCas9-p300	Annotation of Distal gene regulatory elements	61,62
	dCas9-KRAB	Identification of Estrogen receptor enhancers	63
	dCas9-KRAB/ dCas9-VPR	Decipherence of function of Syt1, Bdnf, and Reln in mammalian brain	64,65
	dCas9-VPR Scaffolds SAM	Cell signaling analysis and Generating chimeric receptors	66,67
	SAM	Identification of antiviral factors for Norovirus infection	68
	SAM; SunTag-VP64; dCas9-KRAB	Annotating tumor-related genes involved in cancer: Hells, Egfr, lncRNAs, Myc, Kras-dependent genes	36
Therapeutics	SAM; Scaffold (MCP-p65-HSF1	HIV therapy by activating BST2/tetherin, APOBEC3B and APOBEC3G	43,69
	dCas9-p300	HIV therapy by activating APOBEC3A, APOBEC3B, APOBEC3G, AID	36
	SunTag-VP64; dCas9-VPR; SAM	Reactivating latent HIV in a "shock-and-kill" therapeutic approach	36,44,45,47
	dCas9-KRAB	Repressing TNFR1, IL1R1, IL6st	70–72
	SAM	Neuro- and nephroprotection by activating Klotho gene	73
	dCas9-KRAB	Repressing Pcsk9 to reduce serum cholesterol levels	36
	SAM dCas9-VP160	Generating insulin-producing cells by upregulating Pdx1 or Ins	74,75

Continued

Table 2 Applications of CRISPRi/a approaches in fundamental and therapeutic research.—cont'd

Application domain	CRISPR/dCas system	Application	References
	SAM	Treating DMD by activating Utrophin gene	74
	dCas9	Correcting myotonic dystrophy types 1 and 2 by blocking transcription of expanded microsatellite repeats	36
	dCas9-VP64	Treating obesity by upregulating Sim1; Treating Dravet syndrome by upregulating Scn1a; Treating congenital muscular dystrophy type 1A by upregulating Lama1	50,51,76
	dCas9-DNMT3A dCas9-KRAB dCas9-Ezh2	Repressing Granulin proto-oncogene	52
	dCas9-VP64 dCas9-VPR	Activating tumor suppressors PTEN, CHEK2, DKK3	53–55
	dCas9-VP64	Activating telomere-targeting Cas9 nuclease in cancer cells	77
	SAM	Increased presentation of tumor antigens to immune cells	56
	dCas9-VP64 SAM dCas9-p300	Generating iPSCs by inducing KLF4, LIN28, MYC, OCT4, SOX2	78–80
	dCas9-VPR	Upregulating NANOG to maintain pluripotency	81
	SAM SunTag-VP64	Differentiating stem cells into adipocytes, neural cells and pancreatic cells	82–84
	SAM SunTag-p65-HSF1	Neural reprogramming by activating Neurog2, Ascl1, Neurod1, Dkk1, etc.	21

of the DNA regulatory regions of multiple tumor suppressor gene, including *CDKN2A, RASSF1, HIC1, PTEN*, and *SMARCA2*, using dCas9-DNMT3A system, revolutionized the understanding of gene function in oncogenesis.[88,89] CTCF-linked interaction of the *MYC* proto-oncogene with its super-enhancer was disrupted using dCas9-based methylation, and a new direction was set for the treatment of many oncological diseases.[90] Furthermore, physiological expression of tumor suppressors were normalized by demethylation of *BRCA*[32] and *SARI*[91] promoters. Increased CpG methylation of the unmethylated promoter regions of two specific loci, the BACH and IL6ST, reduced their expression level.[29] Though it seemed that DNA methylation is the only reason behind the induced repression, but later on it was identified that along with the DNA methylation, DNMT3A also induced repression by sterically hindering the transcription machinery of the cell.[28,29] The proliferation of mouse myeloid progenitor cells were enhanced due to the repression Cdkn1a gene by methylation using dCas9-DNMT3A. It was also observed that the efficiency of DNA methylation of the targeted CpG residues were enhanced when CpGs are located downstream of the PAM, within ~50 bp of the sgRNA binding site.[28,29]

To increase the efficiency of the system a combination of dCas9 fusion proteins were administered, in which the KRAB (Krüppel associated box) domain of Kox1, DNMT3A and DNM3L domains were individually fused to respective dCas9, and it resulted in long-term epigenetic silencing of the target genes.[87] Another variant of this technology was designed to improve the methylation efficiency of this system, by coupling dCas9-system with the SunTag system, which recruits multiple effector molecules (e.g., DNMT3A) to the desired site. Interestingly, dCas9-SunTag-DNMT3A was observed to be highly specific, and it does not seem to affect the overall methylation state of the genome[30] and was proposed to be suitable for precisely methylating small portions to large swaths of the genome and subsequently suppress the gene transcription. dCas9-based methylation tools were worthwhile to unravel the relationship between promoter methylation and dysfunction of *desmoplakin* gene expression in the pathogenesis of idiopathic lung fibrosis.[92] Moreover, phospholipid phosphatases associated with calcification of the aortic valve, a pathological condition leading to myocardial infarction was identified to be regulated by methylation of enhancer regions.[93]

Many research and potential therapeutic aspects like, treatment of malignancies by activating tumor suppressors, treatment of genetic diseases, generation or differentiation of stem cells, frequently require demethylation of genomic DNA in order to activate the down regulated genes. DNA demethylation-mediated gene activation was carried out by fusing the catalytic domain of the demethylase TET1 with dCas9, which induced site-specific DNA demethylation of the targeted RANKL gene, silenced through hypermethylation in HEK-293FT cells, resulting in upregulation of the target gene.[32–34] The universality of this epigenome editing strategy was demonstrated in HeLa cells, where expression of MAGEB2 and MMP2, silenced by hypermethylation, was shown to be up-regulated by targeted demethylation. dCas9-TET1 has been used to identify the key role of tumor necrosis factor alpha (*TNFα*) promoter hypermethylation, in the development of nephropathies in diabetes.[94] Later on, the dCas9-SunTag strategy was adapted for targeted DNA demethylation. TET1 was fused to the scFv domain of the dCas9-SunTag system, which carried out demethylation of CpGs successfully and simultaneously upregulated the targeted genes in cancer cell lines, embryonic stem cells, primary neural precursor cells, and *in vivo* in mouse fetuses.[33] New therapeutic techniques has also been developed with the help of dCas9-based methylation and demethylation tools because many diseases such as Parkinson's disease is associated with epigenomic defects. Cell death in an *in vitro* model of Parkinson's disease was observed to be reduced upon hypermethylation of *SNCA* promoter, leading to the establishment of the new therapeutic approach.[95] dCas9-TET1 mediated demethylation of CGG repeats in the promoter region of *FMR1*, has shown to rectify the effects of fragile X syndrome, by activating the gene expression.[96] Likewise, by dCas9-TET1-based demethylation of *Sox1* gene, has resulted in effective programming of neural stem progenitor cells.[97]

An altered variant of this technology was introduced using three components[1]: dCas9 protein[2]; gRNA with two special MS2-hairpins (aptamers); and[3] effector proteins linked to MCP proteins that interact with MS2, where, dCas9 protein in complex with a modified gRNA gets associated with the target genomic region and then TET1-MCP proteins recognizes and binds to MS2 hairpins within gRNA, enabling site-specific demethylation. Since two molecules of MCP can interact with one MS2 hairpin, a single gRNA harboring two MS2 hairpins, can attract up to four units of TET1.[34]

In addition to DNA methylation, posttranslational modifications of histones like acetylation, methylation (lysine and arginine), phosphorylation, ubiquitylation, sumoylation, ADP ribosylation, deamination and proline isomerization,[98] encode essential epigenomic features, associated with the gene expression program of a cell. Formation of heterochromatin and euchromatin largely depends on several epigenetic events like deacetylation of H3K9 and H3K27 histone residues; methylation of H3K9 and H3K27 (H3K9Me3 and H3K27Me3); and methylation of DNA regions wrapped by histones for heterochromatin formation, whereas, methylated H3K4 and H3K79 histone residues and acetylated H3K9 and H3K27 residues marks the transcriptionally active euchromatin.[99–101] These processes are carried out by several effectors, such as histone deacetylases catalyzes the histone deacetylation, SUV39H1 and G9A induces methylation of H3K9,[102,103] whereas, EZH2 methylates H3K27.[104] Factors including MLL and PRDM9 efficiently methylates the H3K4, while histone methyltransferase DOT1L carry out the methylation of H3K79 residues.[105] Histone demethylation is mediated by LSD1[106] and acetylation of H3K9 and H3K27 is carried out by CBP and p300 histone acetyltransferases.[107]

dCas9 has been fused to the catalytic core of the human acetyltransferase p300, which allowed acetylation of histone H3 lysine27 (H3K27), and in turn resulted in activation of targeted genes in human HEK293T cells. Directing dCas9–p300 fusions either to promoter regions or distal regions (proximal or distant enhancers) also activated the gene expression, due to the enrichment of acetylated H3K27 at the corresponding genomic targets.[16] Likewise, the histone acetyltransferase domain of the CREB-binding protein has been fused to dCas9 to build dCas9-CBPHAT and has been used in successful locus-specific acetylation of histones, which in turn upregulated the gene expression when dCas9-CBPHAT was targeted to promoters or proximal and distal enhancer, similar to dCas9-p300.[108] In some cases, a single gene could be transactivated, by targeting the promoters using dCas9-VP64 along with dCas9-p300, where p300 directly catalyses H3K27 acetylation and VP64 recruits subsequent transactivation components.[109]

Likewise, histone demethylase LSD1 coupled with dCas9, targeting the putative enhancers in mouse embryonic stem cells, resulted in the repression of gene expression by decreasing the epigenetic modification level of H3K4Me3 (demethylation) and H3K27Ac (deacetylation) around the target site.[35] Using this strategy, eight pluripotency-specific enhancers were targeted, which resulted in the discovery of four new enhancers regulating the expression of genes, critical for maintaining the embryonic state.[35] dCas9-KRAB were designed to target the Tbx3-enhancer, which resulted in a reduction of H3K27ac and an increase of H3K27me3 and H3K9me3 at the Tbx3 proximal promoter region. In a separate study, the KRAB domain was directed to the HS2 enhancer in the globin locus control region (LCR) in human K562 cells, where H3K9me3 was effectively introduced at the target site.[13] With the advancement of science, diverse dCas9 toolboxes were developed to carry out histone methylations (using histone methyltransferases like DOT1L, SMYD3 and PRDM9) and histone deacetylation (using histone deacetylase HDAC) more efficiently.[110–112]

CRISPR/Cas9 system suffers off target activity that undermines its application as therapeutics. Off-target activity contributes to chromosome instability and severely compromise the utility of CRISPR/Cas-based systems. Though several strategies have been developed to limit the non-specific cleavage, but complete solution of this problem is not yet known. It is noteworthy, that off-target activity of CRISPR-Cas system for epigenetic modifications is not of a major concern, since off-target epigenetic modifications are usually momentary and likely have an effect on transcriptional activity of miss-targeted genes. Matharu et al. observed that there was no off-target epigenetic modifications or non-targeted alterations in mRNA levels in mice, using CRISPRa.[113]

6. dCas9 toolboxes beyond CRISPRi/a

Although the discovery of CRISPR-Cas9 system has revolutionized the concept of gene editing, CRISPR/dCas9 system with its DNA binding dCas9 protein, redesigned from wild Cas9, further extended the scope of application beyond gene editing. This technology allowed us to unravel the characteristics of unknown genes and their role in cellular functions, as described earlier in this chapter. dCas9 alone or in combination with attached effector proteins are commonly used to address target specific gene expression to understand unsolved biological issues related to health and medicine. In this context a number of constructs have been generated for targeted epigenetic modifications in cells.

Apart from the tools related to CRISPRi/a, there are many applications and recent achievements of dCas toolboxes in medical and agricultural fields that are beyond the scope of this chapter. dCas-Cas12a-based multiplex system has been developed for simultaneous modulation of several genes to simplify and understand the complex biological processes with unparalleled accuracy.[114] RNA guided targeted insertion of DNA sequence with dCas associated transposases, is also providing exciting information on recent advances in genome editing research.[115,116] Along with DNA, epigenetic modification of RNA with m6A marker has also recently been described.[117]

Besides, in the field of molecular and cellular imaging, CRISPR/dCas9 tool has been used to develop programmable DNA recognition systems for the visualization of endogenous genomic elements in live cells. Advances in this technology potentially helped to visualize target DNA/RNA sequences, enabling researchers to study chromatin dynamics playing a crucial role in genome organization and transcriptional regulation in living cells.[118–120] The ease of reprogramming the specificity determinant sgRNA and customizing the CRISPR protein to visualize new sequences allowed development of CRISPR-based next generation imaging tool. CRISPR imaging has been employed successfully to visualize repetitive sequences in the genome.[121,122] With the success of CRISPR/Cas system as an imaging tool, great efforts have been made to optimize and develop more efficient system for broader application. Since the first demonstration of CRISPR imaging, one issue of low signal-to-noise ratio due to the freely diffusing dCas9-eGFP molecules, was prevalent, which was solved using the SunTag (SUperNova tag), as it significantly amplifies the fluorescence signal at the target loci.[20,123] This SunTag (SUperNova) system is comprised of a GFP-tagged cognate single-chain variable fragment antibody (scFv) and a 24 copies of short peptide epitope (GCN4) arranged tandemly. Using this system for CRISPR imaging of telomeres resulted in ~20-fold increase in the signal, which made the visualization of genomic loci (two low repeat loci in human embryonic kidney (HEK293T) cells) with lower light illumination.[123] With the gradual development of CRISPR imaging, a novel approach using a split version of super-folder GFP (sfGFP) has been adopted. In this approach, the sfGFP is splitted between the 10th and 11th β-strand (sfGFP1–10 and sfGFP11), resulting in the loss of fluorescence.[124] A tandem array of up to seven repeats of the sfGFP11 epitope (dCas9- sfGFP11x7), can be fused to a protein of interest which can be subsequently recognized by the co-expressed sfGFP1–10 fragment, as upon self-complementation, the chromophore matures and sfGFP regains its fluorescence.[125] It was demonstrated that dCas9- sfGFP11x7 together with VP64-sfGFP1–10, drastically increases the expression level of the targeted CXCR4 gene. In contrast to the SunTag system, which causes background fluorescence due to unbound scFv-GFP molecules, this GFP-derived epitope tag would be particularly promising, since over expression of the non-fluorescent sfGFP1–10 would not result in background signals. Furthermore, an analogous tag has also been developed from super-folder Cherry (sfCherry11), which could also enhance the versatility of this approach.[124]

The conventional CRISPR imaging technique has been upgraded to overcome the inherent drawback of inability to label multiple loci with multiple color simultaneously. To overcome this limitation, orthogonal dCas9 proteins derived from *Streptococcus pyogenes* (Sp-dCas9), *Neisseria meningitidis* (Nm-dCas9) and *Streptococcus thermophilus* (St1-dCas9) were tagged with a different fluorescent protein for use in multiplexed targeting, due to their different PAM recognitions.[126,127] Hence, by co-expressing differently tagged dCas9 orthologs with their cognate sgRNAs simultaneously, each specific for a distinct locus, allowed to resolve the inter- and intra-chromosomal distances between two repetitive sequences in living human cells. Moreover, the small Cas9 ortholog from *Staphylococcus aureus* (Sa-dCas9) were repurposed for multiplexed CRISPR imaging and demonstrated that a combination of Sp-dCas9-eGFP and Sa-dCas9-mCherry is capable to resolve two different loci, which are separated by less than 300 kb.[128]

Although the type II CRISPR/Cas system has been developed to target double-stranded DNA, recently, the scope of CRISPR imaging has been further expanded to visualize endogenous, unmodified RNA,[120] where Cas9 is guided to RNA by providing the PAM in trans by an oligomer (PAMmer), which is partially complementary to the targeted RNA.[129] However, so far, when large quantities of RNA accumulates within stress granules, RNA can only be visualized. Hence, dCas9- sfGFP11-tag fusion is likely to improve the signal-to-noise ratio, thus allowing the tracking of less abundant RNAs. Moreover, an additional Cas protein, Cas13, a member of the Class II type VI system, which specifically targets single-stranded RNA in an crRNA-guided manner, has recently been developed for targeted RNA visualization, as well as RNA knockdown and editing.[130,131] This CRISPR effector, Cas13, can be converted into a catalytically inactive variant (dCas13) by substituting catalytic residues within the HEPN (Higher Eukaryotes and Prokaryotes Nucleotide-binding) domains. Fusion of dCas13 and monomeric superfolder GFP (msfGFP) has been used to track the accumulation of actin beta (ACTB) transcripts in the stress granules in living cells.[130] Also, two CRISPR mediated systems, CAMERA[132] and DOMINO[133] is devised to record cellular events in bacterial and mammalian cells. Collectively, CRISPR imaging is a highly flexible and potential alternative to visualization techniques and is also time- and cost-effective for the chromatin dynamics study *in vivo*.

7. Conclusion

The field of CRISPR is evolving too fast with continuous incorporation of data from recent research, making the tool more efficient and useful not only to edit genome but also for epigenome editing, transcriptional modulation, live cell imaging, etc. However, several major barriers like off-target activity, accessibility, inefficient delivery and instability need to be overcome. Increasing research evidences suggest that targeted epigenetic modification and transcriptional modulation at genome-scale is possible using CRISPR/dCas9 system. The technology has broad prospects in medical field and it is expected that these new technologies will revolutionize the medical treatment for human diseases.

Conflict of interest

The authors declare that there was no conflict of interest.

Acknowledgment

T.A.K. gratefully acknowledges the financial assistance in the form of fellowship (File No.: 09/202(0070)/2017-EMR-I) from Council of Scientific & Industrial Research (CSIR), India.

References

1. Gupta D, Bhattacharjee O, Mandal D, et al. CRISPR-Cas9 system: A new-fangled dawn in gene editing. *Life Sci.* 2019;232:116636. https://doi.org/10.1016/j.lfs.2019.116636.
2. Zetsche B, Gootenberg JS, Abudayyeh OO, et al. Cpf1 is a single RNA-guided endonuclease of a class 2 CRISPR-cas system. *Cell.* 2015;163(3):759–771. https://doi.org/10.1016/j.cell.2015.09.038.
3. Jiang F, Doudna JA. CRISPR–Cas9 structures and mechanisms. *Annu Rev Biophys.* 2017;46(1):505–529. https://doi.org/10.1146/annurev-biophys-062215-010822.
4. Elowitz MB, Leibier S. A synthetic oscillatory network of transcriptional regulators. *Nature.* 2000;403(6767):335–338. https://doi.org/10.1038/35002125.
5. Jackson AL, Bartz SR, Schelter J, et al. Expression profiling reveals off-target gene regulation by RNAi. *Nat Biotechnol.* 2003;21(6):635–637. https://doi.org/10.1038/nbt831.
6. Kampmann M. CRISPRi and CRISPRa screens in mammalian cells for precision biology and medicine. *ACS Chem Biol.* 2018;13(2):406–416. https://doi.org/10.1021/acschembio.7b00657.
7. Qi LS, Larson MH, Gilbert LA, et al. Repurposing CRISPR as an RNA-γuided platform for sequence-specific control of gene expression. *Cell.* 2013;152(5):1173–1183. https://doi.org/10.1016/j.cell.2013.02.022.
8. Gilbert LA, Larson MH, Morsut L, et al. XCRISPR-mediated modular RNA-guided regulation of transcription in eukaryotes. *Cell.* 2013;154(2):442. https://doi.org/10.1016/j.cell.2013.06.044.
9. Konermann S, Brigham MD, Trevino AE, et al. Genome-scale transcriptional activation by an engineered CRISPR-Cas9 complex. *Nature.* 2015;517(7536):583–588. https://doi.org/10.1038/nature14136.
10. Peters JM, Colavin A, Shi H, et al. A comprehensive, CRISPR-based functional analysis of essential genes in bacteria. *Cell.* 2016;165(6):1493–1506. https://doi.org/10.1016/j.cell.2016.05.003.

11. O'Geen H, Bates SL, Carter SS, et al. Ezh2-dCas9 and KRAB-dCas9 enable engineering of epigenetic memory in a context-dependent manner. *Epigenetics Chromatin.* 2019;12(1). https://doi.org/10.1186/s13072-019-0275-8.
12. Gilbert LA, Larson MH, Morsut L, et al. CRISPR-mediated modular RNA-guided regulation of transcription in eukaryotes. *Cell.* 2013;154(2):442. https://doi.org/10.1016/j.cell.2013.06.044.
13. Thakore PI, D'Ippolito AM, Song L, et al. Highly specific epigenome editing by CRISPR-Cas9 repressors for silencing of distal regulatory elements. *Nat Methods.* 2015;12(12):1143–1149. https://doi.org/10.1038/nmeth.3630.
14. Perez-Pinera P, Kocak DD, Vockley CM, et al. RNA-guided gene activation by CRISPR-Cas9-based transcription factors. *Nat Methods.* 2013;10(10):973–976. https://doi.org/10.1038/nmeth.2600.
15. Maeder ML, Linder SJ, Cascio VM, Fu Y, Ho QH, Joung JK. CRISPR RNA-guided activation of endogenous human genes. *Nat Methods.* 2013;10(10):977–979. https://doi.org/10.1038/nmeth.2598.
16. Hilton IB, D'Ippolito AM, Vockley CM, et al. Epigenome editing by a CRISPR-Cas9-based acetyltransferase activates genes from promoters and enhancers. *Nat Biotechnol.* 2015;33(5):510–517. https://doi.org/10.1038/nbt.3199.
17. Cheng AW, Wang H, Yang H, et al. Multiplexed activation of endogenous genes by CRISPR-on, an RNA-guided transcriptional activator system. *Cell Res.* 2013;23(10):1163–1171. https://doi.org/10.1038/cr.2013.122.
18. Chavez A, Scheiman J, Vora S, et al. Highly efficient Cas9-mediated transcriptional programming. *Nat Methods.* 2015;12(4):326–328. https://doi.org/10.1038/nmeth.3312.
19. Fang L, Hung SSC, Yek J, et al. A simple cloning-free method to efficiently induce gene expression using CRISPR/Cas9. *Mol Ther Nucleic Acids.* 2019;14:184–191. https://doi.org/10.1016/j.omtn.2018.11.008.
20. Tanenbaum ME, Gilbert LA, Qi LS, Weissman JS, Vale RD. A protein-tagging system for signal amplification in gene expression and fluorescence imaging. *Cell.* 2014;159(3):635–646. https://doi.org/10.1016/j.cell.2014.09.039.
21. Zhou H, Liu J, Zhou C, et al. In vivo simultaneous transcriptional activation of multiple genes in the brain using CRISPR-dCas9-activator transgenic mice. *Nat Neurosci.* 2018;21(3):440–446. https://doi.org/10.1038/s41593-017-0060-6.
22. Chavez A, Tuttle M, Pruitt BW, et al. Comparison of Cas9 activators in multiple species. *Nat Methods.* 2016;13(7):563–567. https://doi.org/10.1038/nmeth.3871.
23. Wright AV, Sternberg SH, Taylor DW, et al. Rational design of a split-Cas9 enzyme complex. *Proc Natl Acad Sci USA.* 2015;112(10):2984–2989. https://doi.org/10.1073/pnas.1501698112.
24. Zetsche B, Volz SE, Zhang F. A split-Cas9 architecture for inducible genome editing and transcription modulation. *Nat Biotechnol.* 2015;33(2):139–142. https://doi.org/10.1038/nbt.3149.
25. Nihongaki Y, Yamamoto S, Kawano F, Suzuki H, Sato M. CRISPR-Cas9-based photoactivatable transcription system. *Chem Biol.* 2015;22(2):169–174. https://doi.org/10.1016/j.chembiol.2014.12.011.
26. Nihongaki Y, Kawano F, Nakajima T, Sato M. Photoactivatable CRISPR-Cas9 for optogenetic genome editing. *Nat Biotechnol.* 2015;33(7):755–760. https://doi.org/10.1038/nbt.3245.
27. Polstein LR, Gersbach CA. A light-inducible CRISPR-Cas9 system for control of endogenous gene activation. *Nat Chem Biol.* 2015;11(3):198–200. https://doi.org/10.1038/nchembio.1753.
28. McDonald JI, Celik H, Rois LE, et al. Reprogrammable CRISPR/Cas9-based system for inducing sitespecific DNA methylation. *Biol Open.* 2016;5(6):866–874. https://doi.org/10.1242/bio.019067.

29. Vojta A, Dobrinic P, Tadic V, et al. Repurposing the CRISPR-Cas9 system for targeted DNA methylation. *Nucleic Acids Res.* 2016;44(12):5615–5628. https://doi.org/10.1093/nar/gkw159.
30. Huang YH, Su J, Lei Y, et al. DNA epigenome editing using CRISPR-Cas SunTag-directed DNMT3A. *Genome Biol.* 2017;18(1):176. https://doi.org/10.1186/s13059-017-1306-z.
31. Liu XS, Wu H, Ji X, et al. Editing DNA methylation in the mammalian genome. *Cell.* 2016;167(1):233–247.e17. https://doi.org/10.1016/j.cell.2016.08.056.
32. Choudhury SR, Cui Y, Lubecka K, Stefanska B, Irudayaraj J. CRISPR-dCas9 mediated TET1 targeting for selective DNA demethylation at BRCA1 promoter. *Oncotarget.* 2016;7(29):46545–46556. https://doi.org/10.18632/oncotarget.10234.
33. Morita S, Noguchi H, Horii T, et al. Targeted DNA demethylation in vivo using dCas9-peptide repeat and scFv-TET1 catalytic domain fusions. *Nat Biotechnol.* 2016; 34(10):1060–1065. https://doi.org/10.1038/nbt.3658.
34. Xu X, Tao Y, Gao X, et al. A CRISPR-based approach for targeted DNA demethylation. *Cell Discov.* 2016;2. https://doi.org/10.1038/celldisc.2016.9.
35. Kearns NA, Pham H, Tabak B, et al. Functional annotation of native enhancers with a Cas9-histone demethylase fusion. *Nat Methods.* 2015;12(5):401–403. https://doi.org/10.1038/nmeth.3325.
36. Brezgin S, Kostyusheva A, Kostyushev D, Chulanov V. Dead cas systems: types, principles, and applications. *Int J Mol Sci.* 2019;20(23):6041. https://doi.org/10.3390/ijms20236041.
37. Schultz DC, Friedman JR, Rauscher FJ. Targeting histone deacetylase complexes via KRAB-zinc finger proteins: The PHD and bromodomains of KAP-1 form a cooperative unit that recruits a novel isoform of the Mi-2α subunit of NuRD. *Genes Dev.* 2001;15(4):428–443. https://doi.org/10.1101/gad.869501.
38. Gilbert LA, Horlbeck MA, Adamson B, et al. Genome-scale CRISPR-mediated control of gene repression and activation. *Cell.* 2014;159(3):647–661. https://doi.org/10.1016/j.cell.2014.09.029.
39. Jost M, Chen Y, Gilbert LA, et al. Combined CRISPRi/a-based chemical genetic screens reveal that rigosertib is a microtubule-destabilizing agent. *Mol Cell.* 2017;68(1): 210–223.e6. https://doi.org/10.1016/j.molcel.2017.09.012.
40. Mani R, St. Onge RP, Hartman IV JL, Giaever G, Roth FP. Defining genetic interaction. *Proc Natl Acad Sci USA.* 2008;105(9):3461–3466. https://doi.org/10.1073/pnas.0712255105.
41. Du D, Roguev A, Gordon DE, et al. Genetic interaction mapping in mammalian cells using CRISPR interference. *Nat Methods.* 2017;14(6):577–580. https://doi.org/10.1038/nmeth.4286.
42. Boettcher M, Tian R, Blau JA, et al. Dual gene activation and knockout screen reveals directional dependencies in genetic networks. *Nat Biotechnol.* 2018;36(2):170–178. https://doi.org/10.1038/nbt.4062.
43. Bogerd HP, Kornepati AVR, Marshall JB, Kennedy EM, Cullen BR. Specific induction of endogenous viral restriction factors using CRISPR/Cas-derived transcriptional activators. *Proc Natl Acad Sci USA.* 2015;112(52):E7249–E7256. https://doi.org/10.1073/pnas.1516305112.
44. Saayman SM, Lazar DC, Scott TA, et al. Potent and targeted activation of latent HIV-1 using the CRISPR/dCas9 activator complex. *Mol Ther.* 2016;24(3):488–498. https://doi.org/10.1038/mt.2015.202.
45. Limsirichai P, Gaj T, Schaffer DV. CRISPR-mediated activation of latent HIV-1 expression. *Mol Ther.* 2016;24(3):499–507. https://doi.org/10.1038/mt.2015.213.
46. Zhang Y, Yin C, Zhang T, et al. CRISPR/gRNA-directed synergistic activation mediator (SAM) induces specific, persistent and robust reactivation of the HIV-1 latent reservoirs. *Sci Rep.* 2015;5(1):16277. https://doi.org/10.1038/srep16277.

47. Ji H, Jiang Z, Lu P, et al. Specific reactivation of latent HIV-1 by dCas9-SunTag-VP64-mediated guide RNA targeting the HIV-1 promoter. *Mol Ther.* 2016;24(3): 508–521. https://doi.org/10.1038/mt.2016.7.
48. Kostyushev D, Brezgin S, Kostyusheva A, Zarifyan D, Goptar I, Chulanov V. Orthologous CRISPR/Cas9 systems for specific and efficient degradation of covalently closed circular DNA of hepatitis B virus. *Cell Mol Life Sci.* 2019;76(9): 1779–1794. https://doi.org/10.1007/s00018-019-03021-8.
49. Kostyushev D, Kostyusheva A, Brezgin S, et al. Suppressing the NHEJ pathway by DNA-PKcs inhibitor NU7026 prevents degradation of HBV cccDNA cleaved by CRISPR/Cas9. *Sci Rep.* 2019;9(1):111123. https://doi.org/10.1038/s41598-019-38526-6.
50. Matharu N, Rattanasopha S, Tamura S, et al. CRISPR-mediated activation of a promoter or enhancer rescues obesity caused by haploinsufficiency. *Science (80-).* 2019;363 (6424). https://doi.org/10.1126/science.aau0629.
51. Colasante G, Lignani G, Brusco S, et al. dCas9-based Scn1a gene activation restores inhibitory interneuron excitability and attenuates seizures in dravet syndrome mice. *Mol Ther.* 2019. https://doi.org/10.1016/j.ymthe.2019.08.018. Published online.
52. Wang H, Guo R, Du Z, et al. Epigenetic targeting of granulin in hepatoma cells by synthetic CRISPR dCas9 epi-suppressors. *Mol Ther Nucleic Acids.* 2018;11:23–33. https://doi.org/10.1016/j.omtn.2018.01.002.
53. Moses C, Nugent F, Waryah CB, Garcia-Bloj B, Harvey AR, Blancafort P. Activating PTEN tumor suppressor expression with the CRISPR/dCas9 system. *Mol Ther Nucleic Acids.* 2019;14:287–300. https://doi.org/10.1016/j.omtn.2018.12.003.
54. Brauna CJ, Brunoa PM, Horlbeckc MA, Gilbertc LA, Weissmanc JS, Hemanna MT. Versatile in vivo regulation of tumor phenotypes by dCas9-mediated transcriptional perturbation. *Proc Natl Acad Sci USA.* 2016;113(27):E3892–E3900. https://doi.org/10.1073/pnas.1600582113.
55. Kardooni H, Gonzalez-Gualda E, Stylianakis E, Saffaran S, Waxman J, Kypta RM. CRISPR-mediated reactivation of DKK3 expression attenuates TGF-β signaling in prostate cancer. *Cancers (Basel).* 2018;10(6):10. https://doi.org/10.3390/cancers10060165.
56. Wang G, Chow RD, Bai Z, et al. Multiplexed activation of endogenous genes by CRISPRa elicits potent antitumor immunity. *Nat Immunol.* 2019;20(11):1494–1505. https://doi.org/10.1038/s41590-019-0500-4.
57. Bernstein BE, Meissner A, Lander ES. The mammalian epigenome. *Cell.* 2007;128(4): 669–681. https://doi.org/10.1016/j.cell.2007.01.033.
58. Cedar H, Bergman Y. Linking DNA methylation and histone modification: patterns and paradigms. *Nat Rev Genet.* 2009;10(5):295–304. https://doi.org/10.1038/nrg2540.
59. Pulecio J, Verma N, Mejía-Ramírez E, Huangfu D, Raya A. CRISPR/Cas9-based engineering of the epigenome. *Cell Stem Cell.* 2017;21(4):431–447. https://doi.org/10.1016/j.stem.2017.09.006.
60. Kato Y, Kaneda M, Hata K, et al. Role of the Dnmt3 family in de novo methylation of imprinted and repetitive sequences during male germ cell development in the mouse. *Hum Mol Genet.* 2007;16(19):2272–2280. https://doi.org/10.1093/hmg/ddm179.
61. Gasperini M, Hill AJ, McFaline-Figueroa JL, et al. A genome-wide framework for mapping gene regulation via cellular genetic screens. *Cell.* 2019;176(1-2):377–390. e19. https://doi.org/10.1016/j.cell.2018.11.029.
62. Klann TS, Black JB, Chellappan M, et al. CRISPR-Cas9 epigenome editing enables high-throughput screening for functional regulatory elements in the human genome. *Nat Biotechnol.* 2017;35(6):561–568. https://doi.org/10.1038/nbt.3853.
63. Carleton JB, Berrett KC, Gertz J. Multiplex enhancer interference reveals collaborative control of gene regulation by estrogen receptor α-bound enhancers. *Cell Syst.* 2017; 5(4):333–344.e5. https://doi.org/10.1016/j.cels.2017.08.011.

64. Zheng Y, Shen W, Zhang J, et al. CRISPR interference-based specific and efficient gene inactivation in the brain. *Nat Neurosci*. 2018;21(3):447–454. https://doi.org/10.1038/s41593-018-0077-5.
65. Savell KE, Bach SV, Zipperly ME, et al. A neuron-optimized CRISPR/dCas9 activation system for robust and specific gene regulation. *eNeuro*. 2019;6(1). https://doi.org/10.1523/ENEURO.0495-18.2019.
66. Baeumler TA, Ahmed AA, Fulga TA. Engineering synthetic signaling pathways with programmable dCas9-based chimeric receptors. *Cell Rep*. 2017;20(11):2639–2653. https://doi.org/10.1016/j.celrep.2017.08.044.
67. Kipniss NH, Dingal PCDP, Abbott TR, et al. Engineering cell sensing and responses using a GPCR-coupled CRISPR-Cas system. *Nat Commun*. 2017;8(1):1–10. https://doi.org/10.1038/s41467-017-02075-1.
68. Orchard R, Sullender M, Dunlap B, Balce D, Doench J, Virgin H. Identification of anti-norovirus genes in mouse and human cells using genome-wide CRISPR activation screening. *bioRxiv*. 2018. https://doi.org/10.1101/350090. Published online. 350090.
69. Zhang Y, Ozono S, Yao W, et al. CRISPR-mediated activation of endogenous BST-2/tetherin expression inhibits wild-type HIV-1 production. *Sci Rep*. 2019;9(1):1–8. https://doi.org/10.1038/s41598-019-40003-z.
70. Farhang N, Ginley-Hidinger M, Berrett KC, Gertz J, Lawrence B, Bowles RD. Lentiviral CRISPR epigenome editing of inflammatory receptors as a gene therapy strategy for disc degeneration. *Hum Gene Ther*. 2019;30(9):1161–1175. https://doi.org/10.1089/hum.2019.005.
71. Farhang N, Brunger JM, Stover JD, et al. CRISPR-based epigenome editing of cytokine receptors for the promotion of cell survival and tissue deposition in inflammatory environments. *Tissue Eng Part A*. 2017;23(15-16):738–749. https://doi.org/10.1089/ten.tea.2016.0441.
72. Stover JD, Farhang N, Lawrence B, Bowles RD. Multiplex epigenome editing of dorsal root ganglion neuron receptors abolishes redundant interleukin 6, tumor necrosis factor alpha, and interleukin 1β signaling by the degenerative intervertebral disc. *Hum Gene Ther*. 2019;30(9):1147–1160. https://doi.org/10.1089/hum.2019.032.
73. Chen CD, Zeldich E, Li Y, Yuste A, Abraham CR. Activation of the anti-aging and cognition-enhancing gene Klotho by CRISPR-dCas9 transcriptional effector complex. *J Mol Neurosci*. 2018;64(2):175–184. https://doi.org/10.1007/s12031-017-1011-0.
74. Liao HK, Hatanaka F, Araoka T, et al. In vivo target gene activation via CRISPR/Cas9-mediated trans-epigenetic modulation. *Cell*. 2017;171(7):1495–1507.e15. https://doi.org/10.1016/j.cell.2017.10.025.
75. Giménez CA, Ielpi M, Mutto A, Grosembacher L, Argibay P, Pereyra-Bonnet F. CRISPR-on system for the activation of the endogenous human INS gene. *Gene Ther*. 2016;23(6):543–547. https://doi.org/10.1038/gt.2016.28.
76. Kemaladewi DU, Bassi PS, Erwood S, et al. A mutation-independent approach for muscular dystrophy via upregulation of a modifier gene. *Nature*. 2019;572(7767):125–130. https://doi.org/10.1038/s41586-019-1430-x.
77. Dai W, Xu X, Wang D, Wu J, Wang J. Cancer therapy with a CRISPR-assisted telomerase-activating gene expression system. *Oncogene*. 2019;38(21):4110–4124. https://doi.org/10.1038/s41388-019-0707-8.
78. Xiong K, Zhou Y, Blichfeld KA, et al. RNA-guided activation of pluripotency genes in human fibroblasts. *Cell Reprogram*. 2017;19(3):189–198. https://doi.org/10.1089/cell.2017.0006.
79. Hu J, Lei Y, Wong WK, et al. Direct activation of human and mouse Oct4 genes using engineered TALE and Cas9 transcription factors. *Nucleic Acids Res*. 2014;42(7):4375–4390. https://doi.org/10.1093/nar/gku109.

80. Weltner J, Balboa D, Katayama S, et al. Human pluripotent reprogramming with CRISPR activators. *Nat Commun.* 2018;9(1):1–12. https://doi.org/10.1038/s41467-018-05067-x.
81. Guo J, Ma D, Huang R, et al. An inducible CRISPR-ON system for controllable gene activation in human pluripotent stem cells. *Protein Cell.* 2017;8(5):379–393. https://doi.org/10.1007/s13238-016-0360-8.
82. Furuhata Y, Nihongaki Y, Sato M, Yoshimoto K. Control of adipogenic differentiation in mesenchymal stem cells via endogenous gene activation using CRISPR-Cas9. *ACS Synth Biol.* 2017;6(12):2191–2197. https://doi.org/10.1021/acssynbio.7b00246.
83. Liu Y, Yu C, Daley TP, et al. CRISPR activation screens systematically identify factors that drive neuronal fate and reprogramming. *Cell Stem Cell.* 2018;23(5):758–771.e8. https://doi.org/10.1016/j.stem.2018.09.003.
84. Balboa D, Weltner J, Eurola S, Trokovic R, Wartiovaara K, Otonkoski T. Conditionally stabilized dCas9 activator for controlling gene expression in human cell reprogramming and differentiation. *Stem Cell Rep.* 2015;5(3):448–459. https://doi.org/10.1016/j.stemcr.2015.08.001.
85. Chédin F, Lieber MR, Hsieh CL. The DNA methyltransferase-like protein DNMT3L stimulates de novo methylation by Dnmt3a. *Proc Natl Acad Sci USA.* 2002;99(26):16916–16921. https://doi.org/10.1073/pnas.262443999.
86. Guo JU, Su Y, Zhong C, Ming GL, Song H. Hydroxylation of 5-methylcytosine by TET1 promotes active DNA demethylation in the adult brain. *Cell.* 2011;145(3):423–434. https://doi.org/10.1016/j.cell.2011.03.022.
87. Amabile A, Migliara A, Capasso P, et al. Inheritable silencing of endogenous genes by hit-and-run targeted epigenetic editing. *Cell.* 2016;167(1):219–232.e14. https://doi.org/10.1016/j.cell.2016.09.006.
88. Saunderson EA, Stepper P, Gomm JJ, et al. Hit-and-run epigenetic editing prevents senescence entry in primary breast cells from healthy donors. *Nat Commun.* 2017;8(1). https://doi.org/10.1038/s41467-017-01078-2.
89. Wu J, He K, Zhang Y, et al. Inactivation of SMARCA2 by promoter hypermethylation drives lung cancer development. *Gene.* 2019;687:193–199. https://doi.org/10.1016/j.gene.2018.11.032.
90. Schuijers J, Manteiga JC, Weintraub AS, et al. Transcriptional dysregulation of MYC reveals common enhancer-docking mechanism. *Cell Rep.* 2018;23(2):349–360. https://doi.org/10.1016/j.celrep.2018.03.056.
91. Wang Q, Dai L, Wang Y, et al. Targeted demethylation of the SARI promotor impairs colon tumour growth. *Cancer Lett.* 2019;448:132–143. https://doi.org/10.1016/j.canlet.2019.01.040.
92. Qu J, Zhu L, Zhou Z, et al. Reversing mechanoinductive DSP expression by CRISPR/dCas9–mediated epigenome editing. *Am J Respir Crit Care Med.* 2018;198(5):599–609. https://doi.org/10.1164/rccm.201711-2242OC.
93. Mkannez G, Gagné-Ouellet V, Nsaibia MJ, et al. DNA methylation of a PLPP3 MIR transposon-based enhancer promotes an osteogenic programme in calcific aortic valve disease. *Cardiovasc Res.* 2018;114(11):1525–1535. https://doi.org/10.1093/cvr/cvy111.
94. Park J, Guan Y, Sheng X, et al. Functional methylome analysis of human diabetic kidney disease. *JCI Insight.* 2019;4(11). https://doi.org/10.1172/jci.insight.128886.
95. Kantor B, Tagliafierro L, Gu J, et al. Downregulation of SNCA expression by targeted editing of DNA methylation: a potential strategy for precision therapy in PD. *Mol Ther.* 2018;26(11):2638–2649. https://doi.org/10.1016/j.ymthe.2018.08.019.
96. Liu XS, Wu H, Krzisch M, et al. Rescue of fragile X syndrome neurons by DNA methylation editing of the FMR1 gene. *Cell.* 2018;172(5):979–992.e6. https://doi.org/10.1016/j.cell.2018.01.012.

97. Baumann V, Wiesbeck M, Breunig CT, et al. Targeted removal of epigenetic barriers during transcriptional reprogramming. *Nat Commun.* 2019;10(1):1–12. https://doi.org/10.1038/s41467-019-10146-8.
98. Kouzarides T. Chromatin modifications and their function. *Cell.* 2007;128(4): 693–705. https://doi.org/10.1016/j.cell.2007.02.005.
99. Barski A, Cuddapah S, Cui K, et al. High-resolution profiling of histone methylations in the human genome. *Cell.* 2007;129(4):823–837. https://doi.org/10.1016/j.cell.2007.05.009.
100. Schneider R, Bannister AJ, Myers FA, Thorne AW, Crane-Robinson C, Kouzarides T. Histone H3 lysine 4 methylation patterns in higher eukaryotic genes. *Nat Cell Biol.* 2004;6(1):73–77. https://doi.org/10.1038/ncb1076.
101. Creyghton MP, Cheng AW, Welstead GG, et al. Histone H3K27ac separates active from poised enhancers and predicts developmental state. *Proc Natl Acad Sci USA.* 2010;107(50):21931–21936. https://doi.org/10.1073/pnas.1016071107.
102. Shinkai Y, Tachibana M. H3K9 methyltransferase G9a and the related molecule GLP. *Genes Dev.* 2011;25(8):781–788. https://doi.org/10.1101/gad.2027411.
103. Rea S, Eisenhaber F, O'Carroll D, et al. Regulation of chromatin structure by site-specific histone H3 methyltransferases. *Nature.* 2000;406(6796):593–599. https://doi.org/10.1038/35020506.
104. Cao R, Zhang Y. SUZ12 is required for both the histone methyltransferase activity and the silencing function of the EED-EZH2 complex. *Mol Cell.* 2004;15(1):57–67. https://doi.org/10.1016/j.molcel.2004.06.020.
105. Steger DJ, Lefterova MI, Ying L, et al. DOT1L/KMT4 recruitment and H3K79 methylation are ubiquitously coupled with gene transcription in mammalian cells. *Mol Cell Biol.* 2008;28(8):2825–2839. https://doi.org/10.1128/mcb.02076-07.
106. Shi Y, Lan F, Matson C, et al. Histone demethylation mediated by the nuclear amine oxidase homolog LSD1. *Cell.* 2004;119(7):941–953. https://doi.org/10.1016/j.cell.2004.12.012.
107. Pasini D, Malatesta M, Jung HR, et al. Characterization of an antagonistic switch between histone H3 lysine 27 methylation and acetylation in the transcriptional regulation of Polycomb group target genes. *Nucleic Acids Res.* 2010;38(15):4958–4969. https://doi.org/10.1093/nar/gkq244.
108. Cheng AW, Jillette N, Lee P, et al. Casilio: a versatile CRISPR-Cas9-Pumilio hybrid for gene regulation and genomic labeling. *Cell Res.* 2016;26(2):254–257. https://doi.org/10.1038/cr.2016.3.
109. Memedula S, Belmont AS. Sequential recruitment of HAT and SWI/SNF components to condensed chromatin by VP16. *Curr Biol.* 2003;13(3):241–246. https://doi.org/10.1016/S0960-9822(03)00048-4.
110. Cano-Rodriguez D, Gjaltema RAF, Jilderda LJ, et al. Writing of H3K4Me3 overcomes epigenetic silencing in a sustained but context-dependent manner. *Nat Commun.* 2016;7(1):1–11. https://doi.org/10.1038/ncomms12284.
111. Kim JM, Kim K, Schmidt T, et al. Cooperation between SMYD3 and PC4 drives a distinct transcriptional program in cancer cells. *Nucleic Acids Res.* 2015;43(18): 8868–8883. https://doi.org/10.1093/nar/gkv874.
112. Kwon DY, Zhao YT, Lamonica JM, Zhou Z. Locus-specific histone deacetylation using a synthetic CRISPR-Cas9-based HDAC. *Nat Commun.* 2017;8(1):1–8. https://doi.org/10.1038/ncomms15315.
113. Galonska C, Charlton J, Mattei AL, et al. Genome-wide tracking of dCas9-methyltransferase footprints. *Nat Commun.* 2018;9(1):1–9. https://doi.org/10.1038/s41467-017-02708-5.
114. Campa CC, Weisbach NR, Santinha AJ, Incarnato D, Platt RJ. Multiplexed genome engineering by Cas12a and CRISPR arrays encoded on single transcripts. *Nat Methods.* 2019;16(9):887–893. https://doi.org/10.1038/s41592-019-0508-6.

115. Klompe SE, Vo PLH, Halpin-Healy TS, Sternberg SH. Transposon-encoded CRISPR–Cas systems direct RNA-guided DNA integration. *Nature*. 2019;571(7764):219–225. https://doi.org/10.1038/s41586-019-1323-z.
116. Strecker J, Ladha A, Gardner Z, et al. RNA-guided DNA insertion with CRISPR-associated transposases. *Science (80-)*. 2019;364(6448):48–53. https://doi.org/10.1126/science.aax9181.
117. Liu XM, Zhou J, Mao Y, Ji Q, Qian SB. Programmable RNA N 6-methyladenosine editing by CRISPR-Cas9 conjugates. *Nat Chem Biol*. 2019;15(9):865–871. https://doi.org/10.1038/s41589-019-0327-1.
118. Ma H, Tu LC, Naseri A, et al. Multiplexed labeling of genomic loci with dCas9 and engineered sgRNAs using CRISPRainbow. *Nat Biotechnol*. 2016;34(5):528–530. https://doi.org/10.1038/nbt.3526.
119. Duan J, Lu G, Hong Y, et al. Live imaging and tracking of genome regions in CRISPR/dCas9 knock-in mice. *Genome Biol*. 2018;19(1):192. https://doi.org/10.1186/s13059-018-1530-1.
120. Nelles DA, Fang MY, O'Connell MR, et al. Programmable RNA Tracking in Live Cells with CRISPR/Cas9. *Cell*. 2016;165(2):488–496. https://doi.org/10.1016/j.cell.2016.02.054.
121. Anton T, Leonhardt H, Markaki Y. Visualization of genomic loci in living cells with a fluorescent CRISPR/Cas9 system. In: *Methods in molecular biology*. vol 1411. Humana Press Inc; 2016:407–417. https://doi.org/10.1007/978-1-4939-3530-7_25.
122. Chen B, Gilbert LA, Cimini BA, et al. Dynamic imaging of genomic loci in living human cells by an optimized CRISPR/Cas system. *Cell*. 2013;155(7):1479–1491. https://doi.org/10.1016/j.cell.2013.12.001.
123. Ye H, Rong Z, Lin Y. Live cell imaging of genomic loci using dCas9-SunTag system and a bright fluorescent protein. *Protein Cell*. 2017;8(11):853–855. https://doi.org/10.1007/s13238-017-0460-0.
124. Kamiyama D, Sekine S, Barsi-Rhyne B, et al. Versatile protein tagging in cells with split fluorescent protein. *Nat Commun*. 2016;7(1):1–9. https://doi.org/10.1038/ncomms11046.
125. Cabantous S, Terwilliger TC, Waldo GS. Protein tagging and detection with engineered self-assembling fragments of green fluorescent protein. *Nat Biotechnol*. 2005;23(1):102–107. https://doi.org/10.1038/nbt1044.
126. Ma H, Naseri A, Reyes-Gutierrez P, Wolfe SA, Zhang S, Pederson T. Multicolor CRISPR labeling of chromosomal loci in human cells. *Proc Natl Acad Sci USA*. 2015;112(10):3002–3007. https://doi.org/10.1073/pnas.1420024112.
127. Esvelt KM, Mali P, Braff JL, Moosburner M, Yaung SJ, Church GM. Orthogonal Cas9 proteins for RNA-guided gene regulation and editing. *Nat Methods*. 2013;10(11):1116–1123. https://doi.org/10.1038/nmeth.2681.
128. Chen B, Hu J, Almeida R, et al. Expanding the CRISPR imaging toolset with Staphylococcus aureus Cas9 for simultaneous imaging of multiple genomic loci. *Nucleic Acids Res*. 2016;44(8). https://doi.org/10.1093/nar/gkv1533, e75.
129. O'Connell MR, Oakes BL, Sternberg SH, East-Seletsky A, Kaplan M, Doudna JA. Programmable RNA recognition and cleavage by CRISPR/Cas9. *Nature*. 2014;516(7530):263–266. https://doi.org/10.1038/nature13769.
130. Abudayyeh OO, Gootenberg JS, Essletzbichler P, et al. RNA targeting with CRISPR-Cas13. *Nature*. 2017;550(7675):280–284. https://doi.org/10.1038/nature24049.
131. Cox DBT, Gootenberg JS, Abudayyeh OO, et al. RNA editing with CRISPR-Cas13. *Science (80-)*. 2017;358(6366):1019–1027. https://doi.org/10.1126/science.aaq0180.
132. Tang W, Liu DR. Rewritable multi-event analog recording in bacterial and mammalian cells. *Science (80-)*. 2018;360(6385). https://doi.org/10.1126/science.aap8992.
133. Farzadfard F, Gharaei N, Higashikuni Y, Jung G, Cao J, Lu TK. Single-nucleotide--resolution computing and memory in living cells. *Mol Cell*. 2019;75:769–780.e4. https://doi.org/10.1016/j.molcel.2019.07.011.

CHAPTER FIVE

CRISPR-Cas system, antibiotic resistance and virulence in bacteria: Through a common lens

Subhasree Roy†, Sharmi Naha†, Ankur Rao, and Sulagna Basu*
Division of Bacteriology, ICMR-National Institute of Cholera and Enteric Diseases, Scheme XM, Beliaghata, Kolkata, India
*Corresponding author: e-mail addresses: basus.niced@gov.in; supabasu@yahoo.co.in

Contents

† Both authors have contributed equally.

Progress in Molecular Biology and Translational Science, Volume 178
ISSN 1877-1173
https://doi.org/10.1016/bs.pmbts.2020.12.005

Abstract

CRISPR-Cas system, antibiotic resistance and virulence are different modes of survival for the bacteria. CRISPR-Cas is an adaptive immune system that can degrade foreign DNA, antibiotic resistance helps bacteria to evade drugs that can threaten their existence and virulence determinants are offensive tools that can facilitate the establishment of infection by pathogens. This chapter focuses on these three aspects, providing insights about the CRISPR system and resistance mechanisms in brief, followed by understanding the synergistic or antagonistic relationship of resistance and virulence determinants in connection to the CRISPR system. We have addressed the discussion of this evolving topic through specific examples and studies. Different approaches for successful detection of this unique defense system in bacteria and various applications of the CRISPR-Cas systems to show how it can be harnessed to tackle the increasing problem of antibiotic resistance have been put forth. World Health Organization has declared antibiotic resistance as a serious global problem of the 21st century. As antibiotic-resistant bacteria increase their footprint across the globe, newer tools such as the CRISPR-Cas system hold immense promise to tackle this problem.

Abbreviations

ABC	ATP-binding cassette
ABD	antibacterial drone
AMR	antimicrobial resistance
ARGs	antimicrobial resistance genes
AT	adenine and thymine
BLASTn	nucleotide basic local search alignment tool
BLP	bacterial lipoprotein
bp	base pair
bPEI	branched polyethyleneimine
carO	carbapenem-associated resistance protein
Cas/*cas*	CRISPR-associated protein/gene
cat	chloramphenicol acetyltransferase gene
CCs	clonal complexes
CDC	Center for Disease Control and Prevention
CEC	commensal *Escherichia coli*
CFU	colony forming unit
CR	carbapenem-resistant

CRISPR	clustered regularly interspaced short palindromic repeats
crRNAs	CRISPR RNAs
CS	carbapenem-susceptible
CTX-M	cefotaxime β-lactamase
CTX-M-15	cefotaxime β-lactamase-15
DinG	damage inducible gene G
DNA	deoxyribonucleic acid
DRs	direct repeats
DSB	double-strand breaks
EnPEC	intestinal pathogenic *Escherichia coli*
erm	erythromycin ribosomal methylation gene
ESBLs	extended spectrum β-lactamases
ExPEC	extraintestinal pathogenic *Escherichia coli*
FISH	fluorescence *in situ* hybridization
GBS	Guillain-Barré syndrome
GC	guanine and cytosine
gRNA	guide RNA
gyrA	DNA gyrase subunit A gene
gyrB	DNA gyrase subunit B gene
HGT	horizontal gene transfer
H-NS	histone-like nucleoid structuring protein
IMP	imipenemase
IncF	incompatibility group F
IncFIB	incompatibility group FIB
IncHIB	incompatibility group HIB
IS	insertion sequence
KPC	*Klebsiella pneumoniae* carbapenemase
LOS	lipooligosaccharide
MATE	multidrug and toxin extrusion
Mb	megabase
mcr-1	mobilized colistin resistance-1 gene
MFS	major facilitator super
MGEs	mobile genetic elements
MRSA	methicillin-resistant *Staphylococcus aureus*
NDM-1	New Delhi-metallo-β-lactamase-1
OmpC	outer membrane porin C
OmpD	outer membrane porin D
OmpK35	outer membrane porin K35
OmpK36	outer membrane porin K36
OprD	porin D
ORF	open reading frame
OXA-1	oxacillinase-1
OXA-23	oxacillinase-23
OXA-24	oxacillinase-24
OXA-48	oxacillinase-48
OXA-54	oxacillinase-54
PAM	protospacer adjacent motif

parC	DNA topoisomerase IV subunit C gene
parE	DNA topoisomerase IV subunit E gene
PBPs	penicillin-binding proteins
PCR	polymerase chain reaction
PhoE	outer membrane porin PhoE
pre-crRNA	precursor CRISPR RNA
qRT-PCR	real time quantitative reverse transcription polymerase chain reaction
RGN	RNA guided Cas9 nuclease
RNA	ribonucleic acid
RND	resistance-nodulation-division
rpoB	beta subunit of RNA polymerase
rRNA	ribosomal RNA
scaRNA	small CRISPR-Cas-associated RNA
SEVA	Standard European Vector Architecture
SHV	sulfhydryl reagent variable β-lactamase
SRSR	short regularly spaced repeats
STAR	*Staphylococcus aureus* repeat
STs	sequence types
TEM	temoniera β-lactamase
Tet	tetracycline resistance protein
tgt	queuine tRNA-ribosyltransferase
TLR2	Toll-like receptor 2
TLR4	Toll-like receptor 4
Tn	transposon
tracrRNA	trans-activating CRISPR RNA
UPEC	uropathogenic *Escherichia coli*
van	vancomycin-resistance gene
VIM	verona integron encoded metallo-β-lactamase
WT	wild type

1. A place to begin

For all living organisms, survival is a basic instinct and bacteria are no different. Bacteria have devised ingenious methods for survival in various environmental niches. Both the clustered regularly interspaced short palindromic repeat (CRISPR) system and antibiotic resistance are distinct methods that are found in bacterial cells for their survival. CRISPR-Cas is an adaptive immune system of bacteria that contains *cas* (CRISPR *associated*) genes and CRISPR RNA that are complementary to the target nucleic acids of plasmids and viruses, and specifically guide different Cas proteins to degrade the foreign sequences.[1] The system also retains portions of the invading DNA as spacers so as to evade future attack from similar DNA sequences in the form of bacteriophages or bacterial plasmids, and is analogous to the eukaryotic immune system. Not only this, a potential link of the CRISPR-Cas system with bacterial virulence, and pathogenicity have been noted in various studies.[1]

Antibiotic resistance (AMR) mechanisms are also systems to survive under antibiotic pressure. There are various mechanisms in different species which include efflux pumps that can pump out different antibiotics, enzymes that can degrade antibiotics and proteins that modify the antibiotic binding site. These and other mechanisms make their presence felt only when the bacteria is in an environment containing antibiotics. The irrational use of antimicrobials has created selective pressure that has driven the evolution of antibiotic-resistant bacterial populations by acquisition and dissemination of antimicrobial resistance genes (ARGs) in pathogenic as well as in commensal bacteria.[2,3] Antibiotic usage increased globally by 36% between 2000 and 2010 and for carbapenems it reached 45%.[4] The Center for Disease Control and Prevention (CDC) reported that in the USA, nearly 2.8 million people get infected with antibiotic-resistant bacteria, of which more than 35,000 succumb to death (https://www.cdc.gov/drugresistance/about.html). It is also expected that by 2050, around 10 million people will die per year due to antibiotic-resistant pathogens.[5]

Of all different mechanisms of resistance, the most difficult to tackle are the transmissible ARGs. Many of such genes are bla_{NDM}, $bla_{\mathrm{CTX\text{-}M}}$, bla_{KPC}, $bla_{\mathrm{OXA\text{-}48\text{-}like}}$, and *mcr-1* has spread across the globe within a few years.[6,7] ARGs are mainly assembled in mega plasmids which play an important role in horizontal gene transfer (HGT). Mobile genetic elements (MGEs) comprising transposons, and integrons, can easily move between chromosomes and plasmids and thus play an important role in the accumulation and dissemination of ARGs in bacterial population through transposition and recombination mechanisms.[8] A potential strategy to overcome AMR is elimination of these MGEs or ARGs from the bacterial population.

Presence of CRISPR-Cas in multidrug-resistant and virulent bacteria is a paradox that needs detailed investigation. This review focuses on deciphering the possible association of CRISPR-Cas primarily with antibiotic resistance determinants and the MGEs facilitating their transmission, and to some extent virulence in bacteria. The review assembles different approaches for successful detection of this unique defense system in bacteria and further discusses various applications of the CRISPR-Cas systems to show how it can be harnessed to tackle the increasing AMR in bacteria.

With escalating antibiotic resistance, the need to view the CRISPR-Cas system and antibiotic resistance through a common lens, is greater than ever before.

2. Bacterial CRISPR-Cas system

2.1 Presence of CRISPR-Cas in prokaryotes

CRISPRs are interspaced, short, palindromic direct repeat DNA sequences found exclusively in prokaryotic genomes. They assist in providing adaptive immunity from bacteriophage and plasmid DNA.[9–13] Initially discovered in the *E. coli* genome in 1987,[14] these sequences were later found in other prokaryotes.[15] CRISPR loci are found at specific positions in different prokaryotic chromosomes as clusters, with their number varying from one (in *Mycobacterium tuberculosis* and few others) to as many as 15 in *Methanococcus jannaschi*.[15] Since then, it has been found in 87% and 50% of the analyzed archaeal and bacterial genomes, respectively.[16] According to the CRISPR Database in CRISPRFinder website (https://crispr.i2bc.paris-saclay.fr/Server), 87.1% and 45.1% of bacteria and archaeal strains possessed CRISPR-Cas; with 520 and 69 identified and culturable genera of bacteria and archaea, respectively, possessing at least one confirmed CRISPR structure.

2.2 Key elements of this system

The CRISPR system is composed of a CRISPR locus and its upstream *cas* genes encoding Cas proteins.[17] The *cas* genes are very tightly associated to CRISPR loci.[15] The locus further comprises an upstream AT-rich, 100–500 bp long leader sequence functioning as a promoter and a CRISPR array of alternating direct repeats (DRs) that separate similar sizes of variable sequences. These variable sequences are called spacers which are derived from foreign DNA and confer immunity against the invading DNA. Spanning from 21 to 47 bp of varying number, DRs are conserved in strains.[17] Such a unique feature led to their characterization as a new family of short regularly spaced repeats (SRSR), later renamed as CRISPR.[16,17] The spacers and repeats are together transcribed and processed into small CRISPR RNAs (crRNAs).[18]

2.3 The pathway through which CRISPR-Cas system acts

CRISPR-Cas mediated immunity occurs in three steps. The initial adaptation stage involves the uptake of exogenous DNA protospacers

(DNA fragments obtained from non-specific DNA breaks during bacteriophage and plasmid replication within the host; which are capable of being inserted as spacers) and their polarized incorporation in the CRISPR array as novel spacers. Separate adaptation-specific Cas proteins, Cas1 and Cas2, are involved in this step. In the second stage called expression, the array is transcribed into a single pre-crRNA (precursor CRISPR RNA), which is further processed into mature crRNAs (CRISPR RNAs). These RNAs possess a partial repeat and a single spacer sequence essential for targeting the cognate exogenous DNA on subsequent invasion.[18,19] In only type II CRISPR-Cas systems, trans-activating CRISPR RNA (tracrRNA) assists in crRNA processing by complementarily binding to pre-crRNA, forming an RNA duplex. This duplex is later cleaved by RNAse III to give the final tracrRNA-crRNA duplex called single-guide RNA (sgRNA).[20,21] The final interference step involves the interaction between the crRNAs or sgRNAs and effector Cas proteins—the former complementarily binds to exogenous sequence-specific targets, and the latter cleave them through nuclease activity.[18,19] Prior to target binding and cleavage, interaction with a two to four nucleotide sequence upstream of the target called protospacer adjacent motif (PAM) is mediated by the effector Cas proteins. PAMs increase the target specificity and prevent an auto-immune attack on host spacers.[18] Thus, through this entire mechanism involving the adaptive memory of spacers, host CRISPR-Cas systems remove the exogenous DNA from bacteriophages and plasmids.[13]

2.4 Classification of CRISPR-Cas system

CRISPR-Cas systems have been classified hierarchically under class, type and subtype. Two classes exist based on the complexity of the effector module—Class 1, comprising a multi-subunit crRNA-effector complex and Class 2, comprising a single crRNA-effector complex. On the basis of signature *cas* genes present, six types exist (Type I–VI). Types I, III and IV are classified under Class 1; while the others are classified under Class 2.[16,22] Each type possesses a unique signature gene; *cas3* in type I, *cas9* in type II, *cas10* in type III, *csf1* in type IV, *cas12* in type V and *cas13* in type VI. Each type can be classified into subtypes and till date, 34 subtypes have been identified, ranging from two under type IV to 11 under type V.[22,23] Arrangement of *cas* genes, presence of *cas* variants and other associated genes (DinG helicase in type IV-A) primarily define subtypes.[22] Type I CRISPR-Cas system is the most abundant in bacteria and archaea. Particularly in bacteria, subtype I-B, followed by subtypes I-C and I-E, is the most abundant.[16]

CRISPR-Cas has been found to be primarily involved in providing immunity in bacteria. Apart from this, it has been found to be involved in other functions such as virulence in pathogenic bacteria, sporulation and biofilm formation.[1]

3. Bacterial antimicrobial resistance

3.1 The problem of antimicrobial resistance

Antimicrobial resistance (AMR) occurs when a drug loses its ability to inhibit bacterial growth effectively, thereby making the bacteria "resistant" which then continue to multiply in the presence of therapeutic levels of the antibiotics.[24] This means that standard treatments no longer work; the risk of the spread of infection to others is increased; hospital stays are prolonged, economic burden increases; and the risk of death is greater. In fact, with the discovery of each new antibiotic, bacteria have devised techniques to evade the drug.[24,25] When placed in any adverse situation, which could be an environment with antibiotics, bacteria can quickly change themselves and they can also in most cases quickly pass on this change to their progeny or to other bacteria that is in the neighborhood. This change allows them to survive in an antibiotic laden environment. Once such a change is achieved, it can be well maintained by these bacteria particularly in environments where antibiotics are used in large proportions, i.e., health-care setups.

Many efficient antibiotic compounds were introduced and administered for the treatment of life-threatening infections caused by bacterial pathogens. Most antibiotics used for the treatment of bacterial infections may be classified according to their modes of action: (1) antibiotics that interfere with bacterial cell wall synthesis (β-lactams: penicillins, cephalosporins, monobactams, carbapenems; glycopeptides: vancomycin, teicoplanin); (2) antibiotics that inhibit protein synthesis (chloramphenicol, macrolides, clindamycin, linezolid, tetracyclines, aminoglycosides); (3) interfere with nucleic acid synthesis (fluoroquinolones and rifampicin); (4) inhibit metabolic pathways (folic acid analogues, sulfonamides); and (5) antibiotics that cause disruption of bacterial membrane structure (polymyxins and daptomycin).[25]

The irrational use of these antimicrobials has created selective pressure that has driven the evolution of antibiotic-resistant bacterial populations. According to the World Health Organization (WHO), antibiotic-resistant bacteria are a serious global problem to public health of the 21st century (https://www.who.int/news-room/fact-sheets/detail/antimicrobial-resistance). The Infectious Diseases Society of America has identified a group of bacterial pathogen as "ESKAPE" pathogens, associated with most antibiotic-resistant nosocomial infections and encompassing both gram-positive and gram-negative species (*Enterococcus faecium*, *Klebsiella pneumoniae*, *Staphylococcus aureus*, *Pseudomonas aeruginosa*, *Acinetobacter baumannii* and *Enterobacter* spp.).[26] These pathogens have the capability to acquire resistance very quickly, thus they escape antibiotic treatment and survive in the presence of antibiotics.

3.2 Genetic basis of antibiotic resistance

To evade antibiotic attack, bacteria use different genetic strategies which fall into several broad categories including: (i) modifications of the antimicrobial target to decrease the affinity for the drug, (ii) activation of multidrug efflux pumps to expel out the harmful molecules, (iii) a decrease in the drug uptake due to outer membrane impermeability (porin loss), and (iv) drug inactivation or alteration due to the production of different enzymes (Fig. 1).[27,28] Of these, the first three mechanisms are mostly chromosomal or intrinsic and can be passed on by vertical gene transfer (transfer of genetic material from

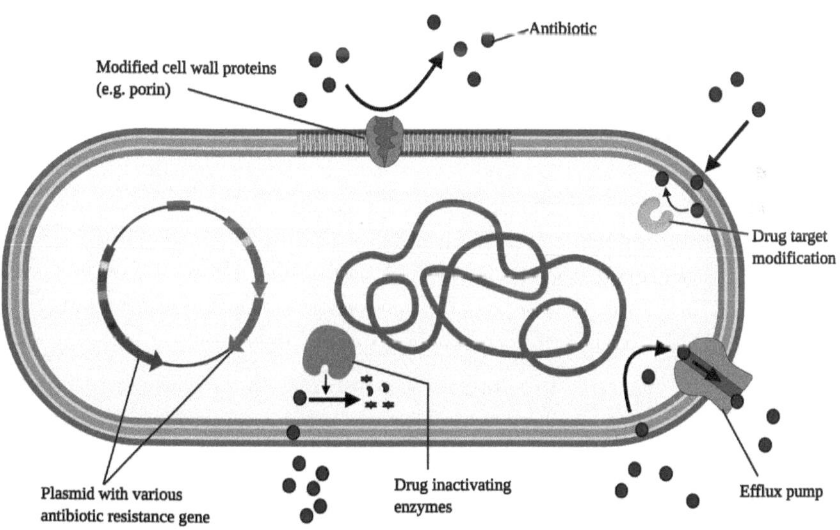

Fig. 1 Schematic diagram highlighting the antimicrobial resistance mechanisms in bacteria: Drug target modification: prevent binding of drugs to the targets; Efflux pumps: expel out structurally unrelated drugs from bacterial cytoplasm; modified cell wall proteins (porins): prevent influx of drug into bacterial cytoplasm due to their loss or reduced expression; Drug inactivating enzymes: hydrolyze the drug to prevent their attack. *Created using BioRender.com.*

a bacterium to its progeny) only. But the last mechanism of enzymatic modification is generally acquired through HGT.[2,28]

3.2.1 Drug target modification

To avoid recognition by antimicrobial agents, bacteria modify their target sites of antibiotic binding. As for example, the mutation of genes encoding penicillin-binding proteins (PBPs) decreases the affinity for all β-lactam antibiotics especially among gram-positive organisms, thereby generating antibiotic-resistant pathogens.[29] Sometimes bacteria evolve new targets with similar functions of the original target but are not inhibited by the antimicrobial molecule due to reduced affinity. Methicillin resistance in *S. aureus* developed due to involvement of an exogenous PBP (PBP2a) and vancomycin resistance in enterococci developed through modifications of the peptidoglycan structure by the *van* (vancomycin-resistance gene) gene clusters. A classic example is amino acid substitutions in the *rpoB* gene (encodes β subunit of RNA polymerase) resulting in decreased affinity of the drug rifampin for its target.[30] Another well-characterized example of drug target modification involves the mechanism of fluoroquinolone resistance due to chromosomal mutations in the genes *gyrA-gyrB* (DNA gyrase subunits A and B genes) and *parC-parE* (DNA topoisomerase IV subunits C and E genes) encoding DNA gyrase and topoisomerase IV, respectively, to which fluoroquinolone binds.[31] Enzymatic alteration of the target is also seen when macrolide resistance occurs by erythromycin ribosomal methylation (*erm*) genes which methylate a specific adenine in the 23S rRNA causing erythromycin resistance.[32] The *erm* genes in clinical strains are present on mobile genetic elements and are widespread among both gram-positive and gram-negative bacteria.

3.2.2 Efflux pumps

Drug-resistant bacteria also use efflux pumps to extrude structurally unrelated drugs from the bacterial cell to prevent its accumulation within the bacterium. Till date, five different families of efflux pumps have been reported in gram-positive and gram-negative bacteria: ABC (ATP-binding cassette family), MFS (major facilitator super family), RND (resistance nodulation-division family), MATE (multidrug and toxin extrusion family), and SMR (small multidrug resistance family).[33] The most important type of efflux pump in gram-negative bacteria is the RND-family pump which is a chromosomally encoded tripartite system consisting of inner membrane protein, periplasmic protein and outer membrane protein. AcrAB-TolC pump in *K. pneumoniae* and MexAB-OprM, MexXY-OprM, MexCD-OprJ, MexEF-OprN efflux pumps in *P. aeruginosa* are RND-family efflux pumps responsible for resistance to different classes of antibiotics including aminoglycosides, fluoroquinolones, cephalosporins and carbapenems.[34] *A. baumannii* also possess RND-type efflux pumps AdeABC, AdeDE, AdeFGH, and AdeIJK which play a role in resistance to fluoroquinolones, erythromycin, aminoglycosides, chloramphenicol, tetracycline, and carbapenems.[34] Though most efflux pumps are borne on the chromosome few can be on plasmids. A classic example of efflux pump harbored in MGEs is MFS-family Tet efflux pump that expels tetracyclines using proton exchange as the source of energy.[35]

3.2.3 Modification of outer membrane permeability

Porins are proteins present in the outer membranes of gram-negative bacteria that form channels to allow the entry of antibiotics. Reduction in the number of porins, decreases the influx of drugs into bacteria allowing them to develop resistance against different antimicrobials.[36] Classic examples of porin-mediated resistance are decreased production of OprD (porin D) in *P. aeruginosa* causing imipenem resistance,[37] loss of CarO (carbapenem-associated resistance protein) in *A. baumannii* causing carbapenem resistance, loss of porins OmpK35 (outer membrane porin K35) and OmpK36 (outer membrane porin K36) in *K. pneumoniae* leading to reduced susceptibility to β-lactams (such as cephalosporins and carbapenems) and loss of porins OmpF (outer membrane porin F), OmpC (outer membrane porin C) and PhoE (outer membrane porin PhoE) in *E. coli* exhibiting resistance.

3.2.4 Enzymatic alteration of drugs

Irreversible inactivation or modification of antibiotics by enzymes such as β-lactamases and aminoglycoside modifying enzymes are well-characterized mechanisms by which bacteria exhibit resistance.[27] Enzymes capable of hydrolyzing the β-lactam ring present in all β-lactam antimicrobials (penicillin, cephalosporins, monobactams, carbacephems, and carbapenems) are termed as β-lactamases such as extended spectrum β-lactamases (ESBLs), carbapenemases, AmpC-type cephalosporinases, and oxacillinases. Each of these enzymes target the active site of the antimicrobials and inhibit its activity.[27] These enzymes are classified into four classes: (i) Ambler class A: ESBLs (CTX-M, TEM, SHV, OXA-1), and *Klebsiella pneumoniae* carbapenemase (KPC) belong to this class. These enzymes have a serine molecule at its active site and are inhibited by β-lactamase inhibitors, (ii) Ambler class B: the enzymes belonging to this class are metallo-β-lactamase (MBL) that possess a metal ion (generally Zn^{2+}) at the active site and can be potentially inhibited by metal chelating inhibitors. The most common metallo-β-lactamases are carbapenemases such as imipenemase (IMP), Verona integron encoded metallo-β-lactamases (VIM), and the newly reported New Delhi-metallo-β-lactamase-1 (NDM-1), (iii) Ambler class C: AmpC-type cephalosporinase, which provide low level resistance to narrow spectrum cephalosporins and cephamycins, but are not inhibited by β-lactamase inhibitors, and (iv) Ambler class D: this group consists of oxacillin hydrolyzing enzymes possessing serine at its active site and are rarely inhibited by any known β-lactamase inhibitors.[27] Of all these enzymes, NDM has spread extensively and has the potential to hydrolyze nearly all β-lactam antibiotics. Apart from NDM, spread of KPC and OXA-48 have also been noted in various parts of the world.[6] In addition to these, CTX-M is another β-lactamase which is apparently present in various geographical settings.[38] Earlier, carbapenems were used for treating serious infections caused by multidrug-resistant bacteria, however this antibiotic is frequently failing us now. Resistance towards this antibiotic has emerged as bacteria have evolved to produce different carbapenemases to hydrolyze this antibiotic.[27] Further, presence of multiple β-lactamases has also been noted in various bacteria which make the bacteria multidrug-resistant. Now, with limited options, colistin (a polycationic peptide), a long forsaken drug has been put into use to treat carbapenem-resistant bacterial infections. Surprisingly, resistance toward this antibiotic also has surfaced due to the emergence of colistin resistance gene, *mcr-1*.[7] Aminoglycoside modifying enzymes cause post-translational methylation of 16S rRNA which alters the antibiotic binding site and make the aminoglycosides ineffective. These enzymes, like other β-lactamases, are present in plasmids or transposons, and have the potential to diffuse among various bacterial groups.[38]

3.3 Acquisition of antibiotic resistance genes

The antibiotic hydrolyzing enzymes discussed above are encoded by genes which are present in genetic materials (plasmids, transposons, etc.) acquired by the bacteria from other resistant bacteria or environment. Acquisition of these genetic material occurs via HGT by which bacteria confers resistance to various antimicrobials.[2] Transfer of genetic material from one bacteria to another is mediated by conjugation, transformation, and transduction. Conjugation is one of the most effective methods of transferring genetic materials among the bacterial community[28] but requires a vehicle for the purpose. Transformation involves transfer of genetic materials directly to other bacteria without the need of any shuttle, while transduction is a process of transfer of genetic material by phage or viral vector.[28] Genetic materials involved in conjugal transfer comprises plasmids, transposons, insertion sequences, and integrons, collectively termed as MGEs.[8] Of these, plasmids are the most common "vehicle" for gene transfer among different bacterial groups. Plasmids are self-replicating extrachromosomal genetic elements which harbor different resistance genes along with genes required for its own propagation.[28] Another important MGE is transposon (Tn), a stretch of DNA possessing antibiotic resistance genes and is a part of plasmids. This MGE has a peculiar characteristic to jump from plasmid to chromosome and vice versa by non-homologous recombination.[8] Hence, Tn also contributes to dispersal of resistance genes. Presence of short transposons with transposition capability is known as insertion sequence (IS) elements.[8] Like Tn, they also possess resistance genes within them and are often associated with Tn. Like other MGEs, integrons can transfer resistance genes to different bacteria and also can capture genes from others.[39] Hence, MGEs play an important role in dissemination of antibiotic resistance genes.

Emergence of AMR pathogens is a serious global concern. It is not confined to clinics anymore but has spread to the environment. As intrinsic or chromosomal resistance can only be transferred by vertical gene transfer, so the spread of intrinsic resistance is limited. Acquired resistance on the other hand is of much concern due to the association of resistance determinants with MGEs. Their containment is of utmost importance right now. To accomplish this, newer antimicrobials need to be developed along with development of newer strategies targeting different MGEs, blocking their transfer or making the genes inactive to combat this increasing resistance and retrieve the efficacy of old drugs.

4. Interference of CRISPR-Cas with bacterial antimicrobial resistance

An essential process in bacterial adaptation to different environments is acquisition of genetic material which include virulence, fitness and antibiotic resistance genes by HGT.[8] To acquire new beneficial traits by HGT, each microbe must allow entry of plasmids carrying resistance or virulence genes that will allow the microbe to survive in a hostile environment. But such genetic acquisitions impose fitness cost on the bacteria while in an environment without any selection pressure. Hence, bacteria try to prevent the

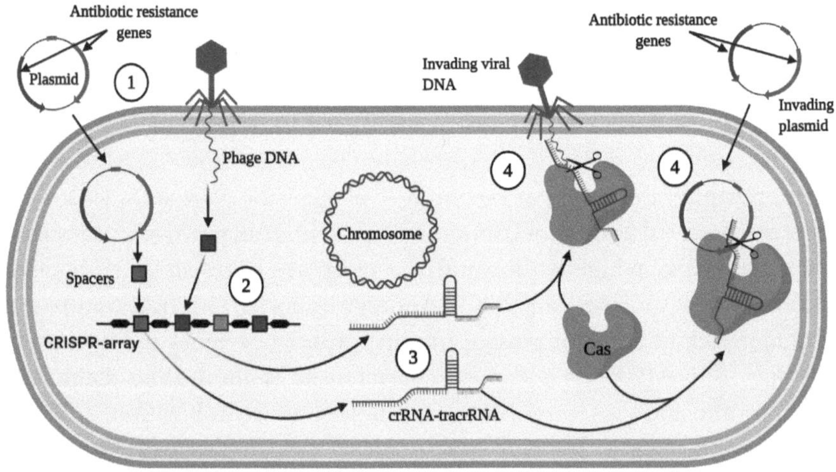

Fig. 2 Schematic diagram featuring CRISPR-Cas activity towards horizontal gene transfer both by plasmids and phages. 1. Insertion of foreign DNA by phage and plasmid into the bacteria, 2. spacer acquisition and incorporation of spacers in CRISPR array, 3. formation of crRNA (CRISPR RNAs) and tracrRNA (trans-activating CRISPR RNAs) complex containing specific spacers, and 4. crRNAs guide *cas* genes to target invading DNA. *Created using BioRender.com.*

entry of genetic elements or try to lose those genetic materials that impose fitness costs on their survival.[40] In this context, bacterial CRISPR-Cas system helps to limit the acquisition of genetic materials via the mode of phage infection, plasmid conjugation, and artificial transformation to provide defense against invading bacteriophages and plasmids (Fig. 2).[12,13,19,41] However, in the presence of antibiotics, ARGs are the saviors and their existence in the bacteria cannot be ignored.[40] The CRISPR-Cas system has been found in both gram-positive and gram-negative bacilli. Different studies have shown significant antagonistic association between CRISPR and antibiotic resistance within bacteria, which has been explored here (Table 1).

4.1 Role of CRISPR-Cas system in drug-resistant gram-positive bacteria

Marraffini and Sontheimer in 2008, were the first to work with *Staphylococcus epidermidis* to show how CRISPR-Cas activity could be used to block plasmid transmission.[41] They identified a spacer sequence (*spc1*) homologous to the nickase (*nes*) gene of conjugative plasmid (pG0400). To test the intervention of plasmid transfer by CRISPR, they introduced nine silent mutations within *nes* gene generating pG0 (mut). Conjugation was performed with these two plasmids in a CRISPR-negative ATCC 12228 *S. epidermidis* and a CRISPR-positive clinical RP62a strain.[59] Conjugation was successful for both the plasmids in ATCC 12228 while only the pG0 (mut) with inactivated *nes* gene was transferred to the CRISPR-positive RP62a. These results indicated that CRISPR interference could prevent plasmid conjugation in a manner that is specified by sequence identity between a spacer and a plasmid target sequence.

Garneau et al. in 2010 showed how absence of CRISPR-Cas system promotes acquisition of plasmid with resistance gene in *Streptococcus thermophilus*.[13] They inactivated *cas5* or *cas7* of plasmid free *S. thermophilus* strain DGCC7710 generating isogenic mutants and then introduced a plasmid, pNT1 into *S. thermophilus* strain DGCC7710. Stability of pNT1 was found in *cas5* mutant only, in comparison to *cas7* mutant, reinforcing the antagonistic role of CRISPR in plasmid stability. Later, they tried to introduce this pNT1 plasmid again into DGCC7710 strain where *cas* genes were not inactivated. But they found the presence of plasmid-insensitive mutants possessing the spacer sequences homologous to the *cat* gene (chloramphenicol resistance gene) of pNT1. These findings indicated that presence of active CRISPR helps to acquire spacers to block the plasmid transformation.

Several authors worked with *Enterococcus faecalis* and demonstrated a similar antagonistic relationship between the CRISPR-Cas system and antibiotic resistance. *E. faecalis*, acquire antibiotic resistance genes by conjugative pheromone-responsive plasmids. Previous studies have shown presence of two type II CRISPR-Cas systems (CRISPR1-Cas and orphan CRISPR2-lacking associated *cas* genes) in *E. faecalis*.[42,60–62] Palmer and Gilmore in 2010, constructed a database with draft genomes of *E. faecalis*. They found a novel CRISPR3 within the genomes with varying occurrences of CRISPR1 also. They noticed a lack of antibiotic resistance genes among strains with complete CRISPRs, and also found absence of CRISPRs in multidrug-resistant clonal complexes (CCs) such as CC2, CC8, and CC9. The authors proposed that homologous recombination and excision of CRISPR-Cas loci occurred at two different conserved unique sequences (52 and 205 bp) that led to loss of CRISPR-Cas loci (CRISPR1 and CRISPR3) among all antibiotic-resistant *E. faecalis*.[42] Following the work of Palmer and Gilmore,[42] Price et al. in 2016 showed that absence of CRISPR-Cas in multidrug-resistant *E. faecalis* make them immunocompromised, and provide an advantage to adapt to new antibiotics by rapidly acquiring new genes.[43] Firstly, they showed that *cas9* of CRISPR1 was required for functionality of orphan CRISPR2 locus. Secondly, absence of ~620 kb of horizontally acquired

Table 1 Details of CRISPR-Cas system and their association with antimicrobial resistance and virulence

Organisms	CRISPR locus and *cas* genes	Association with antimicrobial resistance or virulence	Study
		Antimicrobial resistance	
S. epidermidis	CRISPR-associated genes (*cas1*,*cas2*, and *cas6*) and spacer sequence (*spc1*)	CRISPR locus provides immunity against plasmid conjugation.	Marraffini and Sontheimer, 2008[41]
S. thermophilus	CRISPR1-Cas5	Active CRISPR-Cas helps to acquire spacers blocking plasmid transformation.	Garneau *et al.*, 2010[13]
E. faecalis	Type II CRISPR-Cas systems (CRISPR1-Cas and CRISPR3-Cas)	Strains with complete CRISPRs lack antibiotic resistance genes, and CRISPR-Cas is uniformly absent in multidrug-resistant CC2, CC8, and CC9 clonal complexes (CCs).	Palmer and Gilmore, 2010[44]
E. faecalis	Novel CRISPR3 along with CRISPR1 and CRISPR2	Orphan CRISPR2 becomes functional in presence of CRISPR1-Cas9; presence of CRISPR3-Cas locus with restriction modification system provides immunity to the strains from acquisition of new resistance genes and different mobile genetic elements.	Price *et al.*, 2016[47]
E. coli	Type I CRISPR-Cas systems (subtypes I-E and I-F)	No significant association of CRISPR-*cas* genes (both subtypes I-E and I-F) with the presence/absence of replicons and integrons detected.	Touchon *et al.*, 2012[48]
E. coli	CRISPR1 and CRISPR2 (subtype I-E), and CRISPR3 and CRISPR4 (subtype I-F)	Absence of CRISPR loci makes UPEC more susceptible to infection by phages or plasmids and allows them to adapt more quickly to various environments.	Dang *et al.*, 2013[49]

A. baumannii	Type I CRISPR-Cas systems (subtype I-F {I-Fa and I-Fb})	Presence of numerous strain-specific unique spacers identified, some are critical for conjugation machinery. Interference of CRISPR in maintaining genome diversity and transfer of conjugative elements.	Touchon *et al.*, 2014[50]
A. baumannii	CRISPR1-Cas9	Bacteria with CRISPR arrays had a very low number of plasmids and cas9-like genes along with restriction-modification system prohibited entry of new plasmids into the bacteria.	Mangas *et al.*, 2019[52]
K. pneumoniae	Type I CRISPR-Cas systems (subtypes I-E and I-E*)	Presence of type I-E* CRISPR limits acquisition of plasmids and exogenous DNA and absence of CRISPR in multidrug-resistant CC258	Li *et al.*, 2018[53]
K. pneumoniae	Type I CRISPR-Cas system (subtype I-E*)	Presence of type I-E* CRISPR inhibits acquisition of carbapenem resistance plasmids and helps in maintaining sensitivity to antimicrobials.	Mackow *et al.*, 2019[56]
K. pneumoniae	Type IV CRISPR-Cas system (plasmid) and type I CRISPR-Cas systems (subtypes I-E and I-E* {chromosome})	Bacteria with chromosomal type I-E CRISPR tend to have more plasmids when compared to bacteria with type I-E* CRISPR. CRISPR carrying plasmids had IncFIB & IncHIB replicons.	Kamruzzaman and Iredell, 2020[57]
		Virulence	
S. pneumoniae	CRISPR1 locus from *S. pyogenes*	Presence of CRISPR can prevent emergence of virulence and loss of CRISPR activity or mutations led to emergence of new pathogenic strains.	Bikard *et al.*,2012[61]
C. jejuni	Type II CRISPR-Csn1 system	CRISPR-Cas marker gene, *csn1* enhanced adherence, invasion, and translocation for proper manifestation in host cells.	Louwen *et al.*, 2013[69]

Continued

Table 1 Details of CRISPR-Cas system and their association with antimicrobial resistance and virulence—cont'd

Organisms	CRISPR locus and *cas* genes	Association with antimicrobial resistance or virulence	Study
C. jejuni	Type II CRISPR-Cas9 system	Cas9 increased production of biofilm, adhesion and invasion, prolonged survivability, high motility and increased production of cytotoxins.	Shabbir *et al.*, 2018[73]
F. novicida	Type II CRISPR-Cas locus (*cas1*, *cas2*, *cas4*), and effector phases (*cas9*, crRNA, tracrRNA)	CRISPR-Cas-associated RNA (scaRNA), *cas9* and tracrRNA were necessary to repress bacterial lipoprotein (BLP), in order to overcome host innate immune response.	Sampson and Weiss, 2013[74]
P. aeruginosa	CRISPR core genes *cas1* and *cas3* and Ypest CRISPR subtype genes *csy1* to *csy4*	Disruption of CRISPR loci restores biofilm formation and motility in DMS3 lysogens.	Zegans *et al.*, 2009[76]
P. aeruginosa	Type I CRISPR-Cas3 system (subtype I-F)	CRISPR-Cas system represses bacterial quorum-sensing regulator, *LasR*, to inhibit the recognition by toll-like receptor 4 (TLR4) of the host cell, thereby increasing pathogenicity of the bacteria.	Li *et al.*, 2016[79]
P. aeruginosa	Type I CRISPR-Cas3 system (subtype I-F)	CRISPR-Cas immunity was repressed by a KinB/AlgB regulated biofilm-activating pathway.	Borges *et al.*, 2020[80]
E. coli	Type I CRISPR-Cas systems (subtypes I-E and I-F)	Fewer repeats and loss of CRISPR activity leads to increased virulence in the bacteria.	García-Gutiérrez E *et al.*, 2015[85]

genetic element in susceptible *E. faecalis* possessing CRISPR3-Cas locus along with the presence of restriction-modification system was noted. Lack of CRISPR3-Cas and restriction-modification system in multidrug-resistant *E. faecalis* reduced their genome defense against clinically relevant mobile elements, making them immunocompromised. Thus, both these authors highlighted the critical importance of the *E. faecalis* CRISPR-Cas loci in HGT as well as in decreased antimicrobial resistance.[42,43]

4.2 Role of CRISPR-Cas system in drug-resistant gram-negative bacteria

Apart from gram-positive pathogens, study of CRISPR was also conducted in different gram-negative pathogens. Touchon et al. in 2012 tried to establish the role of CRISPR in the spread of plasmids and antibiotic resistance in *E. coli.*[44] They characterized 263 natural *E. coli* from human and animal hosts in terms of CRISPRs, replicons, class 1 integrons and further studied the association of CRISPR with the presence/absence of plasmids and mobile antibiotic resistance determinants. Presence of type I CRISPR with subtype I-E and I-E* was noted among most of the *E. coli* isolates. They found IncF plasmids and class I integron to be the prevalent mobile genetic elements in these strains. Strains possessing IncF plasmids, had more *cas* subtype I-E systems, while susceptible strains harbored subtype I-F systems. *cas* genes were found to be prevalent in all resistant as well as in susceptible *E. coli* strains irrespective of integron and plasmids. Thus, authors did not find any meaningful association of CRISPR activity and the presence/absence of the *cas* genes with the plasmids, integrons or antibiotic resistance.

Contrary to the finding of Touchon et al.,[44] Dang et al. in 2013, observed that absence of CRISPR loci make uropathogenic *E. coli* (UPEC) more susceptible to infection by phages or plasmids and allow them to adapt more quickly to various environments.[45] Authors analyzed 81 pairs of UPEC and fecal *E. coli* (isolated from women with urinary tract infection) in terms of number of CRISPR loci and spacers, and sensitivity to antibiotics. They detected CRISPR1/CRISPR2 (subtype I-E), and CRISPR3/CRISPR4 (subtype I-F) among the fecal *E. coli* isolates and not in UPEC. While the presence of fewer spacers and repeats among UPEC probably made them more resistant and adaptable to the ever-changing environment compared to the fecal isolates. Absence of CRISPR and spacers in UPEC isolates facilitated acquisition of conjugative plasmids and phages, which made the uropathogens more resistant to various antimicrobials. Hence, authors successfully established a strong link between absence of CRISPR and increased resistance.

After the study with *E. coli*, Touchon et al. in 2014, analyzed the presence of CRISPR-Cas system in *Acinetobacter baumannii*.[46] They analyzed 131 complete genomes of *A. baumannii* retrieved from various genome databases and identified two I-F subtypes (I-Fa and I-Fb) in most of the *A. baumannii* genomes, as reported previously.[63] They identified numerous unique, strain-specific spacers matching with 831 complete phage and 3861 complete plasmid sequences when compared to sequences in NCBI database. Presence of 47 spacers matching different resistance genes of plasmids were also detected. Few spacers were homologous to various virulence and mobile elements. They concluded that CRISPR-Cas in Acinetobacter plays an important role in maintaining genome diversity and transfer of conjugative elements. An extension to the work of Touchon et al.,[46] was carried forward by Mangas et al. in 2019, wherein they exhibited association of CRISPR-Cas with the absence of plasmid.[47] They conducted genome analysis of 2500 strains and detected 2 groups, the first group possessed a larger number of resistance genes while the second group had fewer resistance genes. Presence of CRISPR was high in the latter group. They detected *cas9*-like gene which interacted with other *cas* genes restricting new plasmid entry into the bacteria. On the other hand, some strains of this group lacked *cas* genes and possessed additional type I restriction-modification system known to block plasmid entry. Like other authors, they also showed an inverse relationship of CRISPR with antibiotic resistance.

Li et al. in 2018 had reported the role of CRISPR-Cas in antibiotic susceptibility among *Klebsiella pneumoniae* strains.[48] They analyzed 97 strains from genome database and 176 clinical strains, and detected two types of CRISPR (type I-E and I-E*), of which type I-E* was the prevalent one. Not only type I-E* harboring isolates have fewer plasmids and phage elements, but also possess spacers targeting several regions of these mobile elements. Again, susceptibility to various antimicrobials was high in them when compared to type I-E and CRISPR-Cas negative isolates. Hence, they hypothesized that the possibility of limited acquisition of plasmids and exogenous DNA is mediated by type I-E* CRISPR. Again, they statistically confirmed a strong association of CRISPR with certain sequence types (STs) such as type I-E in ST34/ST45/ST66/ST67/ST147/ST273/ST383/ST392/ST941 whereas type I-E* within ST14/ST15/ST23/ST28/ST111/ST374/ST493/ST505.[48] But no CRISPR was found in ST258 and ST11 belonging to the largest multidrug-resistant clonal complex CC258,[64,65] indicating absence of CRISPR-Cas in multidrug-resistant CC258 *K. pneumoniae*.

Contrary to the previous findings, Mackow et al. in 2019 found few ST258 *K. pneumoniae* possessing CRISPRs.[49] They analyzed 304 clinical *K. pneumoniae* both carbapenem-susceptible (CS-*K. pneumoniae*) and carbapenem-resistant (CR-*K. pneumoniae*) *K. pneumoniae* for the presence and function of CRISPR-Cas system. CRISPR possessing ST258 *K. pneumoniae* were sensitive to carbapenems whereas the majority of ST258 without CRISPRs were resistant to carbapenems. To understand the role of CRISPR in preventing plasmid acquisition, transformation was carried out with a plasmid possessing carbapenem-resistant determinant, bla_{KPC} (*Klebsiella pneumoniae* carbapenemase). The recipient strain was carbapenem sensitive and carried a type I-E* CRISPR. Presence of spacers targeting bla_{KPC} in the recipient strain, hindered the transformation. The same experiment when carried out with a recipient strain that lacked the same spacer sequence showed successful transformation. Hence, this study proposed that the lack of CRISPR-Cas facilitates acquisition of carbapenem-resistant plasmids, while those having CRISPR-Cas are protected from acquiring these plasmids.

Until 2019, presence of CRISPR-Cas system was found in the chromosome, but Kamruzzaman and Iredell in 2020 identified for the first time the presence of type IV CRISPR-Cas system in large sized (200–430 kb) antibiotic-resistant plasmids IncFIB and IncHIB replicons of *K. pneumoniae*.[50] They examined 699 complete plasmid sequences from *K. pneumoniae* and 217 *K. pneumoniae* genome sequences from the GenBank for the presence of CRISPR-Cas systems, their types, acquired spacers and relationship between presence or absence of CRISPR and plasmids. Out of 217 *K. pneumoniae* genomes assessed, they found plasmids in 85% cases, while 37% *K. pneumoniae* harbored putative CRISPR-Cas system. They identified the CRISPR array in 5% of plasmid. Presence of unique spacers homologous to various conjugal transfer genes and transcription regulators were detected among CRISPR-positive plasmids of *K. pneumoniae*. Also, they found a lower GC percentage in CRISPR-positive plasmids compared to CRISPR-negative plasmids. An increased GC percentage indicates acquisition of foreign DNA in the plasmid.[66] In this study, lower GC indicated restriction to the entry of foreign DNA by CRISPR. They found that bacteria with chromosomal type I-E CRISPR tend to have more plasmids when compared to bacteria with type I-E* CRISPR as reported previously.[48]

Newire et al. in 2019 also detected the presence of a novel type IV-B CRISPR-Cas on IncH1IB/IncFIB plasmids with multiple putative transposable elements among multidrug-resistant *K. pneumoniae*.[67] But they did not find any CRISPR in the chromosome. They detected presence of spacer array with a single repeat at the upstream of *cas* loci matching regions of conjugal transfer genes indicating role of this type IV-B CRISPR-Cas in plasmid incompatibility.[67] Presence of CRISPR in plasmids indicated restriction of plasmid mobility and resistance among the strains.

To summarize, the presence of CRISPR was detected in gram-positive as well as in gram-negative organisms. Although most studies suggest an anticorrelation of the CRISPR-Cas system with respect to antimicrobial resistance in bacteria, very few reports have also found that antibiotic-resistant bacteria possess the CRISPR-Cas system. A probable reason for this might be absence of spacers homologous to the resistance genes for the time being, which they might acquire with subsequent generation and will eventually get rid of the resistance genes later on. Studies commendably pointed the CRISPR-Cas system as a defense mechanism in bacteria against invading phages and clinically relevant MGEs. Absence of this system would make the bacteria immunocompromised, and hence, will give rise to a generation of antibiotic-resistant bacteria rendering new antibiotics ineffective.

5. Association of CRISPR-Cas with bacterial virulence

Virulence is the ability of an organism to infect the host and cause a disease thereby assisting the bacterium to colonize within the host.[68] The impact of CRISPRs on the emergence of virulence is not well documented. However, several studies have shown that the CRISPR-Cas system has a role in virulence apart from defending bacteria against invaders and also preventing acquisition of MGEs.[1] Authors have investigated the relationship of CRISPR (antagonistic or synergistic) with virulence among different gram-positive and gram-negative bacilli (Table 1).

5.1 Interference of CRISPR-Cas with virulence in gram-positive bacteria

Bikard et al. in 2012 have tested the activity of CRISPR-Cas loci against the emergence of virulence in the human pathogen (gram positive) *Streptococcus pneumoniae*.[51] The pneumococcus unveils a unique ability to incorporate new genetic material via HGT, particularly through direct uptake of exogenous DNA, or DNA transformation.[69,70] To prevent the recognition by the host immune system, pneumococci change their surface polysaccharide capsule by capsule switching mechanism. This helps them to evade recognition by capsule-specific antibodies of the host, and establish a successful infection.[70,71] As none of the available *S. pneumoniae* genomes harbor CRISPR loci,[72] authors introduced CRISPR sequences from related *Streptococcus pyogenes* into *S. pneumoniae*. They found that presence of CRISPR with appropriate sequence-specific spacers (homologous to capsule switching genes) can prevent transfer of capsular genes. This result indicated that presence of CRISPR inhibits virulence in *S. pneumoniae*. They also found that the transfer of CRISPR-Cas systems from *S. pyogenes* into *S. pneumoniae* containing specific targets (capsular genes) led to the death of the transformants. In accordance with other studies, these findings also indicated that CRISPR loci and their targets cannot coexist in the same cell. In addition, their findings also showed that loss of CRISPR activity or mutation of the target sequence upon acquisition of resistance genes led to emergence of new pathogenic strains.[73,74]

5.2 Interference of CRISPR-Cas with virulence in gram-negative bacteria

Several studies have also reported how the CRISPR system works among gram-negative pathogens to produce enhanced virulence. Few pathogens like *Campylobacter jejuni*, and *Haemophilus influenzae*, are known to cause acute or chronic damage to the host or are linked to (post infectious) complications such as Guillain-Barré syndrome (GBS). GBS is a post infectious disease in which the human peripheral nervous system is affected.[52,75–77] These pathogens harbor type II CRISPR-Cas systems. They express sialylated lipooligosaccharide (LOS) structures on its cell envelope.[78] Sialylation of the cell envelope helps the pathogens to adhere, invade, and translocate across epithelial cells and to evade host immune responses. Louwen et al. in 2013 presented a potential link between pathogenicity of *C. jejuni* and presence of type II CRISPR-Cas marker gene, *csn1*, among GBS-causing isolates.[52] They inactivated *csn1* gene in three *C. jejuni* sialyl transferase (*cst*-II)-positive GBS isolates and one *cst*-II-negative isolate. Comparison between these two groups was done against their isogenic *Δcsn1* mutants by infection assay using Caco-2 cells. They found the isogenic Δ*csn1* mutant of the *cst*-II positive GBS isolates primarily showed decreased adherence, invasion, and translocation in Caco-2 monolayers compared to *cst*-II negative isolates. *csn1* mutants of the *cst*-II positive GBS isolates were unable to cause any damage to the epithelial cells. Hence, it is clear that CRISPR-Cas provides virulence advantage to *C. jejuni* for proper manifestation in host cells.

Shabbir et al. again in 2018 proposed a potential involvement of type II CRISPR-Cas9 in increasing the virulence of *C. jejuni*.[53] They performed biofilm assay, adhesion/invasion assay, intracellular survivability assay, cytotoxin assay and motility assay with WT *C. jejuni* NCTC11168, *cas9* mutant and complemented strains. They found increased production of biofilm, adhesion/invasion, prolonged survivability, high motility and cytotoxic factors in WT and complemented strains when compared to Δ*cas9* mutants. They clearly pointed to the involvement of CRISPR-Cas with the virulence of *C. jejuni*.

A study led by Sampson and Weiss in 2013 on *Francisella novicida* showed similar correlation between CRISPR and virulence as reported by previous authors.[54] *F. novicida* genomes possess type II CRISPR-Cas locus comprising of full-length forms of all the necessary components for the adaptation (*cas1*, *cas2*, *cas4*) and effector phases (*cas9*, crRNA, tracrRNA-also required for crRNA processing and interaction of the crRNA with Cas9) for targeting foreign DNA. *F. novicida* uses scaRNA, *cas9* and tracrRNA to repress bacterial lipoprotein (BLP), which activates proinflammatory immune response factors such as Toll-like receptor 2 (TLR-2) in host cells. Thus, to survive within the host, this pathogen needs to prevent the expression of BLP leading to inhibition of recognition by TLR2.[79] Deletion of *cas9*, tracrRNA, and scaRNA resulted in an increased level of BLP by 100-fold. This led to increased recognition by TLR-2 among the CRISPR deleted strains, thereby causing bacterial clearance by host immune cells. This finding further reinforces the role of CRISPR-Cas in regulation of gene expression to evade immune response and promote virulence leading to bacterial survival in hostile environments.

Zegans et al. in 2009 and Palmer & Whiteley in 2011, demonstrated that the disruption of CRISPR locus increased the biofilm formation and motility of *Pseudomonas aeruginosa* lysogens.[55,80] *P. aeruginosa* possessed CRISPR core genes *cas1* and *cas3* along with Ypest CRISPR subtype genes *csy1* to *csy4* bracketed by two CRISPR spacer arrays, CRISPR1 and CRISPR2. Infection of *P. aeruginosa* by phage DMS3 hampered their biofilm or swarming motility, a phenotype that helped *Pseudomonas* to overcome phagocytosis by mammalian white blood cells. They infected *Pseudomonas* PA14[81] with phage DMS3 which produced PA14 DMS3 lysogens. Biofilm production and swarming motility were found to be repressed among the lysogens. To know the role of CRISPR in biofilm production and swarming motility among lysogenized strains, CRISPR2 was deleted. The biofilm formation in DMS3 lysogens increased. Similar phenotype was noted when *csy3* or *csy4* were deleted or when a transposon was inserted in *cas3*, *csy1*, or *csy2*. These results strongly supported that biofilm formation and swarming depends on the presence of the *P. aeruginosa* PA14 CRISPR region (Ypest subtype CRISPR locus).[55,80]

Li et al. in 2016 established the synergistic role of CRISPR with virulence.[56] They identified a novel role of Type I CRISPR-Cas3 system in evading host defense by *P. aeruginosa*, PA14. They found type I CRISPR-Cas system targeted the bacterial quorum-sensing regulator, LasR, to inhibit the recognition by Toll-like receptor-4 (TLR-4), thereby weakening the proinflammatory responses of the host. They established this both by *in vivo* and *in vitro* experiments. They infected C57/DL/6J mice with WT PA14 and PA14 mutants with complete deletion of the CRISPR-Cas region (ΔTCR). They found increased expression of type 3 secretory system and LasR in the mutant strain which was reversed upon complementation with the CRISPR-Cas system components. Mice lethality was found to be high when infected with wild type or the complements of CRISPR-Cas (where decreased expression of LasR was noted) but not with mutants of CRISPR-Cas (with increased expression of LasR). LasR expression was decreased in WT and complemented mutants, resulting in inhibition of recognition by TLR-4 leading to bacterial survival in the host. When they infected mouse macrophage cell line (MH-S) with both WT and ΔTCR mutant strains, an elevated expression of TLR-4 was detected in case of ΔTCR mutant infection compared to WT-infected cells. This again indicated bacterial survival within the host was decreased in case of mutants where CRISPR locus was deleted. This finding highlighted the role of the CRISPR-Cas system in regulation of LasR and evasion of the host immune system by the bacteria.[56]

Borges et al. in 2020 on the other hand, established a negative connection between CRISPR and biofilm production among *P. aeruginosa*. They showed that CRISPR-Cas immunity was repressed by a biofilm-activating pathway in *P. aeruginosa*.[57] KinB/AlgB is a well-known two-component system of *P. aeruginosa* where KinB acts as a sensor kinase/phosphatase and AlgB acts as a response regulator. It is known from previous literature that this system controls the biofilm production in *P. aeruginosa*.[82] Absence of KinB function (lack of KinB phosphatase activity) leads to accumulation of phosphorylated form of the response regulator AlgB (P-AlgB). Enhancement of P-AlgB helps to liberate a sigma factor, AlgU, which positively regulates many factors involved in biofilm production.[83–85] Thus, accumulation of P-AlgB, increases biofilm production. On the contrary, the authors showed in their study that accumulation of P-AlgB stimulates the activity of CRISPR-Cas repressors AlgU, AlgR, and AmrZ resulting in CRISPR-Cas repression.

Therefore, they successfully established that in *P. aeruginosa*, surface-association/biofilm formation and CRISPR-Cas expression are inversely correlated.

García-Gutiérrez et al. in 2015 showed a negative connection between the numbers of CRISPR units and the presence of virulence traits.[58] They worked with 124 *E. coli* strains (commensal, intestinal, extraintestinal) and two strains of *E. fergusonii* ATCC 35469 and *E. albertii* TW07627. They analyzed the strains in terms of pathotypes, and CRISPRs. Like others, they also noted the presence of type I-E and I-F CRISPR among the isolates. They detected fewer CRISPR repeats in extraintestinal *E. coli* (ExPEC) than commensal *E. coli* (CEC) followed by intestinal *E. coli* (EnPEC), and loss of *cas* genes from type I-E CEC and EnPEC can make them more pathogenic like ExPEC. Also the numbers of *cas* genes were less in the pathogenic type I-F *E. coli* than I-E types, suggesting insignificant impact on pathogenicity of I-F in comparison to I-E. Also, they found a lower number of repeats in strains with increased virulence traits, i.e., strains with only one virulence trait harbor more repeats and vice versa. Hence, they postulated that lower repeats and loss of CRISPR activity leads to increased virulence of *E. coli*. Although, most of the previously mentioned studies showed positive correlation between CRISPR and virulence, however, Borges et al. and García-Gutiérrez et al. established antagonistic role of CRISPR with virulence in their studies.

The examples discussed in this section reveal a love-hate relationship between bacterial pathogens and their CRISPR-Cas systems. The role of these systems in modulating virulence is certainly a new and exciting area of research. Taken together, these examples demonstrate the remarkable versatility of CRISPR-Cas systems and warrant future investigations focused on their roles that extend beyond the canonical disruption of invading phage or plasmids. Understanding of the mechanisms leading to the control of virulence via CRISPR-Cas systems will provide a deeper view into the regulation of gene expression and the means by which bacteria respond to environmental changes. Future work explaining how these systems contribute to bacterial pathogenesis will facilitate identification of bacterial host defense strategies during infection.

6. Detection of CRISPR-Cas system in bacteria

Detection of CRISPR-Cas can be done either by using bioinformatics tools or by studying expression of *cas* genes by real time quantitative reverse transcription PCR (qRT-PCR) assays.

6.1 Bioinformatics approach

Dedicated user-friendly websites like CRISPRFinder[86] recently updated to CRISPRCasFinder[87] and CRISPRone[88] are helpful in bioinformatics analysis. For detection using these sites, FASTA format of whole genome sequences as well as Sanger sequences obtained after normal PCR detection of CRISPR loci and *cas* genes from bacterial isolates can be used as input. Conventional PCR is used to amplify specific CRISPR loci and *cas* genes from bacterial isolates, which are verified by the given websites. Normally, amplification of the universal *cas1* gene as well as signature *cas* genes indicate presence of CRISPR loci; use of CRISPR array specific primers can further validate the results.

Before any proper bioinformatics analysis was developed, older programs like Tandem Repeat Finder, LUNA, Patscan, PYGRAM, REPFind, CRT and PILER-CR were used to detect CRISPR sequences in prokaryotic genomes. However, many disadvantages prevailed, since these could not eliminate background data completely for efficient CRISPR array detection, as well as lack boundaries defining direct repeat (DR) consensus. Omission of truncated DRs, repeated identical CRISPR identification with different DRs and lack of distinguishing false CRISPRs from true ones were also observed. Moreover, technical expertise and different hardware were required for different programs, along with time-consuming compiling processes. Till then, no user-friendly website or software had been developed.[86]

The above need was fulfilled by the development of the first user-friendly website, CRISPRFinder (https://crispr.i2bc.paris-saclay.fr/), by Grissa et al.[86] It overcame the previous major deficiencies of hardware specificity, difficult programming language, as well as failure in defining DR boundaries. For using this website, DNA sequence of maximum 67 Mb length in FASTA format is pasted in the input field. An algorithm locates CRISPRs through the input sequence with DRs having 23–55 bp size separated by variable sequences of a size in between 0.6 and 2.5 times the length of the DRs. On detecting CRISPR sequence, two types of results are generated—the presence of confirmed CRISPR sequences, possessing three or more identical DRs with variable spacers, and questionable CRISPR sequences, having two identical DRs flanking a variable sequence. Each detected sequence has their DRs and their unique spacers coded in yellow and other different colors, respectively. Advantages of CRISPRFinder include single-base resolution and defining of DR sequences, extraction of flanking sequences and spacers for further analysis, as well as compilation-free uploading. Presence of associated databases in the website for quicker analysis had made it an indispensable tool for studying CRISPR.[86]

Although CRISPRFinder could demarcate consensus DRs as well as detect even the shortest CRISPR; Zhang and Ye discovered that it could not differentiate between true CRISPRs and false CRISPRs.[88] Thus, they created a website named CRISPRone (http://omics.informatics.indiana.edu/), which could differentiate the two on the basis of (i) presence of strictly unique spacers and (ii) presence of *cas* genes in the vicinity of CRISPR array. These criteria could be met only by the true CRISPR sequences. False CRISPRs were further classified and annotated as tandem repeats, simple repeats, *Staphylococcus aureus* specific STAR elements, and unknown false CRISPRs. Also, CRISPRone, unlike CRISPRFinder, described questionable CRISPRs as those possessing predicted CRISPR array(s), but lacking *cas* genes in the sequence. The advantages of this website include greater detection sensitivity than CRISPRFinder, as well as display of orientation of CRISPR-Cas sequences with different shapes and colors for CRISPR loci, arrays, and various *cas* genes up to subtype level.[88] These advantages have made CRISPRone verify CRISPRFinder results, and thus, supplement them.[50]

In 2018, CRISPRFinder was upgraded to CRISPRCasFinder (https://crisprcas.i2bc.paris-saclay.fr/) by integrating CasFinder, a separate website previously dedicated to detect Cas proteins and *cas* genes.[87] For input, CRISPRCasFinder can accept a maximum 50 Mb size of FASTA DNA sequence files, including a maximum of 100 sequences. In comparison to its decade-old predecessor, CRISPRCasFinder distinguishes true CRISPRs from false ones using a parameter called evidence level, which is dependent on the number of spacers, levels of DR conservation based on DR entropy values, and overall spacer identity percentage that refers to the degree of similarity, or, uniqueness of the spacers. These levels range from the lowest value 1 to the highest value 4. Putative CRISPR values with lowest evidence value possess less than 4 spacers; those in level 2 have $\geq$4 spacers with DR entropy values $<$70. If level 2 CRISPRs have the entropy values $>$70, then they are CRISPRs of evidence level 3; the level 3 ones with spacer similarity percentage $<$8% are level 4 CRISPRs. Evidence levels 1 and 2 are usually indicative of false positives due to low DR consensus and spacer number; putative levels 3 and 4 CRISPRs are considered potential CRISPRs. Apart from this primary function, CRISPRCasFinder also identifies orientation of CRISPR arrays on the basis of a dataset of consensus DRs as well as

AT% of the 100 bp flanking regions of the array for leader sequence detection. Based on the sequence similarity with known Cas sequences and updated CRISPR-Cas classification, CRISPRCasFinder identifies putative Cas proteins using three levels of detection—the sensitive general level for maximum Cas search, the intermediate typing level and the stringent subtyping level for precise identification. CRISPRCasFinder offers a number of advantages over the previously developed sites, including CRISPRone. Unlike other sites, it gives a clear and robust summary of the putative CRISPR properties, facilitating quicker and better analysis. CRISPRCasFinder also offers a unique feature of comparing shorter evidence level 1 arrays with larger valid level 4 arrays within the same genome; thus giving room to validate these short arrays as proper CRISPRs provided their consensus DRs matched. As CRISPRone based Cas protein search is comparatively less stringent causing detection of too many potential proteins; CRISPRCasFinder with its general-to-specific three level detection possesses an advantage over it.[87]

Thus, the start in use of CRISPRFinder followed by its upgradation to CRISPRCasFinder and the development of a more efficient site, CRISPRone, have established simple bioinformatics methods for CRISPR-Cas based studies, whose examples are discussed below.

CRISPRFinder has been primarily used for detecting CRISPR-Cas systems in *E. coli*,[89] *K. pneumoniae*,[48,50,90,91] *E. faecalis*,[42] and *E. faecium*.[42] It has also been used to verify conventional PCR results for CRISPR-Cas detection[8,42,48,49] as well as detect a new subtype in *K. pneumoniae*.[91] CRISPRone has been used mainly to verify CRISPR results for detection in *K. pneumoniae*.[50] Spacers from detected CRISPRs were checked for homology using BLASTn.[42,50,90–92]

6.2 Wet lab based approach

The functionality of CRISPR-Cas systems has been found to be associated with certain factors. Increased activity of these systems has been correlated with increased number of unique spacers and DRs in the array, greater spacer homology with various mobile genetic elements (MGEs) and increased expression of *cas* genes.[89,93,94] Conventional PCR can only detect presence of genes, but cannot quantify their expression. A bioinformatics analysis can provide a rough prediction of function on the basis of the first two factors. However, it cannot predict the functionality as a whole because it cannot determine *cas* gene expression—the most important factor dictating the overall activity of the system. Moreover, expression of *cas* genes can be downregulated by certain global transcriptional repressors like H-NS (Histone-Like Nucleoid Structuring Protein).[95,96] Hence, qRT-PCR is used for quantifying expression of *cas* genes, thus, the overall system functionality.[92,96]

Post CRISPR-Cas identification, qRT-PCR has been primarily used in studying the expression of *cas1* and *csy1* genes, thereby detecting the functional status of subtype I-F system in *E. coli* as well as its association with antimicrobial resistance.[92,96] Thus, it has proved to be helpful in analyzing the functionality of CRISPR-Cas systems.

Bioinformatics analysis has been useful in detecting the presence of CRISPR-Cas systems along with validation of conventional PCR CRISPR-Cas detection. An additional functionality analysis using qRT-PCR can further increase the specificity of CRISPR-Cas identification.

7. Applications of CRISPR-Cas for controlling antimicrobial resistance and virulence

The CRISPR-Cas system has different applications among which five approaches in relation to AMR and virulence have been discussed (Fig. 3).

7.1 Use of CRISPR-Cas system to prevent plasmid transfer (MGEs) among bacteria via conjugation, thus restricting HGT and AMR dissemination

Increasing evidence of the use of CRISPR-Cas in preventing acquisition of MGEs conferring resistance to different antibiotic classes has been documented in recent years. Such a system opens new avenues for therapeutic application of bacteria's own defense against themselves to inhibit resistance acquisition and spread.

The first study of using CRISPR in blocking plasmid transfer (conjugation) was performed by Marraffini and Sontheimer in 2008.[41] The authors used CRISPR spacers (*spc1*) homologous to nickase gene, present in all conjugative plasmids of Staphylococcal strains, to block HGT in *S. epidermidis*. They demonstrated that CRISPR loci have the potential to block multiple routes of HGT and can contain spread of antibiotic resistance in pathogenic bacteria (discussed in Section 4.1). Garneau et al., while working with *S. thermophilus* also showed involvement of CRISPR in blocking plasmid transformation by introducing pNT1 into a CRISPR-positive strain and found plasmid-insensitive transformants (discussed in Section 4.1).[13]

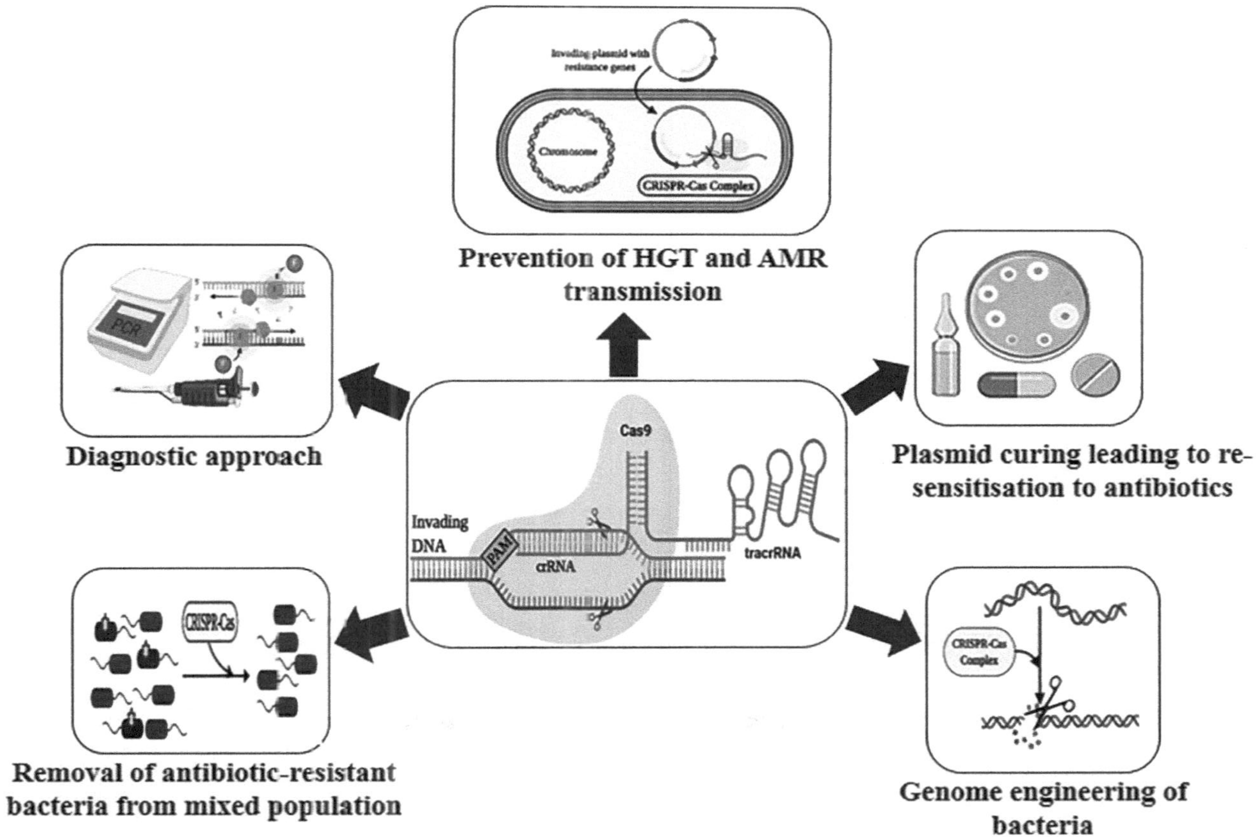

Fig. 3 Schematic diagram highlighting different approaches of CRISPR-Cas system application in the field of antimicrobial resistance and virulence. *Created using BioRender.com.*

Role of CRISPR in blocking conjugation in *E. faecalis* was explained by Price et al.[43] They used a susceptible *E. faecalis* as recipient for conjugation whose CRISPR3-Cas9 was deleted and found an increased acquisition of plasmid. This finding suggested the role of CRISPR3-Cas9 in transfer and acquisition of plasmids in *E. faecalis* (discussed in Section 4.1).

7.2 Role of CRISPR system in plasmid curing from drug-resistant pathogens and re-sensitization of bacteria to different antimicrobials

Blocking the entry of plasmids possessing ARGs is one mode of restricting the transmission of resistance whereas eliminating such plasmids from drug-resistant strains can be another approach. Elimination of plasmids or plasmid curing from bacteria enables the study of plasmid functionality in drug resistance but more importantly can re-sensitize the bacteria against particular antimicrobials, thereby improving effective treatment protocol.

Yosef et al. in 2015 had shown a genetic strategy to kill antibiotic-resistant bacteria by sensitizing the bacteria to antibiotics.[97] They used genetically modified lytic phages to deliver a programmable CRISPR-Cas system which contained antibiotic resistance genes as spacers from antibiotic-resistant plasmid of the target bacteria. Authors created a transferable CRISPR-Cas system using CRISPR cascade genes (*cse1*, *cse2*, *cas7*, *cas5*, and *cas6e*) and *cas3* of the *E. coli* subtype I-E CRISPR system. This engineered CRISPR system was introduced into lambda phages through homologous recombination, along with CRISPR spacers targeting $bla_{\text{NDM-1}}$ and $bla_{\text{CTX-M-15}}$ immediately downstream of the *cas* genes. Observations showed that lysogenized bacteria with CRISPR-Cas possessed a greater fitness compared to the plasmid bearing ones. Transformation of these lysogenized bacteria again, with plasmids possessing $bla_{\text{NDM-1}}$ and $bla_{\text{CTX-M-15}}$, was effectively blocked leading to plasmid curing. Hence, they exhibited that the engineered CRISPR-Cas system could be easily transferred to bacteria by lysogenization to target and destroy plasmids encoding resistance genes.[97]

Cao et al. in 2017 had used CRISPR-Cas9 system to eliminate high-copy number plasmids in *Zymomonas mobilis* ZM4 strain,[98] which harbor five native plasmids lacking antimicrobial resistance genes.[99] They showed that in absence of any selective marker gene (resistance genes) within the plasmids, CRISPR-Cas9 system could be used to cure the plasmids easily by targeting their replicase genes. They constructed two plasmids: pSUZM2a-Cas9, used to express *cas9* gene cloned from *S. pyogenes*; and pUC-T7sgRNA with a T7 promoter used for synthesis of sgRNA (targeting replicase gene). These two plasmids were co-transformed in a *Z. mobilis* ZM4 which possessed high-copy plasmids pZZM402 and pZZM403 (with more than 30 copies). Their results showed successful curing of these two plasmids by knock-out of replicase gene using CRISPR-Cas9:sgRNA system.[98]

Another novel method of plasmid elimination using CRISPR-Cas was developed by Lauritsen et al., in 2018, where they constructed CRISPR-Cas9 based plasmid curing system (pFREE) which could specifically target major plasmid replicons including the Standard European Vector Architecture (SEVA) vectors.[100] This system was efficient in curing two sets of SEVA vectors possessing different replicons. For this, they inserted genes encoding *cas9*, tracrRNA and gRNAs targeting the conserved regions of two replicons along with other essential modules into pFREE. Transformation of resistant bacteria with pFREE showed increased curing efficiencies (40%–100%) for most plasmids employed. Thus, this study established a single-step method for curing multiple plasmids in a broad host range.[100,101]

Recently, the spread of colistin resistance by plasmid mediated gene, *mcr-1* has compounded the problem of antimicrobial resistance. In view of preventing spread of colistin resistance, Sun et al. in 2017 designed a plasmid encoding *cas9* and sgRNA targeting *mcr-1*, and transformed them in *E. coli* NJ-15-3 harboring *mcr-1* gene. They noted elimination of *mcr-1* in transformed cells along with increased sensitization to colistin.[102] With an objective of tackling conjugative plasmids bearing *mcr-1*; unlike the previous study, Dong et al. in 2019, modified a conjugative plasmid pFSEC-01 with CRISPR-Cas9 system targeting *mcr-1* generating pMob-Cas9. This plasmid (pMob-Cas9) was transformed into *E. coli* strain MG1655, serving as the donor for conjugation with *E. coli* strain SHP45, bearing the *mcr-1* containing plasmid pHNSHP45. They found loss of *mcr-1* plasmid among the recipient, making them susceptible to colistin.[103] Both the studies have pinpointed the efficacy of CRISPR-Cas9 system in elimination of resistance gene bearing plasmids and re-sensitization to antimicrobials.[102,103]

7.3 Use of CRISPR system for genome engineering of bacteria with/without killing them

A prerequisite to genome editing using the CRISPR-Cas system is the introduction of double-strand breaks (DSBs) within the target DNA by Cas9 nuclease complexed with sgRNA. For genome editing in cells, CRISPR-Cas9 is delivered in different ways: Cas9 and sgRNA encoded by the same or separate DNA expression vectors; delivery of *in vitro* transcribed Cas9 mRNA and sgRNA, or, delivery as RNP (ribonucleoprotein) complex of Cas9 recombinant protein and sgRNA. Several studies are discussed in this section to show the efficacy of CRISPR-mediated genome engineering of bacteria for their greater utility.

Jiang et al. in 2013 used CRISPR-Cas9 system to cleave a genomic target preventing phage infection in *Streptococcus pneumoniae* as well as to make the bacteria susceptible to any antimicrobials by introducing mutations in various genomic sites of the bacteria.[104] The CRISPR-Cas system which was detected in *S. pneumoniae* strain crR6 contains a dual-RNA:Cas9 gene for cleavage of a target sequence in bacteriophage φ8232.5. They introduced the target sequence with/without mutated PAM region in the chromosomal locus (*strA*) of two *S. pneumoniae* strains. Transformation was carried out in these two strains with genomic DNA from crR6 (containing dual-RNA:Cas9 and wild-type *strA* locus). Transformants were retrieved which were the product of a double recombination event that replaced φ8232.5 target with the wild-type *srtA* locus resulting in prevention of cell death. Transformants were retrieved with 10-fold higher efficiency in case of mutation in PAM of target sequence within the *srtA* locus. They integrated the erythromycin resistance gene (*ermAM*) in the *srtA* locus, and introduced a premature stop codon using dual-RNA:Cas9-mediated editing. This resulted in sensitization of the strain toward erythromycin.

So et al. in 2017 had devised a new method called iterative genome engineering to perform efficient genome editing in *Bacillus subtilis*.[105] The authors had transformed the wild-type strain 168, with two plasmids. One plasmid, pHCas9, contained *cas9* gene; another plasmid, derived from pAD123, had both the sgRNA targeting the gene of interest, and the homologous donor DNA, containing the desired mutation in that gene to be inserted by homology directed repair. Using this method, almost 100% efficient deletion of 38 kb non-essential plipastatin operon (*pps*) was done to create space for incorporation of essential genes (that will be useful for industrial and food grade applications) without disturbing the bacterial chromosomal integrity. Apart from generating deletions, efficient point mutation in *sigE* (a gene involved in sporulation pathway) was also done to assess change in bacterial morphology. To further facilitate observation of altered morphology, insertion of *gfp* (green fluorescence protein encoding gene) within *sigE* was carried out. This study established how genetic engineering could be used to increase utility of a strain.[105]

Liu et al. in 2017 had used a similar approach for efficient genome editing in *Staphylococcus aureus*.[106] An *E. coli/S. aureus* shuttle vector pKOR was selected for insertion of *cas9* gene, which was placed under the constitutive promoter $P_{xyl/tet}$, giving rise to the plasmid pLQ-$P_{xyl/tet}$-cas9. The same plasmid was modified by the insertion of the sgRNA targeting *tgt* gene (queuine tRNA-ribosyltransferase), and the homologous donor DNA (50 bp *tgt* deletion) giving rise to the new construct pLQ-$P_{xyl/tet}$-cas9-P_{spac}-sgRNA. This plasmid was used for homologous recombination to generate 50 bp deletion of *tgt* or removal of the whole 1.1 kb *tgt* open reading frame (ORF). To investigate knock-in generation efficiency by the plasmid, erythromycin resistance gene *ermA* (1.2 kb) was inserted in the plasmid containing the 50 bp *tgt* deletion. The plasmid had exhibited similar transformation efficiency with that of the one responsible for 50 bp of *tgt* deletion. Thus, a single plasmid construct containing both *cas9* and sgRNA was capable of generating deletion and knock-in mutants.[106]

Ram et al. in 2018 used the CRISPR system to genetically modify the pathogenicity islands (SaPIs) of *S. aureus* to reduce infection caused by this pathogen.[107] They developed antibacterial drones (ABDs) by replacing the toxin genes of SaPIs with antibacterial cargos to specifically target the infecting bacteria in the animal host either by killing or de-activating them. They constructed two ABDs namely ABD2003 (containing CRISPR-Cas9) and ABD2006 (containing CRISPR-dCas9). *In vivo* experiment with ABD2003 showed double stranded break in *agrA* gene (a non-essential global gene regulator of staphylococcal virulence) resulting in protection toward abscess formation by killing the bacteria. On the other hand, ABD2006 blocked expression of Agr operon resulting in blocking of abscess formation without killing the bacteria. *In vitro* experiments with bacteria isolated from ABD2006 treatment showed hampered hemolytic activity. These findings reinforced that genetic engineering using CRISPR can be used both to kill bacteria as well as to reduce pathogenicity among them.[107]

7.4 Use of CRISPR arrays to remove specific virulent or antibiotic-resistant bacteria from a mixed population by targeting their antibiotic resistance or virulence genes

Several authors have used CRISPR-Cas as a tool to kill the antibiotic-resistant or virulent strains in a heterogeneous bacterial population allowing non-pathogenic strains to overgrow the pathogens.

Gomaa et al. in 2014 showed that CRISPR-Cas systems can be employed in a sequence-specific manner to remove the individual bacterial strains and species from a diverse microbial population.[108] They used the *E. coli* K-12 type I-E CRISPR system as a model. The experiment had addressed different aspects of targeting bacteria—removal of *E. coli* K12, *E. coli* BL21 and *Salmonella enterica* in pure culture as well as from mixed culture, using different spacers for particular strains and species. They used three plasmids: pCasA-E & pCas3 for inducible expression of all six *cas* genes and pCRISPR for sequential insertion of engineered spacer sequences. The authors designed the spacer using PAM-flanking different target sequences (genes) for different strains (such as *fucA* for K-12, *ogr* for BL21, and *mviM* for *S. enterica*).[108] They transformed the pCRISPR- containing type I system with the engineered spacer sequences into the pure and mixed culture and found removal of these individual strains from a mixed culture by selectively using different spacers against different strains.[108,109]

Bikard et al. in 2014 used reprogrammed *cas9* nuclease to specifically kill the virulent strain of *Staphylococcus aureus*.[110] They used a Cas9 phagemid as a sequence-specific antimicrobial to target methicillin-resistant *S. aureus* (MRSA) in a mixed population. The CRISPR-Cas system was programmed to target the methicillin resistance gene, *mecA*, of a clinical MRSA isolate lysogenized with the mentioned phagemid. Treatment with the phagemid::*mecA* had successfully reduced the proportion of MRSA from 50% to 0.4%. Thus, the Cas9 phagemid proved to be efficient in killing specific Staphylococci from a mixed population.[110] On the other hand, Kang et al. in 2018 first used a vector-free nanocomplex made up of polymer-derivatized Cas9 protein and sgRNA targeting the *mecA* gene of MRSA.[111] They covalently linked Cas9 sulfhydryl groups with a cationic polymer, branched polyethyleneimine (bPEI), a common gene delivery carrier. They mixed Cas9 bound bPEI with the sgRNA targeting *mecA*, forming small nanocomplexes. Uptake of these nanocomplexes by the cells decreased the cell growth thus showing removal of MRSA by sequence-specific targeting using CRISPR-Cas9-nanocomplexes.[111]

Citorik et al. in 2014 had used transmissible programmable CRISPR-Cas systems as a platform to induce site-specific cleavage resulting in cytotoxicity, in activating toxin-antitoxin systems, re-sensitize bacterial populations to antibiotics resulting in removal of a particular bacterial group.[112] In this study, they delivered a designed RNA guided cas9 nuclease (RGN) by bacteriophage into carbapenem-resistant *E. coli* to create specific cleavage against plasmid (harboring carbapenem-resistance gene, $bla_{\mathrm{NDM-1}}$ and β-lactamase resistance gene, $bla_{\mathrm{SHV-18}}$) or chromosomal target sequences (quinolone resistance gene, $gyrA_{\mathrm{D87G}}$). They showed that delivery of RGN targeting chromosomal sequences lead to cytotoxic effect whereas presence of plasmid sequence within RGN leads to cell death or plasmid loss depending on the presence or absence of toxin-antitoxin (TA) systems, respectively. Again, phagemid treatment is found as a viable alternative apart from traditional antimicrobial agents.[112]

Kim et al. in 2016 described CRISPR-Cas9 system as a genome editing tool in sensitizing several antibiotic-resistant bacteria toward versatile antibiotics especially to kill extended spectrum β-lactamase (ESBL)-producing *E. coli*.[113] They developed CRISPR-Cas9 plasmid possessing spacer sequences from each bla_{TEM} and bla_{SHV} ESBL genes that were conserved among most ESBL mutants to specifically target these ESBL genes. Transformation of bacterial cells with CRISPR-Cas9 plasmid containing the target sequences for ESBL genes resulted in the abolition of β-lactamase production thereby sensitizing them to a variety of antimicrobials to which they previously showed resistance.[113]

Wang et al. in 2019 constructed a CRISPR-Cas9 plasmid (pMBLcas9) and used it to clone single-guide RNAs (sgRNAs) for plasmid curing.[114] They developed this plasmid to sensitize clinical isolates of *E. coli* toward multiple antibiotics. They transferred the recombinant plasmid pMBLcas9-sgRNA by conjugation into two clinical isolates of *E. coli* (14EC003 and 14EC007). Native plasmids were successfully eliminated in a step wise manner from the *E. coli* isolates and sensitized the cells to polymyxins and carbenicillin. In addition, strain 14EC033 also carried two *mcr-1*-possessing plasmids among which one plasmid was cured using a single *mcr-1* sgRNA and *mcr-1* gene was deleted from another plasmid to generate a mutant plasmid. Transposase-mediated (IS5) recombination was responsible for *mcr-1* gene deletion instead of plasmid curing.[114]

Tagliaferri et al. in 2020 had utilized the Type II CRISPR-Cas9 system to control AMR; first in a non-AMR *E. coli* BL21 model strain; followed by three other clinical Enterobacteriaceae samples—*E. coli* 189A, *Enterobacter hormaechei* 4962 *and Klebsiella variicola* 68AI isolated from bacteremia patients.[115] The model and the clinical strains were transformed with a high-copy plasmid PSB1A2 containing the bla_{TEM-1} gene and another plasmid PSB1C3 containing CRISPR-Cas9 system programmed to specifically target the above gene. It was found that CRISPR-Cas9 successfully reversed AMR in the model strain. However, entire plasmid loss was not observed. In the case of clinical isolate *E. coli* 189A, complete plasmid clearance, phenotypic reversal and re-sensitization to several antibiotics including third-generation cephalosporins was found. On the contrary, *E. hormaechei*

4962 *and K. variicola* 68AI showed some plasmid retention, partial phenotypic reversal and re-sensitization. Such phenomena had occurred in these two isolates due to deletions in CRISPR-Cas9 locus, as well as presence of other resistance genes not targeted by the system.[115]

Removal of antibiotic-resistant bacteria from a population of *E. faecalis* was documented by Hullahalli et al. in 2017.[116] They constructed plasmid pCR2 with introduction of the entire CRISPR2 locus of a multidrug-resistant *E. faecalis* V583 in pKH12. Later they manipulated pCR2 to target erythromycin resistance gene (*ermB*) generating pCR2-*ermB*. Both the plasmids (pCR2 and pCR2-*ermB*) were introduced in *E. faecalis* CK111SSp (pCF10–101).[117] Conjugation was carried out with CK111SSp as donor and V583 and V649 (a derivative of V583 with *cas9*, promoter and tracrRNA inserted in it) as recipient. A reduction in conjugation frequency was observed with pCR2-*ermB* relative to pCR2 in both V583 and V649. Hence, their experiment demonstrated that delivery of CRISPR systems via conjugation eradicates prevalence of drug-resistant bacteria from *E. faecalis* populations.[116]

7.5 Diagnostic approach for detection of antibiotic resistance genes using CRISPR system

The versatile CRISPR system has also been used as a diagnostic tool given the fact that timely detection of resistance determinants in patient specimens can lead to correct use of antibiotics. Characteristics of a good diagnostic tool lies with the accuracy, sensitivity, specificity and early recognition of resistance allowing implementation of proper infection control and health measures. The CRISPR-Cas system can be used to serve such a motto. It is not only affordable, but also portable and can be carried out in various clinical settings. Various authors have used CRISPR-Cas9 to develop diagnostic tools for detection of infectious diseases.

Müller et al. in 2016 employed such an approach showing a combined effect of CRISPR system with optical DNA mapping to identify bacterial antibiotic resistance genes.[118] They initially used gRNAs specific to the sequence of CTX-M-15 (a β-lactamase), a fluorescent dye (YOYO-1)[119] and netropsin (a polyamide that binds to AT-rich regions of DNA selectively). A varying fluorescence emission unique to each DNA segment, were recorded like a barcode with lower emission from AT-rich region than GC region. They found a double-strand break at the same position along the barcode indicating the presence of target sequence. By using a cocktail of gRNAs, targeting of multiple resistance genes was possible along with the plasmids harboring them. This helped to distinguish between plasmids producing different variants of CTX-M along with carbapenemases such as KPC and NDM-1. Their findings indicated the role of the CRISPR-Cas system to track bacterial resistance within hours, to effectively control spread and implement efficient treatment protocols.[118]

Similarly, Guk et al. in 2017 and Quan et al. in 2019, developed unique diagnostic tools using the CRISPR system for rapid detection of resistance genes in the bacterial population of *S. aureus*.[120,121] Guk et al. used CRISPR-Cas9 with fluorescence *in situ* hybridization (FISH) technique to detect methicillin-resistant *S. aureus* (MRSA).[120] sgRNA-dCas9 was coupled with a SYBR green I fluorescent probe to specifically target the *mecA* gene of MRSA, without causing any DNA break, hence, making it competent for FISH. Ten-fold increase in fluorescence intensity of MRSA was noted when compared to methicillin-susceptible *Staphylococcus aureus*. Cell concentration as low as 10 CFU/mL could be detected by this method. The technique was able to track any resistance genes within a short span of time, was cost effective, and sensitive in detection of various pathogens.[120] On the other hand, Quan et al.,[121] used a computational tool, FLASHit to construct gRNAs targeting resistance genes of *S. aureus* (both chromosomal and plasmid). They digested genomic DNA by *cas9* complexed with a set of gRNAs, and ligated universal adapters to ensure amplification of target sequences compared to other genomic sequences. Afterward, next-generation sequencing (FLASH-NGS) of the templates was done and up to five orders of magnitude of enrichment was achieved for any particular target sequences. They also theorized highly multiplexed detection of antimicrobial resistance genes with precision and sequence identity by this CRISPR-based diagnostic tool FLASH. Hence, both the studies focused on use of CRISPR as an effective tool for rapid detection of resistance genes with lower volume of bacterial concentration.[121]

The capability of CRISPR-Cas immunity to inhibit the transfer of drug resistance and virulence genes into pathogens can reduce the potential of their evolution. CRISPR-Cas systems can be repurposed to regulate gene expression and enhance pathogenesis. The utility of these systems, as explained by these findings mentioned (Table 2), are of interest to those in both academia and industries and finally for the welfare of society. In the future, more avenues will open up enhancing control of bacterial resistance and virulence using CRISPR-Cas.

Table 2 Applications of CRISPR-Cas system in the field of antimicrobial resistance and virulence.

Sl No.	Organism	MGE/gene target	Delivery	Strategy and result	Study
A. CRISPR-Cas system prevents plasmid transfer (MGE) among bacteria via conjugation, thus prevent HGT and AMR dissemination					
1.	*S. epidermidis*	pG0400 plasmid and nickase gene	Conjugation and electroporation	Blocking plasmid transfer by conjugation and transformation with CRISPR system possessing *spc1* CRISPR spacer targeting nickase gene	Marraffini and Sontheimer, 2008[41]
2.	*S. thermophilus*	Chloramphenicol resistance gene (*cat*) of pNT1	Transformation	Blocking plasmid transformation in the presence of active *cas* genes	Garneau *et al.*, 2010[13]
3.	*E. faecalis*	pAM714	Conjugation	Blocking of conjugal transfer of erythromycin-resistant plasmid in presence of CRISPR3-Cas	Price *et al.*, 2016[47]
B. Role of CRISPR system in plasmid curing					
4.	*E. coli*	$bla_{\text{NDM-1}}$ and $bla_{\text{CTX-M-15}}$ causing carbapenem and cephalosporin resistance respectively	Lysogenisation and transformation	Temperate phage encoding CRISPR-Cas had specifically cured the plasmids bearing the carbapenem resistance genes or ESBL genes from cells favouring only susceptible cell survival.	Yosef *et al.*, 2015[97]
5.	*Z. mobilis*	pZZM402 and pZZM403	Electroporation	Effective curing of native plasmids by engineering CRISPR-Cas9 plasmid targeting plasmid replicase genes.	Cao *et al.*, 2017[98]
6.	*E. coli*, *P. putida*	Various replicons	Transformation	Effective plasmid curing using a single-step pFREE CRISPR-Cas9 plasmid programmed to target conserved regions of replicon.	Lauritsen *et al.*, 2017[101]
7.	*E. coli*	Plasmids bearing colistin resistance gene, *mcr-1*	Electroporation	Elimination of *mcr-1* and sensitisation to colistin using CRISPR-Cas9 plasmid.	Sun *et al.*, 2017[102]
8.	*E. coli*	Plasmids bearing colistin resistance gene, *mcr-1*	Transformation and conjugation	Effective plasmid curing and sensitisation to colistin using a host-independent conjugative plasmid pMob-Cas9.	Dong *et al.*, 2019[103]

C. CRISPR system could be used for genome engineering of bacteria with/ without killing them					
9.	*S. pneumoniae*	A target sequence of bacteriophage and erythromycin resistance gene (*ermAM*)	Transformation	Using dual-RNA: Cas9-mediated editing to cleave a genomic target preventing phage infection in bacteria and re-sensitisation towards antimicrobials by incorporating mutations within a particular genomic site.	Jiang *et al.*, 2013[104]
10.	*B. subtilis*	*pps* operon, *sigE*	Transformation	Effective incorporation of various mutations in chromosomal genomes using newly devised CRISPR-Cas9 based iterative genome engineering to increase industrial utility of bacteria.	So *et al.*, 2017[105]
11.	*S. aureus*	Queuine tRNA-ribosyltransferase (*tgt*) gene	Transformation	Effective and specific genome editing of target loci using pLQ-$p_{xyl/tet}$-cas9-p_{spac}-sgRNA plasmid generating deletion and knock-in mutants using single plasmid.	Liu *et al.*, 2017[106]
12.	*S. aureus*	*agrA* gene and expression of its operon system	Transformation, subcutaneous and intraperitoneal injection in mice model	Using CRISPR-Cas9 incorporated pathogenicity islands called antibacterial drones (ABDs) targeting specific genes responsible for virulence. Antibacterial drones (ABDs) efficiently blocked *agr* expression and hamper virulence potential.	Ram *et al.*, 2018[107]
D. CRISPR arrays have been used to remove specific virulent or antibiotic-resistant bacteria from a population by targeting their antibiotic resistance or virulence genes					
13.	*S. enterica* *E. coli*	*mviM and groL* *ftsA*, *asd*, *msbA*, *nusB*, *fucP*, *groL*, *ogr and arpA*	Transformation	Three plasmids: pCasA-E, pCas3 and pCRISPR for targeting chromosomal regions were introduced which resulted in deletion of targeted genes and removal of specific bacterial species from a mixed population.	Gomaa *et al.*, 2014[108]

Continued

Table 2 Applications of CRISPR-Cas system in the field of antimicrobial resistance and virulence.—cont'd

SI No.	Organism	MGE/gene target	Delivery	Strategy and result	Study
14.	*S. aureus*	Methicillin resistance (*mecA*) gene	Transduction	Phagemid CRISPR-Cas9 system had selectively killed methicillin-resistant strains within a mixed population	Bikard *et al.*, 2014[110]
15.	*S. aureus*	*mecA*	Nanoparticle delivery	Vector-free Cas9-bPEI-sgRNA nanocomplex delivery increased cellular uptake and reduced growth of cells by sequence-specific targeting without side-effects.	Kang *et al.*, 2017[111]
16.	*E. coli*	bla_{NDM-1}, bla_{SHV-18}, $gyrA_{D87G}$ causing quinolone resistance	Conjugation and transduction	Phagemid CRISPR-Cas9 killed specifically the cells bearing the plasmid-borne genes as well as quinolone resistant ones, without causing off-target effects.	Citorik *et al*, 2014[112]
17.	*E. coli*	ESBL genes (bla_{TEM}, bla_{SHV})	Transformation	Introduction of CRISPR-Cas9 plasmid targeting ESBL genes which abolished targeted resistance genes and re-sensitised the bacteria to various antibiotics.	Kim *et al.*, 2016[113]
18.	*E. coli*	Curing of multiple plasmids possessing polymyxins and carbenicillin resistance genes	Conjugation	Construction of a CRISPR-Cas9 plasmid (pMBLcas9) that possessed single-guide RNAs (sgRNAs) for plasmid curing.	Wang *et al.*, 2019[114]
		Deletion of *mcr-1* gene		Transposase-mediated (IS5) recombination was responsible for *mcr-1* gene deletion instead of plasmid curing.	

19.	*E. coli*, *E. hormaechei*, *K. variicola*	bla_{TEM-1}	Transformation	CRISPR-Cas9 system bearing pSB1C3 plasmid was used which reversed resistant phenotype by curing bla_{TEM-1} bearing plasmid and permitted reuse of traditional antibiotics.	Tagliaferri *et al.*, 2020[115]
20.	*E. faecalis*	pTEF1 and pAM771	Conjugation and electroporation	Manipulation of CRISPR2 locus for targeting antibiotic-resistant plasmids which decreased conjugation frequency of plasmids harboring resistance genes	Hullahalli *et al.*, 2017[116]
E. Diagnostic approach for detection of antibiotic resistance genes using CRISPR system					
21.	Gram-negative bacteria	$bla_{CTX-M-15, CTX-M-14, KPC, NDM-1}$	gRNAs specific to the target, a fluorescent dye (YOYO-1) and netropsin	Optical DNA mapping mediated CRISPR system detected presence of resistance genes within hours, also the plasmids harboring them by making DSB and matching break position along the bar codes.	Müller *et al.*, 2016[118]
22.	Methicilin-resistant *S. aureus*	*mecA* gene	Fluorescent tagging of genomic DNA of MRSA and their cell lysate with dCas9/gRNA SGI	Using CRISPR-mediated DNA FISH system, presence of *mecA* gene producing *S. aureus* can be detected within 30 mins even in low bacterial load of 10 CFU/ml.	Guk *et al.*, 2017[120]
23.	*S. aureus*	Various resistance genes of *S. aureus*	CRISPR-Cas9 complexed with gRNAs mediated amplification of target sequences followed by FLASH-Next Gen Sequencing	Targeting multiple resistance genes in a single shot with CRISPR-Cas9 mediated FLASH-NGS which exhibited 5 orders of magnitude of enrichment and detection with minimal amount.	Quan J *et al.*, 2019[121]

8. A place to reflect

Recognizing problems, acknowledging them and finding ways to solve them has taken the human race forward. It has been well understood that AMR pathogens are a major public health concern worldwide. Again, virulent strains with increased ability to persist in hospitals as well as community are worrisome. Several clones of bacteria which are both resistant and virulent have also emerged. With no promising antibiotics in the pipeline, the solution to this problem will be multi-pronged and the strategic use of the CRISPR system to fight against antibiotic-resistant and virulent pathogens may prove to be an extremely important arsenal.

Several elegant studies discussed in this chapter have shown that this bacterial adaptive immune system has immense potential and can be harnessed to bring some respite to the problem of antibiotic resistance. With increasing whole genome analysis of bacteria, the presence of CRISPR in diverse species, their types and subtypes have been delineated. However, sequences are just the blue-print, the interplay of resistance/virulence with the CRISPR in natural settings or in the laboratory needs further probing. Once this understanding increases, its use in eradicating antibiotic resistance will also advance. As antibiotic resistance is also an evolving, changing, multi-faceted landscape in bacterial existence, keeping-up in defeating it will not be easy. With multiple resistance mechanisms working together in various combinations in gram-negative and gram-positive bacteria, antibiotic resistance will surely be difficult to solve but the initial emphasis should be on restricting the transmissible resistance determinants. The bacterial MGEs act as the vehicles in transporting the resistance/virulence genes quickly and efficiently to large diverse populations of bacteria, increasing the capability of bacteria in invading newer geographies. The promise that certain studies have already shown in eradicating pathogenic bacterial population, rapid detection of resistance and the revival of efficacy of antibiotics to which bacteria showed resistance earlier, is the silver lining behind this grave problem. With the use of CRISPRs, it will be beating bacteria at their own game.

Acknowledgments

The authors acknowledge the Council of Scientific and Industrial Research (CSIR), India for fellowship to Dr. Subhasree Roy and Indian Council of Medical Research (ICMR), India for fellowship to Sharmi Naha and Ankur Rao.

Conflict of interest

The authors declare no conflict of interest.

References

1. Louwen R, Staals RHJ, Endtz HP, van Baarlen P, van der Oost J. The role of CRISPR-Cas systems in virulence of pathogenic bacteria. *Microbiol Mol Biol Rev.* 2014;78:74–88. https://doi.org/10.1128/mmbr.00039-13.
2. Wright GD. The antibiotic resistome: the nexus of chemical and genetic diversity. *Nat Rev Microbiol.* 2007;5:175–186. https://doi.org/10.1038/nrmicro1614.
3. World Health Organization. *Antimicrobial resistance: global report on surveillance.* https://www.who.int/antimicrobial-resistance/publications/surveillancereport/en/. Published 2014 Apr. Accessed on 2020 Aug 29.
4. Van Boeckel TP, Gandra S, Ashok A, et al. Global antibiotic consumption 2000 to 2010: an analysis of national pharmaceutical sales [published correction appears in *Lancet Infect Dis.* 2017;17:897]. *Lancet Infect Dis.* 2014;14:742–750. https://doi.org/10.1016/S1473-3099(14)70780-7.
5. World Health Organization, Regional Office for South-East Asia. *Antimicrobial resistance and its containment in India.* http://origin.searo.who.int/india/topics/antimicrobial_resistance/amr_containment.pdf. Published 2016 Nov. Accessed on 2020 August 13.
6. Lee CR, Lee JH, Park KS, Kim YB, Jeong BC, Lee SH. Global dissemination of carbapenemase-producing *Klebsiella pneumoniae*: epidemiology, genetic context, treatment options, and detection methods. *Front Microbiol.* 2016;7:895. https://doi.org/10.3389/fmicb.2016.00895.
7. Liu YY, Wang Y, Walsh TR, et al. Emergence of plasmid-mediated colistin resistance mechanism MCR-1 in animals and human beings in China: a microbiological and molecular biological study. *Lancet Infect Dis.* 2016;16:161–168. https://doi.org/10.1016/S1473-3099(15)00424-7.
8. Frost LS, Leplae R, Summers AO, Toussaint A. Mobile genetic elements: the agents of open source evolution. *Nat Rev Microbiol.* 2005;3:722–732. https://doi.org/10.1038/nrmicro1235.
9. Bolotin A, Quinquis B, Sorokin A, Ehrlich SD. Clustered regularly interspaced short palindrome repeats (CRISPRs) have spacers of extrachromosomal origin. *Microbiology.* 2005;151:2551–2561. https://doi.org/10.1099/mic.0.28048-0.
10. Mojica FJ, Diez-Villaseñor C, Garcia-Martinez J, Soria E. Intervening sequences of regularly spaced prokaryotic repeats derive from foreign genetic elements. *J Mol Evol.* 2005;60:174–182. https://doi.org/10.1007/s00239-004-0046-3.
11. Pourcel C, Salvignol G, Vergnaud G. CRISPR elements in *Yersinia pestis* acquire new repeats by preferential uptake of bacteriophage DNA, and provide additional tools for evolutionary studies. *Microbiology.* 2005;151:653–663. https://doi.org/10.1099/mic.0.27437-0.
12. Barrangou R, Fremaux C, Deveau H, Richards M, Boyaval P. CRISPR provides acquired resistance against viruses in prokaryotes. *Science.* 2007;315:1709–1712. https://doi.org/10.1126/science.1138140.
13. Garneau JE, Dupuis MÈ, Villion M, et al. The CRISPR/Cas bacterial immune system cleaves bacteriophage and plasmid DNA. *Nature.* 2010;468:67–71. https://doi.org/10.1038/nature09523.
14. Ishino Y, Shinagawa H, Makino K, Amemura M, Nakata A. Nucleotide sequence of the *iap* gene, responsible for alkaline phosphatase isozyme conversion in *Escherichia coli*, and identification of the gene product. *J Bacteriol.* 1987;169:5429–5433. https://doi.org/10.1128/jb.169.12.5429-5433.1987.

15. Mojica FJM, Díez-Villaseñor C, Soria E, Juez G. Biological significance of a family of regularly spaced repeats in the genomes of Archaea, bacteria and mitochondria. *Mol Microbiol.* 2000;36:244–246. https://doi.org/10.1046/j.1365-2958.2000.01838.x.
16. Makarova KS, Wolf YI, Alkhnbashi OS, et al. An updated evolutionary classification of CRISPR–Cas systems. *Nat Rev Microbiol.* 2015;13:722–736. https://doi.org/10.1038/nrmicro3569.
17. Jansen R, van Embden JDA, Gaastra W, Schouls LM. Identification of genes that are associated with DNA repeats in prokaryotes. *Mol Microbiol.* 2002;43:1565–1575. https://doi.org/10.1046/j.1365-2958.2002.02839.x.
18. Marraffini LA. CRISPR-Cas immunity in prokaryotes. *Nature.* 2015;526:55–61. https://doi.org/10.1038/nature15386.
19. Barrangou R. Diversity of CRISPR-Cas immune systems and molecular machines. *Genome Biol.* 2015;16:247. https://doi.org/10.1186/s13059-015-0816-9.
20. Jinek M, Chylinski K, Fonfara I, Hauer M, Doudna JA, Charpentier E. A programmable dual-RNA-guided DNA endonuclease in adaptive bacterial immunity. *Science.* 2012;337:816–821. https://doi.org/10.1126/science.1225829.
21. Chylinski K, Le Rhun A, Charpentier E. The tracrRNA and Cas9 families of type II CRISPR-Cas immunity systems. *RNA Biol.* 2013;10:726–737. https://doi.org/10.4161/rna.24321.
22. Makarova KS, Wolf YI, Koonin EV. Classification and nomenclature of CRISPR-Cas systems: where from here? *CRISPR J.* 2018;1:325–336. https://doi.org/10.1089/crispr.2018.0033.
23. Koonin EV, Makarova KS, Zhang F. Diversity, classification and evolution of CRISPR-Cas systems. *Curr Opin Microbiol.* 2017;37:67–78. https://doi.org/10.1016/j.mib.2017.05.008.
24. Zaman SB, Hussain MA, Nye R, Mehta V, Mamun KT, Hossain N. A review on antibiotic resistance: alarm bells are ringing. *Cureus.* 2017;9, e1403. https://doi.org/10.7759/cureus.1403.
25. Neu HC. The crisis in antibiotic resistance. *Science.* 1992;257:1064–1073. https://doi.org/10.1126/science.257.5073.1064.
26. Boucher HW, Talbot GH, Bradley JS, et al. Bad bugs, no drugs: no ESKAPE! An update from the Infectious Diseases Society of America. *Clin Infect Dis.* 2009;48:1–12. https://doi.org/10.1086/595011.
27. Queenan AM, Bush K. Carbapenemases: the versatile beta-lactamases. *Clin Microbiol Rev.* 2007;20:440–458. https://doi.org/10.1128/CMR.00001-07.
28. Peterson E, Kaur P. Antibiotic resistance mechanisms in bacteria: relationships between resistance determinants of antibiotic producers, environmental bacteria, and clinical pathogens. *Front Microbiol.* 2018;9:2928. https://doi.org/10.3389/fmicb.2018.02928.
29. Tang SS, Apisarnthanarak A, Hsu LY. Mechanisms of β-lactam antimicrobial resistance and epidemiology of major community- and healthcare-associated multidrug-resistant bacteria. *Adv Drug Deliv Rev.* 2014;78:3–13. https://doi.org/10.1016/j.addr.2014.08.003.
30. Floss HG, Yu TW. Rifamycin-mode of action, resistance, and biosynthesis. *Chem Rev.* 2005;105:621–632. https://doi.org/10.1021/cr030112j.
31. Hooper DC. Fluoroquinolone resistance among Gram-positive cocci. *Lancet Infect Dis.* 2002;2:530–538. https://doi.org/10.1016/s1473-3099(02)00369-9.
32. Weisblum B. Erythromycin resistance by ribosome modification. *Antimicrob Agents Chemother.* 1995;39:577–585. https://doi.org/10.1128/aac.39.3.577.
33. Sun J, Deng Z, Yan A. Bacterial multidrug efflux pumps: mechanisms, physiology and pharmacological exploitations [published correction appears in *Biochem Biophys Res Commun.*2015;465:165]. *Biochem Biophys Res Commun.* 2014;453:254–267. https://doi.org/10.1016/j.bbrc.2014.05.090.

34. Nikaido H, Pagès JM. Broad-specificity efflux pumps and their role in multidrug resistance of Gram-negative bacteria. *FEMS Microbiol Rev.* 2012;36:340–363. https://doi.org/10.1111/j.1574-6976.2011.00290.x.
35. Roberts MC. Update on acquired tetracycline resistance genes. *FEMS Microbiol Lett.* 2005;245:195–203. https://doi.org/10.1016/j.femsle.2005.02.034.
36. Nikaido H. Molecular basis of bacterial outer membrane permeability revisited. *Microbiol Mol Biol Rev.* 2003;67:593–656. https://doi.org/10.1128/mmbr.67.4.593-656.2003.
37. Fukuoka T, Ohya S, Narita T, et al. Activity of the carbapenem panipenem and role of the OprD (D2) protein in its diffusion through the *Pseudomonas aeruginosa* outer membrane. *Antimicrob Agents Chemother.* 1993;37:322–327. https://doi.org/10.1128/aac.37.2.322.
38. Reygaert WC. An overview of the antimicrobial resistance mechanisms of bacteria. *AIMS Microbiol.* 2018;4:482–501. https://doi.org/10.3934/microbiol.2018.3.482.
39. Gillings MR. Integrons: past, present, and future. *Microbiol Mol Biol Rev.* 2014;78:257–277. https://doi.org/10.1128/mmbr.00056-13.
40. Levin BR. Nasty viruses, costly plasmids, population dynamics, and the conditions for establishing and maintaining CRISPR-mediated adaptive immunity in bacteria. *PLoS Genet.* 2010;6:e1001171. https://doi.org/10.1371/journal.pgen.1001171.
41. Marraffini LA, Sontheimer EJ. CRISPR interference limits horizontal gene transfer in staphylococci by targeting DNA. *Science.* 2008;322:1843–1845. https://doi.org/10.1126/science.1165771.
42. Palmer KL, Gilmore MS. Multidrug-resistant enterococci lack CRISPR-Cas. *mBio.* 2010;1:e00227–10. https://doi.org/10.1128/mBio.00227-10.
43. Price VJ, Huo W, Sharifi A, Palmer KL. CRISPR-Cas and restriction-modification act additively against conjugative antibiotic resistance plasmid transfer in *Enterococcus faecalis. mSphere.* 2016;1:e00064-16. https://doi.org/10.1128/mSphere.00064-16.
44. Touchon M, Charpentier S, Pognard D, et al. Antibiotic resistance plasmids spread among natural isolates of *Escherichia coli* in spite of CRISPR elements. *Microbiology (Reading).* 2012;158:2997–3004. https://doi.org/10.1099/mic.0.060814-0.
45. Dang TN, Zhang L, Zöllner S, et al. Uropathogenic *Escherichia coli* are less likely than paired fecal *E. coli* to have CRISPR loci. *Infect Genet Evol.* 2013;19:212–218. https://doi.org/10.1016/j.meegid.2013.07.017.
46. Touchon M, Cury J, Yoon EJ, et al. The genomic diversification of the whole *Acinetobacter* genus: origins, mechanisms, and consequences. *Genome Biol Evol.* 2014;6: 2866–2882. https://doi.org/10.1093/gbe/evu225.
47. Mangas EL, Rubio A, Álvarez-Marín R, et al. Pangenome of *Acinetobacter baumannii* uncovers two groups of genomes, one of them with genes involved in CRISPR/Cas defense systems associated with the absence of plasmids and exclusive genes for biofilm formation. *Microb Genom.* 2019;5:e000309. https://doi.org/10.1099/mgen.0.000309.
48. Li HY, Kao CY, Lin WH, et al. Characterization of CRISPR-Cas systems in clinical *Klebsiella pneumoniae* isolates uncovers its potential association with antibiotic susceptibility. *Front Microbiol.* 2018;9:1595. https://doi.org/10.3389/fmicb.2018.01595.
49. Mackow NA, Shen J, Adnan M, Khan AS, Fries BC, Diago-Navarro E. CRISPR-Cas influences the acquisition of antibiotic resistance in *Klebsiella pneumoniae. PLoS One.* 2019;14:e0225131. https://doi.org/10.1371/journal.pone.0225131.
50. Kamruzzaman M, Iredell JR. CRISPR-cas system in antibiotic resistance plasmids in *Klebsiella pneumoniae. Front Microbiol.* 2020;10:2934. https://doi.org/10.3389/fmicb.2019.02934.
51. Bikard D, Hatoum-Aslan A, Mucida D, Marraffini LA. CRISPR interference can prevent natural transformation and virulence acquisition during *in vivo* bacterial infection. *Cell Host Microbe.* 2012;12:177–186. https://doi.org/10.1016/j.chom.2012.06.003.

52. Louwen R, Horst-Kreft D, de Boer AG, et al. A novel link between *Campylobacter jejuni* bacteriophage defense, virulence and Guillain-Barré syndrome. *Eur J Clin Microbiol Infect Dis.* 2013;32:207–226. https://doi.org/10.1007/s10096-012-1733-4.
53. Shabbir MAB, Tang Y, Xu Z, et al. The involvement of the *cas9* gene in virulence of *Campylobacter jejuni. Front Cell Infect Microbiol.* 2018;8:285. https://doi.org/10.3389/fcimb.2018.00285.
54. Sampson TR, Weiss DS. Cas9-dependent endogenous gene regulation is required for bacterial virulence. *Biochem Soc Trans.* 2013;41:1407–1411. https://doi.org/10.1042/BST20130163.
55. Zegans ME, Wagner JC, Cady KC, Murphy DM, Hammond JH, O'Toole GA. Interaction between bacteriophage DMS3 and host CRISPR region inhibits group behaviors of *Pseudomonas aeruginosa. J Bacteriol.* 2009;191:210–219. https://doi.org/10.1128/JB.00797-08.
56. Li R, Fang L, Tan S, et al. Type I CRISPR-Cas targets endogenous genes and regulates virulence to evade mammalian host immunity. *Cell Res.* 2016;26:1273–1287. https://doi.org/10.1038/cr.2016.135.
57. Borges AL, Castro B, Govindarajan S, Solvik T, Escalante V, Bondy-Denomy J. Bacterial alginate regulators and phage homologs repress CRISPR-Cas immunity. *Nat Microbiol.* 2020;5:679–687. https://doi.org/10.1038/s41564-020-0691-3.
58. García-Gutiérrez E, Almendros C, Mojica FJ, Guzmán NM, García-Martínez J. CRISPR content correlates with the pathogenic potential of *Escherichia coli* [published correction appears in PLoS One. 2015;10:e0134138]. *PLoS One.* 2015;10:e0131935. https://doi.org/10.1371/journal.pone.0131935.
59. Gill SR, Fouts DE, Archer GL, et al. Insights on evolution of virulence and resistance from the complete genome analysis of an early methicillin-resistant *Staphylococcus aureus* strain and a biofilm-producing methicillin-resistant *Staphylococcus epidermidis* strain. *J Bacteriol.* 2005;187:2426–2438. https://doi.org/10.1128/JB.187.7.2426-2438.2005.
60. Bourgogne A, Garsin DA, Qin X, et al. Large scale variation in *Enterococcus faecalis* illustrated by the genome analysis of strain OG1RF. *Genome Biol.* 2008;9:R110. https://doi.org/10.1186/gb-2008-9-7-r110.
61. Lindenstrauss AG, Pavlovic M, Bringmann A, Behr J, Ehrmann MA, Vogel RF. Comparison of genotypic and phenotypic cluster analyses of virulence determinants and possible role of CRISPR elements towards their incidence in *Enterococcus faecalis* and *Enterococcus faecium. Syst Appl Microbiol.* 2011;34:553–560. https://doi.org/10.1016/j.syapm.2011.05.002.
62. Hullahalli K, Rodrigues M, Schmidt BD, Li X, Bhardwaj P, Palmer KL. Comparative analysis of the orphan CRISPR2 locus in 242 *Enterococcus faecalis* strains. *PLoS One.* 2015;10:e0138890. https://doi.org/10.1371/journal.pone.0138890.
63. Hauck Y, Soler C, Jault P, et al. Diversity of *Acinetobacter baumannii* in four French military hospitals, as assessed by multiple locus variable number of tandem repeats analysis. *PLoS One.* 2012;7:e44597. https://doi.org/10.1371/journal.pone.0044597.
64. Bowers JR, Kitchel B, Driebe EM, et al. Genomic analysis of the emergence and rapid global dissemination of the clonal group 258 *Klebsiella pneumoniae* pandemic. *PLoS One.* 2015;10:e0133727. https://doi.org/10.1371/journal.pone.0133727.
65. Tseng IL, Liu YM, Wang SJ, et al. Emergence of carbapenemase producing *Klebsiella pneumoniae* and Spread of KPC-2 and KPC-17 in Taiwan: a nationwide study from 2011 to 2013. *PLoS One.* 2015;10:e0138471. https://doi.org/10.1371/journal.pone.0138471.
66. Mann S, Chen YP. Bacterial genomic G+C composition-eliciting environmental adaptation. *Genomics.* 2010;95:7–15. https://doi.org/10.1016/j.ygeno.2009.09.002.

67. Enas Newire AA, Samina J, Virve E, Adam R. Identification of a type IV CRISPR-Cas system located exclusively on IncHI1B/IncFIB plasmids in Enterobacteriaceae. *bioRxiv*. 2019, [Preprints]. https://doi.org/10.1101/536375.
68. Sharma AK, Dhasmana N, Dubey N, et al. Bacterial virulence factors: secreted for survival. *Indian J Microbiol*. 2017;57:1–10. https://doi.org/10.1007/s12088-016-0625-1.
69. Avery OT, Macleod CM, McCarty M. Studies on the chemical nature of the substance inducing transformation of pneumococcal types: Induction of transformation by a desoxyribonucleic acid fraction isolated from *Pneumococcus* type III. *J Exp Med*. 1944;79:137–158. https://doi.org/10.1084/jem.79.2.137.
70. Griffith F. The significance of pneumococcal types. *J Hyg (Lond)*. 1928;27:113–159. https://doi.org/10.1017/s0022172400031879.
71. Croucher NJ, Harris SR, Fraser C, et al. Rapid pneumococcal evolution in response to clinical interventions. *Science*. 2011;331:430–434. https://doi.org/10.1126/science.1198545.
72. Grissa I, Vergnaud G, Pourcel C. The CRISPRdb database and tools to display CRISPRs and to generate dictionaries of spacers and repeats. *BMC Bioinf*. 2007;8:172. https://doi.org/10.1186/1471-2105-8-172.
73. Edgar R, Qimron U. The *Escherichia coli* CRISPR system protects from λ lysogenization, lysogens, and prophage induction. *J Bacteriol*. 2010;192:6291–6294. https://doi.org/10.1128/JB.00644-10.
74. Gudbergsdottir S, Deng L, Chen Z, et al. Dynamic properties of the *Sulfolobus* CRISPR/Cas and CRISPR/Cmr systems when challenged with vector-borne viral and plasmid genes and protospacers. *Mol Microbiol*. 2011;79:35–49. https://doi.org/10.1111/j.1365-2958.2010.07452.x.
75. Haft DH, Selengut J, Mongodin EF, Nelson KE. A guild of 45 CRISPR-associated (Cas) protein families and multiple CRISPR/Cas subtypes exist in prokaryotic genomes. *PLoS Comput Biol*. 2005;1:e60. https://doi.org/10.1371/journal.pcbi.0010060.
76. Tomb JF, White O, Kerlavage AR, et al. The complete genome sequence of the gastric pathogen *Helicobacter pylori* [published correction appears in Nature 1997;389:412]. *Nature*. 1997;388:539–547. https://doi.org/10.1038/41483.
77. van Doorn PA, Ruts L, Jacobs BC. Clinical features, pathogenesis, and treatment of Guillain-Barré syndrome. *Lancet Neurol*. 2008;7:939–950. https://doi.org/10.1016/S1474-4422(08)70215-1.
78. Severi E, Hood DW, Thomas GH. Sialic acid utilization by bacterial pathogens. *Microbiology*. 2007;153:2817–2822. https://doi.org/10.1099/mic.0.2007/009480-0.
79. Jones CL, Napier BA, Sampson TR, Llewellyn AC, Schroeder MR, Weiss DS. Subversion of host recognition and defense systems by *Francisella* spp. *Microbiol Mol Biol Rev*. 2012;762:383–404. https://doi.org/10.1128/mmbr.05027-11.
80. Palmer KL, Whiteley M. DMS3-42: the secret to CRISPR-dependent biofilm inhibition in *Pseudomonas aeruginosa*. *J Bacteriol*. 2011;193:3431–3432. https://doi.org/10.1128/JB.05066-11.
81. Budzik JM, Rosche WA, Rietsch A, O'Toole GA. Isolation and characterization of a generalized transducing phage for *Pseudomonas aeruginosa* strains PAO1 and PA14. *J Bacteriol*. 2004;186:3270–3273. https://doi.org/10.1128/jb.186.10.3270-3273.2004.
82. Damron FH, Owings JP, Okkotsu Y, et al. Analysis of the *Pseudomonas aeruginosa* regulon controlled by the sensor kinase KinB and sigma factor RpoN. *J Bacteriol*. 2012;194:1317–1330. https://doi.org/10.1128/JB.06105-11.
83. Tart AH, Blanks MJ, Wozniak DJ. The AlgT-dependent transcriptional regulator AmrZ (AlgZ) inhibits flagellum biosynthesis in mucoid, nonmotile *Pseudomonas aeruginosa* cystic fibrosis isolates. *J Bacteriol*. 2006;188:6483–6489. https://doi.org/10.1128/JB.00636-06.

84. Baynham PJ, Wozniak DJ. Identification and characterization of AlgZ, an AlgT-dependent DNA-binding protein required for *Pseudomonas aeruginosa* algD transcription. *Mol Microbiol.* 1996;22:97–108. https://doi.org/10.1111/j.1365-2958.1996.tb02659.x.
85. Wozniak DJ, Ohman DE. Transcriptional analysis of the *Pseudomonas aeruginosa* genes *algR*, *algB*, and *algD* reveals a hierarchy of alginate gene expression which is modulated by algT. *J Bacteriol.* 1994;176:6007–6014. https://doi.org/10.1128/jb.176.19.6007-6014.1994.
86. Grissa I, Vergnaud G, Pourcel C. CRISPRFinder: a web tool to identify clustered regularly interspaced short palindromic repeats. *Nucleic Acids Res.* 2007;35:W52–W57. https://doi.org/10.1093/nar/gkm360.
87. Couvin D, Bernheim A, Toffano-Nioche C, et al. CRISPRCasFinder, an update of CRISPRFinder, includes a portable version, enhanced performance and integrates search for Cas proteins. *Nucleic Acids Res.* 2018;46:W246–W251. https://doi.org/10.1093/nar/gky425.
88. Zhang Q, Ye Y. Not all predicted CRISPR-Cas systems are equal: isolated *cas* genes and classes of CRISPR like elements. *BMC Bioinf.* 2017;18:92. https://doi.org/10.1186/s12859-017-1512-4.
89. Touchon M, Charpentier S, Clermont O, Rocha EP, Denamur E, Branger C. CRISPR distribution within the *Escherichia coli* species is not suggestive of immunity-associated diversifying selection. *J Bacteriol.* 2011;193:2460–2467. https://doi.org/10.1128/JB.01307-10.
90. Ostria-Hernández ML, Sánchez-Vallejo CJ, Ibarra JA, Castro-Escarpulli G. Survey of clustered regularly interspaced short palindromic repeats and their associated Cas proteins (CRISPR/Cas) systems in multiple sequenced strains of *Klebsiella pneumoniae*. *BMC Res Notes.* 2015;8:332. https://doi.org/10.1186/s13104-015-1285-7.
91. Shen J, Lv L, Wang X, Xiu Z, Chen G. Comparative analysis of CRISPR-Cas systems in *Klebsiella* genomes. *J Basic Microbiol.* 2017;57:325–336. https://doi.org/10.1002/jobm.201600589.
92. Aydin S, Personne Y, Newire E, et al. Presence of Type I-F CRISPR/Cas systems is associated with antimicrobial susceptibility in *Escherichia coli*. *J Antimicrob Chemother.* 2017;72:2213–2218. https://doi.org/10.1093/jac/dkx137.
93. Touchon M, Rocha EP. The small, slow and specialized CRISPR and anti-CRISPR of *Escherichia* and *Salmonella*. *PLoS One.* 2010;5:e11126. https://doi.org/10.1371/journal.pone.0011126.
94. Horvath P, Barrangou R. CRISPR/Cas, the immune system of bacteria and archaea. *Science.* 2010;327:167–170. https://doi.org/10.1126/science.1179555.
95. Westra ER, Pul U, Heidrich N, et al. H-NS-mediated repression of CRISPR-based immunity in *Escherichia coli* K12 can be relieved by the transcription activator LeuO. *Mol Microbiol.* 2010;77:1380–1393. https://doi.org/10.1111/j.1365-2958.2010.07315.x.
96. Almendros C, Guzmán NM, Díez-Villaseñor C, García-Martínez J, Mojica FJ. Target motifs affecting natural immunity by a constitutive CRISPR-Cas system in *Escherichia coli*. *PLoS One.* 2012;7:e50797. https://doi.org/10.1371/journal.pone.0050797.
97. Yosef I, Manor M, Kiro R, Qimron U. Temperate and lytic bacteriophages programmed to sensitize and kill antibiotic-resistant bacteria. *Proc Natl Acad Sci U S A.* 2015;112:7267–7272. https://doi.org/10.1073/pnas.1500107112.
98. Cao QH, Shao HH, Qiu H, Li T, Zhang YZ, Tan XM. Using the CRISPR/Cas9 system to eliminate native plasmids of *Zymomonas mobilis* ZM4. *Biosci Biotechnol Biochem.* 2017;81:453–459. https://doi.org/10.1080/09168451.2016.1189312.
99. Cao QH, Li T, Shao HH, Tan XM, Zhang YZ. Three new shuttle vectors for heterologous expression in *Zymomonas mobilis*. *Electron J Biotechnol.* 2016;19:33–40. https://doi.org/10.1016/j.ejbt.2015.11.004.

100. Lauritsen I, Kim SH, Porse A, Nørholm MHH. Standardized cloning and curing of plasmids. *Methods Mol Biol*. 1772;2018:469–476. https://doi.org/10.1007/978-1-4939-7795-6_28.
101. Lauritsen I, Porse A, Sommer MOA, Nørholm MHH. A versatile one-step CRISPR-Cas9 based approach to plasmid-curing. *Microb Cell Fact*. 2017;16:135. https://doi.org/10.1186/s12934-017-0748-z.
102. Sun L, He T, Zhang L, et al. Generation of newly discovered resistance gene *mcr-1* knockout in *Escherichia coli* using the CRISPR/Cas9 system. *J Microbiol Biotechnol*. 2017;27:1276–1280. https://doi.org/10.4014/jmb.1611.11021.
103. Dong H, Xiang H, Mu D, Wang D, Wang T. Exploiting a conjugative CRISPR/Cas9 system to eliminate plasmid harboring the *mcr-1* gene from *Escherichia coli*. *Int J Antimicrob Agents*. 2019;53:1–8. https://doi.org/10.1016/j.ijantimicag.2018.09.017.
104. Jiang W, Bikard D, Cox D, et al. RNA-guided editing of bacterial genomes using CRISPR-Cas systems. *Nat Biotechnol*. 2013;31:233–239. https://doi.org/10.1038/nbt.2508.
105. So Y, Park SY, Park EH, et al. A highly efficient CRISPR-Cas9-mediated large genomic deletion in *Bacillus subtilis*. *Front Microbiol*. 2017;8:1167. https://doi.org/10.3389/fmicb.2017.01167.
106. Liu Q, Jiang Y, Shao L, et al. CRISPR/Cas9-based efficient genome editing in *Staphylococcus aureus*. *Acta Biochim Biophys Sin*. 2017;49:764–770. https://doi.org/10.1093/abbs/gmx074.
107. Ram G, Ross HF, Novick RP, Rodriguez-Pagan I, Jiang D. Conversion of staphylococcal pathogenicity islands to CRISPR-carrying antibacterial agents that cure infections in mice. *Nat Biotechnol*. 2018;36:971–976. https://doi.org/10.1038/nbt.4203.
108. Gomaa AA, Klumpe HE, Luo ML, Selle K, Barrangou R, Beisel CL. Programmable removal of bacterial strains by use of genome-targeting CRISPR-Cas systems. *mBio*. 2014;5:e00928-13. https://doi.org/10.1128/mBio.00928-13.
109. Gholizadeh P, Köse Ş, Dao S, et al. How CRISPR-Cas system could be used to combat antimicrobial resistance. *Infect Drug Resist*. 2020;13:1111–1121. https://doi.org/10.2147/IDR.S247271.
110. Bikard D, Euler CW, Jiang W, et al. Exploiting CRISPR-Cas nucleases to produce sequence-specific antimicrobials. *Nat Biotechnol*. 2014;32:1146–1150. https://doi.org/10.1038/nbt.3043.
111. Kang YK, Kwon K, Ryu JS, Lee HN, Park C, Chung HJ. Nonviral genome editing based on a polymer-derivatized CRISPR nanocomplex for targeting bacterial pathogens and antibiotic resistance [published correction appears in Bioconjug Chem. 2018;29:3936]. *Bioconjug Chem*. 2017;28:957–967. https://doi.org/10.1021/acs.bioconjchem.6b00676.
112. Citorik RJ, Mimee M, Lu TK. Sequence-specific antimicrobials using efficiently delivered RNA guided nucleases. *Nat Biotechnol*. 2014;32:1141–1145. https://doi.org/10.1038/nbt.3011.
113. Kim JS, Cho DH, Park M, et al. CRISPR/Cas9-mediated re-sensitization of antibiotic-resistant *Escherichia coli* harboring extended-spectrum β-lactamases. *J Microbiol Biotechnol*. 2016;26:394–401. https://doi.org/10.4014/jmb.1508.08080.
114. Wang P, He D, Li B, et al. Eliminating *mcr-1*-harboring plasmids in clinical isolates using the CRISPR/Cas9 system. *J Antimicrob Chemother*. 2019;74:2559–2565. https://doi.org/10.1093/jac/dkz246.
115. Tagliaferri TL, Guimarães NR, Pereira MPM, et al. Exploring the potential of CRISPR-Cas9 under challenging conditions: facing high-copy plasmids and counteracting beta-lactam resistance in clinical strains of Enterobacteriaceae. *Front Microbiol*. 2020;11:578. https://doi.org/10.3389/fmicb.2020.00578.

116. Hullahalli K, Rodrigues M, Palmer KL. Exploiting CRISPR-Cas to manipulate *Enterococcus faecalis* populations. *eLife*. 2017;6, e26664. https://doi.org/10.7554/eLife.26664.
117. Kristich CJ, Chandler JR, Dunny GM. Development of a host-genotype-independent counterselectable marker and a high-frequency conjugative delivery system and their use in genetic analysis of *Enterococcus faecalis*. *Plasmid*. 2007;57:131–144. https://doi.org/10.1016/j.plasmid.2006.08.003.
118. Müller V, Rajer F, Frykholm K, et al. Direct identification of antibiotic resistance genes on single plasmid molecules using CRISPR/Cas9 in combination with optical DNA mapping. *Sci Rep*. 2016;6:37938. https://doi.org/10.1038/srep37938.
119. Nyberg LK, Persson F, Berg J, et al. A single-step competitive binding assay for mapping of single DNA molecules. *Biochem Biophys Res Commun*. 2012;417:404–408. https://doi.org/10.1016/j.bbrc.2011.11.128.
120. Guk K, Keem JO, Hwang SG, et al. A facile, rapid and sensitive detection of MRSA using a CRISPR-mediated DNA FISH method, antibody-like dCas9/sgRNA complex. *Biosens Bioelectron*. 2017;95:67–71. https://doi.org/10.1016/j.bios.2017.04.016.
121. Quan J, Langelier C, Kuchta A, et al. FLASH: a next-generation CRISPR diagnostic for multiplexed detection of antimicrobial resistance sequences. *Nucleic Acids Res*. 2019;47:e83. https://doi.org/10.1093/nar/gkz418.

CHAPTER SIX

Trends in CRISPR-Cas9 technology application in cancer

Subhadip Mukhopadhyay[a,*] and Sujit Kumar Bhutia[b,*]
[a]Department of Radiation Oncology, Laura and Isaac Perlmutter Cancer Center, NYU Medical School, New York, NY, United States
[b]Department of Life Science, National Institute of Technology, Rourkela, Odisha, India
[*]Corresponding authors: e-mail address: subhadip.mukhopadhyay@nyulangone.org; sujitb@nitrkl.ac.in

Contents

Abstract

The evolution of the CRISPR-Cas9 technology in cancer research has tremendous potential to shape the future of oncology. Although this gene-editing tool's pre-clinical progress is into its nascent stage, there are many unanswered questions regarding health benefits and therapy precision using CRISPR. The application of CRISPR is highly specific, economically sustainable, and is a high throughput technique, but on the other hand, its application involves measured risk of countering the toxic immune response of Cas protein, off target effects, limitation of delivering the edited cells back into cancer patients. The current chapter highlights the possibilities and perils of the present-day CRISPR engineering in cancer that should highlight CRISPR translation to therapy.

1. Introduction

Tumors are metabolically reprogrammed pseudo-organs autonomously nurtured in a microenvironment, which helps them to evade host immune surveillance and cancer therapy. Increase in cancer mortality places it as a leading cause of health disorders. As of 2020, cancer has become the primary cause of death in many countries.[1] With the present rate of cancer incidence, it is predicted that by 2040 there would be 37 million cases

Progress in Molecular Biology and Translational Science, Volume 178
ISSN 1877-1173
https://doi.org/10.1016/bs.pmbts.2020.12.004

worldwide. The current situation demands rapid development of smart and economically sustainable therapeutic interventions.

Clustered Regularly Interspaced Short Palindromic Repeat (CRISPR) emerged back in the late 1980s, and since then, it has revolutionized as a genetic engineering tool with tremendous potential for cancer therapy.[2] In a prokaryotic system, CRISPR forms a critical component of an adaptive immunity system that applies different CRISPR associated (Cas) endonuclease to kill invading phages. To explain in simple terms, CRISPR RNAs guide the Cas nuclease to exogenous genetic content using sequence homology of particular species conserved genetic stretch known as an adjacent protospacer motif (PAM). The CRISPR detects the foreign DNA and silence it by binding to it and cleaving it. Double-stranded DNA damage repairs often make errors and end up making "knockout" as well as "knock-in" version of alleles. Research shows different modifications of the Cas9 version provide the scientists with the ability to activate the gene expression instead of cleaving it. CRISPR interference (CRISPRi) and activation (CRISPRa) explore different transcriptional fused regulators to inhibit or induce gene transcription when Cas9 is particularly directed towards the transcription initiation site of different respective genes.[3] Not only, these tools allow us to study the function of the genes, but they also help us analyze and correct different errors in human genes that ultimately lead to diseases.

CRISPR-Cas9 machinery has been extensively applied in specific genetic editing mechanisms from cellular to the whole organism level of microorganisms, plants, and animals. The CRISPR technique's utility involves transcriptional control, analysis of genetic and epigenetic expression, running genome-wide screens, and imaging chromosomes.[4] It is also being utilized as a critical tool in rapid diagnostic kits. The efficacy, promising outputs, and the simplistic approach of RNA–DNA hybridization of CRISPR-Cas9 technology provides end-user benefits over alternative genome editing methods of protein-DNA hybridization involving zinc-finger nucleases (ZFNs) or transcription activator-like effector nuclease (TALEN) proteins.

2. Genome-wide CRISPR-Cas9 library screening in cancer cells

Comprehensive understanding of the molecular cancer signatures in patients become easier after unraveling the CRISPR-Cas9 mediated genome editing and reprogramming. This segment elaborates on the genome wide screening application using CRISPR-Cas9 technology and its potential benefits. Exploring this path shall lead to the therapeutic clues of cancer translational research. It often happens that it becomes impossible to figure out which genes are functioning in a tumor. Often the genes that physiologically drive a cancer help to promote metastasis and therapy resistance. *In vivo* or *in vitro* CRISPR screen aims to target these different annotated genes' ability to regulate growth and metastatic potential in a tumor. In this method, genome-wide loss of function phenotype is studied by selectively knocking out genes using a single guide RNA (sgRNA) library designed on the cancer patient's gene signatures. The analysis identifies the genes by next-generation sequencing in a particular cancer cell undergoing therapy *versus* the control (Fig. 1). The benefits of sgRNA library over shRNA-based screens include the advantage of uniform sgRNA size for targeting the entire genome, low off-target effects, low false positive negative reads. Moreover, it can be easily multiplexed with high throughput assay, quicker to get a readout, economical to practice on genome-wide studies, and can be used with a large-sized library. For example, a lentiviral sgRNA

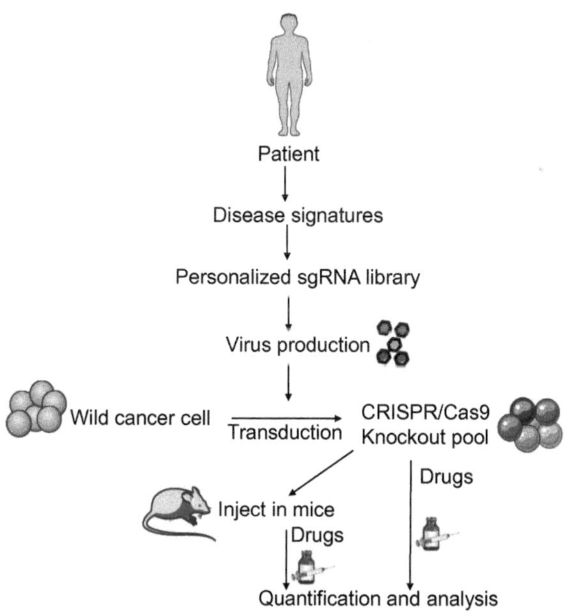

Fig. 1 Schematic representation of CRISPR-Cas9-mediated genome engineering application strategy in preclinical cancer models.

library has been developed to screen for resistance to 6-thioguanine and etoposide to identify DNA damage repair pathway and cyclin-dependent kinase 6 proteins, respectively.[5] Interestingly, Konermann et al. constructed a library comprising 70,290 guides to identify proteins giving resistance to vemurafenib (inhibitor of BRAF) in A375 cells, indicating the presence of BRAF independent MAPK pathway or MAPK independent orthogonal pathway.[6] Although the *in vitro* CRISPR screening in cancer cells reveals some significant targets modulating drug resistance and tumorigenesis, it becomes difficult to validate the physiological significance of key hits of the screening effect of the tumor microenvironment.[7] Furthermore, Chen et al. pioneered the first *in vivo* CRISPR screen study and identified that different sgRNAs involved in lung cancer and late-stage primary tumors target specific genes whose loss of function led to tumor proliferation.[8] Interestingly, transduction of genome-scale CRISPR-Cas9 knockout (GeCKO) lentiviral library A (hGeCKOa) was designed to target 19,050 genes in SMMC7721 HCC cell, which led to the identification of ADAMTSL3 and PTEN as a potential candidate to target for HCC treatment.[9] Besides, high-resolution CRISPR screening has successfully reported important genes involved in controlling cellular signaling, differentiation, growth, and regulation of important survival processes in cancer.[10–12] Kiessling et al. showed that the highest-ranking sgRNA targeting EGFR is the driver of EGFR-mutant lung cancer HCC-827 cells. Interestingly, CHP-212 neuroblastoma cells bearing NRAS mutation exhibited NRAS and MAP2K1 (MEK1) as the highest sgRNA target hit.[13]

Kelkar et al. recently reported a doxycycline-dependent "self-inactivating CRISPR (SiC)" that allows temporal regulation of on and off-targets, both *in vitro* and *in vivo*.[14] Likewise, Jubair et al. reported the *in vivo* CRISPR-Cas9 targeted therapy for cervical carcinoma. They showed complete survival and tumor elimination in treated animals where CRISPR-Cas9 was targeted against E7 oncogene.[15] Similarly, CRISPR-Cas9 mediated genetic editing of oncogenic microRNA-10b using viral vectors was shown to have a therapeutic advantage in malignant gliomas.[16] Deletion of NRAS in CRISPR-inducible embryonal rhabdomyosarcoma (ERMS) xenograft models led to regression of tumor through cell death and myogenic differentiation.[17] Moreover, it showed that the CRISPR screen successfully identified BCR and ABL proteins' lethal role in chronic myelogenous leukemia. In this line, the lethal effect of two other proteins, namely KRAS and PIK3CA, were also elucidated in the colorectal cancer cells.[18] In Fig. 2 we show the major targets of CRISPR/Cas application in cancer as discussed in this chapter.

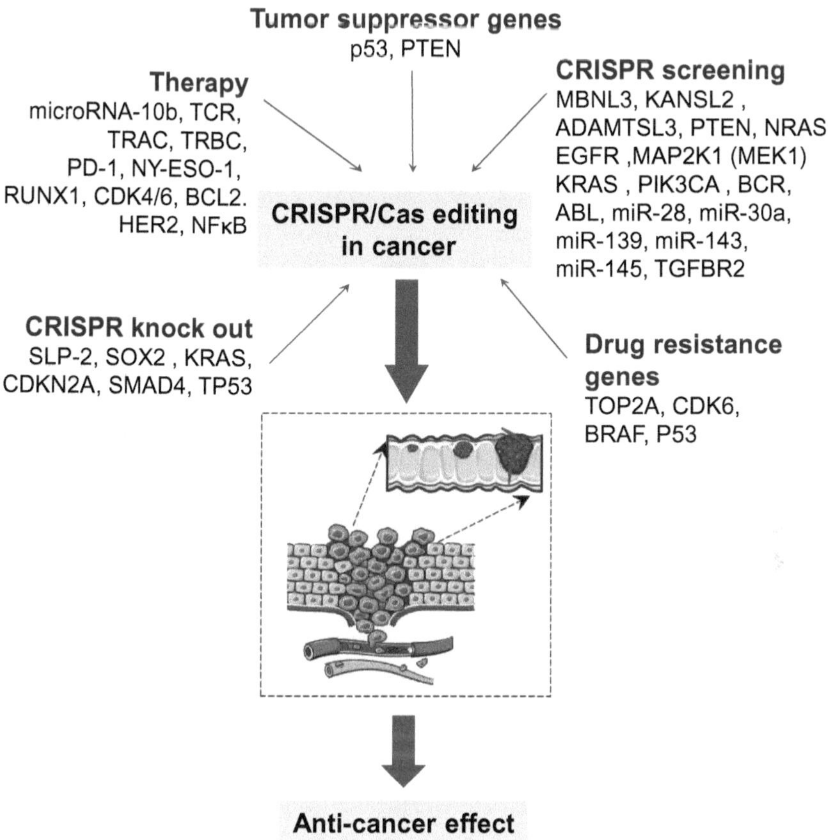

Fig. 2 Major target genes of CRISPR-Cas9 application in cancer.

3. Potential clinical application of CRISPR in cancer therapy

One of the biggest challenges of cancer therapy is identifying and targeting faulty molecules that support tumor growth and proliferation. In this segment of the chapter, we highlight the potential application of CRISPR in cancer therapy. The emergence of the CRISPR-Cas9 mediated *ex vivo* engineering of T cells is called adoptive T cell therapy, which has the capability to pin-point multi-targets in T cells to enhance onco-immunotherapy in a safe way. Stadtmauer et al. pioneered the phase-I CRISPR-Cas9 based T cell clinical trial to validate the safety and achievability of multiplex CRISPR-Cas9 editing to engineer T cells in refractory cancer patients.[19] Guide RNA that targets endogenous T cell receptor (TCR) chains and programmed cell death protein 1 (PD-1; PDCD1) were introduced with ribonucleoproteins (RNPs) of Cas9 protein in T cells isolated from patients. Engineered T-cells were also accompanied by cancer-specific TCR transgene (NY-ESO-1), which was introduced to increase tumor cells' recognition power.[20] Furthermore, Lu et al. highlighted the outcomes of a phase-I trial (ClinicalTrials.gov NCT02793856) of CRISPR-Cas9 based PD-1-edited T cells in grade 1 or 2 of non-small-cell lung cancer (NSCLC) patients to be safe and efficacious for future therapy.[21] *Ex vivo* PD-1-edited T cells exhibited a median overall progression-free survival of 42.6 weeks.[21] In this line, directing a CD19 specific chimeric antigen receptors (CARs) to the T-cell receptor α constant (TRAC) locus with CRISPR-Cas9 enhanced tumor rejection. It revealed the ability of genetic editing mediated immunotherapy in acute lymphoblastic leukemia.[22]

Hou et al. explored the involvement of TP53 gene in the advancement of PI3K inhibitor resistance by deleting key region of p53 (96–393 aa) using CRISPR-Cas technology in lung adenocarcinoma cells to show the importance of p53 interaction with PI3K isoform genes.[23] Wild type form of p53 expressing NSCLC showed non-protective autophagy while the CRISPR-Cas9 mediated p53 null cells exhibited cytoprotective autophagy after cisplatin treatment.[24] Furthermore, it was reported that CRISPR-Cas mediated editing of tumor suppressor genes (e.g. p53, PTEN), as well as oncogenes, demonstrated their therapeutic significance of liver cancer.[25] Again, the mechanism behind clear cell renal cell carcinoma cancer victims expressing a high level of RUNX1/2 depicted poor survival advantage was revealed utilizing CRISPR-Cas9 technology. RUNX1-CRISPR-deleted cells expressed tumorigenic extracellular architecture remodeling that involved STMN3, SERPINH1, and EPHRIN signal axis.[26] Furthermore, RNAi and CRISPR screening study involving genomic, epigenetic, and molecular profiles across 14 different cancer from 5200 patient cohort revealed downregulation of miRNA (miR-28,-30a,-139,-143,-145) during tumorigenesis.[27] Similarly, CRISPR-Cas9 facilitated knockdown of PRK2 in triple-negative breast cancer (TNBC) diminishes cellular proliferation and spheroid formation.[28] TNBC patients have a history of tumor recurrence. Interestingly therapeutically targeting CDK4/6 and BCL2 signaling improves tumor response. In this regard, a CRISPR-Cas9 screen identified that ABT-199 could alleviate the deficiency of Palbociclib.[29] ABT-199 and anti-PD1 therapy was found to extend breast tumor response. CRISPR-Cas9 system directed against HER2 oncogene led to inhibition of breast cancer progression and highlighted a pathway distinct from the known clinical drug Herceptin.[30] CRISPR-Cas9 knock out of nuclear factor κB (NFκB) pathway highlighted therapeutic advantage in breast cancer treated with tamoxifen.[31]

Georgiadis et al. elucidated T cell engineering's therapeutic aspect in leukemia by developing lentivirus-mediated chimeric antigen receptor (CAR) vector with CRISPR guides to delete endogenous TCRαβ-cells. Application of Terminal-TRAC (TT) CAR T cells exhibited non-HLA-matched CAR cells' potential for B cell malignancy.[32] To determine the signaling pathways involved in rendering resistance to FLT3 inhibitor AC220, an unbiased genome-wide CRISPR-pooled library was used to screen novel genes whose loss of function conferred resistance to AC220 in AML. Interestingly, it identified SPRY3, an intracellular inhibitor of FGF signaling, and GSK3, Wnt signaling antagonist, as the major mechanism of resistance to AC220.[33] Moreover, CRISPR screening also identified MBNL3 and KANSL2 as regulators of the high invasiveness of pancreatic cancer, demonstrating the future potential of the presently uncurable pancreatic cancer cases. Another important aspect is the inhibition of the PD-1/PD-L1 signaling axis that leads to tumor regression; however, secondary resistance owing to immune selection pressure drives the development of resistant tumor clones. Williams et al. showed the significance of Ifngr2, Jak1 genes for rendering vulnerability to T cell-mediated killing *in vitro* by utilizing a CRISPR screen involving melanoma cells.[34] Applying the CRISPR Cas editing, Guo et al. reported that the second intron of IncRNA, PVT1, enhances MYC transcription regardless of BRD4 recruitment in leukemia.[35] Another interesting report by Liu et al. showed that enhanced SLP-2 expression exhibited a positive correlation with tumor progression in colorectal cancer (CRC). Implementation of CRISPR-Cas9 mediated deletion of stomatin-like protein 2 (SLP-2) knockout showed abrogation of CRC proliferation. Compound library screening effectively identified PIM1i SGI-1776 and JAK2i TG-101348 as SLP-2-linked drugs. Combinatorial therapies with these two drugs produce a synergistic effect in CRC therapy, highlighting a potential clinical use.[36]

The development of organoid cancer model holds the key to the cancer research system. In this regard, CRISPR-CAS9 based screening strategy was used to identify TGFBR2 as the most important tumor suppressor in human colon organoids with APC −/−;KRAS G12D.[37] Hai et al. utilized CRISPR-Cas9 technology to delete several tumor suppressor genes in lung organoids originating from Cre-dependent SOX2 knock-in mice. They explored the clinical potential and immunological efficacy associated with the combination of PD-1 blockers and inhibition of WEE1 in both *in vivo* and patient-derived cell lines *in vitro*.[38] Prostate cancer patient-derived organoids exhibiting knockout by CRISPR-Cas9 or expression of SLFN11 after were treated with platinum-based chemotherapy.[39] Benedetta et al. recently identified a simple genome editing method called CRISPR-Cas9-mediated homology-independent organoid transgenesis (CRISPR-HOT) that can be used to develop organoids that have knock-in of different genes.[40] Interestingly, CRISPR-Cas9 editing of different PDAC driver genes like p53, KRAS, led to Wnt niche independent PDAC ecosystem development.[41] Despite the incredible potential of the CRISPR-Cas9 in the biotechnological application, important to regulate the dose and exposure of CRISPR-Cas9, so Cas9 was fused with a small molecule-assisted shut-off (SMASh) peptide which consists of a domain for protease, and a hepatitis C virus (HCV) derived degron domain.[42] Though different chemotherapeutic drugs try to achieve drug specificity by targeting the genetic vulnerabilities, loss-of-function screens often struggle to distinguish the key function of genes driving cancer. So, Neggers et al. reported about CRISPRres, a CRISPR-Cas-based fast genetic screening approach that can be employed to target mutations in genes, essentially involved in driving drug resistance.[43] Improvement of genetic diagnostics has led us to figure out genes involved in cancer prevention and tumorigenesis. A CRISPR-based diagnostic system called Specific High Sensitivity Enzymatic Reporter UnLOCKing (SHERLOCK) has been established, which uses Cas13a, an RNA-guided RNase, to produce non-specific single-stranded DNA (ssDNA) trans-cleavage product as a collateral effect.[44] Another system DNA endonuclease targeted CRISPR trans reporter (DETECTR) uses Cas12a and functions similar to Cas13a. Additionally, another enzyme, recombinase polymerase amplification (RPA), was used to economically identify different HPV types, lung carcinomas in a rapid manner.[45]

4. Quality control and bioinformatics of CRISPR guide design

The massive expansion of the application of CRISPR-Cas9 genome editing becomes a tool among the scientific community. It involves rigorous quality control and proper bioinformatics analysis of the CRISPR guides to ensure target specificity and reduced off-target effects. Different well recognized online, open knowledge resource sharing bioinformatics resources are being used for sgRNAs designing purpose.[46] The algorithms governing on-target efficiency of sgRNA are derivative of big-scale sequence data analysis that integrates the composition as well binding site and relative position within a gene model.[47] Implementation of different approaches are followed to predict off-target effects by *in silico* integration of synthetic oligos in double-stranded breaks in living cells or *in vitro* discovery of Cas9 cut sites by sequencing cleaved genetic content.[48] Tables 1 and 2 shows a list of bioinformatics resources routinely used for CRISPR associated experiment design.[47,49–71]

Table 1 Bioinformatic tools for CRISPR analysis.

Tool	Features	URL	References
CHOPCHOP v3	Selecting target sites for CRISPR/Cas9	https://chopchop.cbu.uib.no/	49
Cas-OFFinder	Searches for potential off-target sites of Cas9 RNA-guided endonucleases	http://www.rgenome.net/cas-offinder/	50
CRISPRscan	Select the best gRNAs	https://www.crisprscan.org/	51
E-CRISP	Design of CRISPR constructs	http://www.e-crisp.org/E-CRISP/	52
DeskGen	Effective and affordable guides, with fewer false negatives	https://www.deskgen.com/landing/	53
GPP sgRNA Designer/ CRISPick	Ranks and picks candidate CRISPRko sgRNA sequences for the targets provided, while attempting to maximize on-target activity and minimizing off-target activity	https://portals.broadinstitute.org/gpp/ public/analysis-tools/sgrna-design	47
sgRNA Scorer 1.0	Identify putative sites with minimal off target effects	http://crispr.med.harvard.edu/sgRNAScorer	54
CRISPOR	Design CRISPR guides with off-target and efficiency predictions	http://crispor.tefor.net/	55
Cas-Designer	Quick guide-RNA designer for CRISPR/Cas derived RGENs	http://www.rgenome.net/cas-designer/	56
WU-CRISPR	gRNA Designer for CRISPR/Cas9 Knockout	http://crispr.wustl.edu/	57
sgRNACas9	Designing CRISPR sgRNA and Evaluating Potential Off-Target Cleavage Sites	http://www.biootools.com/col.jsp?id=103	58
CRISPR-DO	Designed for sgRNA design for site specific knock out experiment	http://cistrome.org/crispr/	59
GuideScan	Software for Improved Single and Paired CRISPR Guide RNA Design	http://www.guidescan.com/	60

Continued

Table 1 Bioinformatic tools for CRISPR analysis.—cont'd

Tool	Features	URL	References
GT-Scan	Find targets with minimal similar sequences (potential off-targets) elsewhere in the genome and rank the targets by their uniqueness	https://gt-scan.csiro.au/	61
FlashFry	Fast and flexible command-line tool for characterizing large numbers of potential CRISPR target sequences	http://mckennalab.org/FlashFry/	62
CRISPR-ERA	Fast and Comprehensive Guide RNA Design Tool for Genome Editing, Repression and Activation	http://crispr-era.stanford.edu/	63
CasOT	Genome-Wide Cas9/gRNA Off-Target Searching Tool	http://eendb.zfgenetics.org/casot/	64
CRISPR MultiTargeter	Tool for automatic searches of CRISPR guide RNA targets	http://www.multicrispr.net/	65
CRISPR-P	Design of highly efficient sgRNA with minimal off-target effects in plants	http://crispr.hzau.edu.cn/CRISPR2/	66
CRISPRdirect	Software for Designing CRISPR/Cas Guide RNA With Reduced Off-Target Sites	https://crispr.dbcls.jp/	67
CRISPRseek	Design of target-specific guide RNAs in CRISPR-Cas9, genome-editing systems	http://www.bioconductor.org/packages/release/bioc/html/CRISPRseek.html	68
ZiFitTargeter	Identifies potential target sites in DNA sequences for several DNA binding platforms	http://zifit.partners.org/ZiFiT/	69
CCTOP	CRISPR/Cas9 target online predictor, Off target prediction	http://crispr.cos.uni-heidelberg.de/	70
CRISPRcloud	Secure cloud-based pipeline for CRISPR pooled screen deconvolution	https://crispr.nrihub.org/	71

Table 2 CRISPR resources.

Zhang lab	https://zlab.bio/resources-2
ADDGENE	https://www.addgene.org/crispr/
Synthego	https://www.synthego.com/products/bioinformatics/crispr-design-tool
Horizon inspired cell solutions	https://horizondiscovery.com/en/products/tools/CRISPR-Design-Tool
ATUM	https://www.atum.bio/eCommerce/cas9/input
Benchling	https://www.benchling.com/crispr/
WGE—CRISPR design tool	https://www.sanger.ac.uk/science/tools/wge
ThermoFisher	https://www.thermofisher.com/us/en/home/life-science/genome-editing/geneart-crispr.html
Sigma Aldrich	https://www.sigmaaldrich.com/catalog/product/sigma/crispr?lang=en&region=US
Genscript	https://www.genscript.com/gRNA-design-tool.html
IDT	https://www.idtdna.com/pages/products/crispr-genome-editing

5. Perils of CRISPR-Cas technology

Despite the increasing use of CRISPR-Cas9 technology, its safety and efficiency are important concerns, demanding holistic studies on safety concerns. Off-target alterations hinder the utilization of the CRISPR-Cas9 system in the clinical scenario. CRISPR-Cas9 gene editing is not free of error, and that exists 15% chance of time.[72] In this segment, we elucidate the need for safety concern and the problems that might erupt if proper precautionary measures are not followed during the use of CRISPR-Cas9 in cancer research. A fault in CRISPR-Cas9 may result in off-target cleavage that may result in indel mutation, translocation ultimately making the future of edited cancer cell treatment challenging. Moreover, the most crucial aspect of regulating this genome tool's efficacy is understanding the fitness of CRISPR edited cells, the immune reaction evoking the delivery methods, and the maneuvering of this technology.[73] Biased activities of off-target CRISPR-based genomic cleavage can now be limited by using available online tools such as Cas-OFFinder etc (as listed in Tables 1 and 2). Methods like T7EI assay, Sanger sequencing are routinely used previously to identify mutations at the potential off-target site; however, they involve a more time-consuming process making it less productive with low sensitivity and specificity mutation composition. To beat this shortcoming, scientists have started to perform high-throughput sequencing (HTS) to understand the off-target sites following PCR-based amplification in a rapid, accurate, and sensitive manner compared to Sanger sequencing. Besides, this potential off-target can also be determined unbiasedly to detect undesired cleavages using methods like whole-genome sequencing (WGS) and cross-linking chromatin immunoprecipitation of endonuclease-mutant Cas9 (ChIP-dCas9).[74] Interestingly, genome wide unbiased identification of DNA strand break (DSB) enabled by sequencing (GUIDE-seq) followed by the discovery of *in situ* Cas off-targets and verification by sequencing (DISCOVER-seq) help to eliminate the off-targeting effects of CRISPR-Cas9 technology.[74]

The effectiveness of CRISPR Cas9 application varies by cell type and nature as the nonhomologous end-joining (NHEJ)-mediated and homologous directed repair (HDR)-mediated DNA strand break (DSB) repair. An attempt to escalate HDR efficiency can be achieved by combining nanocarriers or viral vectors with CRISPR-Cas9 components with HDR templates.[74,75] The problem with viral delivery is to pack large genetic materials comprising gRNA within the viral delivery unit. The non-viral vector promises better dose and timing regulation with decreased off-target effects with low potential side effects of tumorigenesis. One such promising carrier for sgRNA delivery to the cancer cell is the exosomes. Recently, Kim et al. documented the preferential uptake of low immunogenic exosomes loaded with Cas9 and PARP-1 sgRNA to induce apoptosis in cancer cells.[76] To improve such exosome carriers containing sgRNA/Cas complex, Ye et al. loaded them with fluorescent GFP to identify the non-small lung cancer cells for targeted delivery with desired knockout effects.[77] Recently, CRISPR operated a stoplight system for functional intercellular RNA exchange (CROSS-FIRE) highlighted a sensitive system involving CRISPR-Cas9 reporter involving exosome-mediated delivery of sgRNA at single cell resolution.[78]

Another burning issue ahead of CRISPR based therapeutics in oncology is the fitness of edited cells.[79] The overall health of the highly efficient CRISPR edited cells demands a sustainable threshold as a key to promoting this field's clinical translation. The limitation might involve an *ex vivo* genome editing therapy, which may be time-consuming, expensive, and low throughput. A significant drawback of CRISPR-Cas9 therapy is the high immunogenicity of bacterial-derived Cas9 protein in humans.[80] Cas9 can be presented to the T cells by the major histocompatibility complex (MHC) molecules to elicit a pro-inflammatory immune response that has the potential to destruct the host cells.[81]

6. Conclusion and future perspective

The rapid evolution of cancer diagnosis and therapy has led to the development of CRISPR-Cas9 based gene editing as a highly specific technology with the potential to improve human health. Various gene-editing tools have been used for HIV, immune-related diseases, and cancer; the fundamentals of life science involved in this mechanism need to be understood better for safe therapeutic application. Although CRISPR has immense potential, before applying this in patients, some trade-offs have to be considered for the acceptable risk for significant benefit. It is still unanswered how to counterbalance the toxic immunogenic potential of Cas9 without inducing off-target gene cleavage. There must be further pre-clinical analysis before CRISPR is deployed as a common clinical tool ensuring maximum specificity to target the diseased cells for improvement of cancer therapy.

Acknowledgment

Figures were developed using Servier Medical Art by Servier (https://smart.servier.com/) is licensed under a Creative Commons Attribution 3.0 Unported License.

References

1. Siegel RL, Miller KD, Jemal A. Cancer statistics, 2020. *CA Cancer J Clin*. 2020;70(1):7–30. https://doi.org/10.3322/caac.21590.
2. Pennisi E. The CRISPR craze. *Science*. 2013;341(6148):833–836. https://doi.org/10.1126/science.341.6148.833.
3. Kampmann M. CRISPRi and CRISPRa screens in mammalian cells for precision biology and medicine. *ACS Chem Biol*. 2018;13(2):406–416. https://doi.org/10.1021/acschembio.7b00657.
4. Barrangou R, Doudna JA. Applications of CRISPR technologies in research and beyond. *Nat Biotechnol*. 2016;34(9):933–941. https://doi.org/10.1038/nbt.3659.
5. Wang T, Wei JJ, Sabatini DM, Lander ES. Genetic screens in human cells using the CRISPR-Cas9 system. *Science*. 2014;343(6166):80–84. https://doi.org/10.1126/science.1246981.
6. Konermann S, Brigham MD, Trevino AE, et al. Genome-scale transcriptional activation by an engineered CRISPR-Cas9 complex. *Nature*. 2015;517(7536):583–588. https://doi.org/10.1038/nature14136.

7. Quail DF, Joyce JA. Microenvironmental regulation of tumor progression and metastasis. *Nat Med.* 2013;19(11):1423–1437. https://doi.org/10.1038/nm.3394.
8. Chen S, Sanjana NE, Zheng K, et al. Genome-wide CRISPR screen in a mouse model of tumor growth and metastasis. *Cell.* 2015;160(6):1246–1260. https://doi.org/10.1016/j.cell.2015.02.038.
9. Zhou X, Li R, Jing R, Zuo B, Zheng Q. Genome-wide CRISPR knockout screens identify ADAMTSL3 and PTEN genes as suppressors of HCC proliferation and metastasis, respectively. *J Cancer Res Clin Oncol.* 2020;146(6):1509–1521. https://doi.org/10.1007/s00432-020-03207-9.
10. Hart T, Chandrashekhar M, Aregger M, et al. High-resolution CRISPR screens reveal fitness genes and genotype-specific Cancer liabilities. *Cell.* 2015;163(6):1515–1526. https://doi.org/10.1016/j.cell.2015.11.015.
11. Wang T, Birsoy K, Hughes NW, et al. Identification and characterization of essential genes in the human genome. *Science.* 2015;350(6264):1096–1101. https://doi.org/10.1126/science.aac7041.
12. Shi J, Zhao Y, Wang K, et al. Cleavage of GSDMD by inflammatory caspases determines pyroptotic cell death. *Nature.* 2015;526(7575):660–665. https://doi.org/10.1038/nature15514.
13. Kiessling MK, Schuierer S, Stertz S, et al. Identification of oncogenic driver mutations by genome-wide CRISPR-Cas9 dropout screening. *BMC Genomics.* 2016;17(1):723. https://doi.org/10.1186/s12864-016-3042-2.
14. Kelkar A, Zhu Y, Groth T, et al. Doxycycline-dependent self-inactivation of CRISPR-Cas9 to temporally regulate on- and off-target editing. *Mol Ther.* 2020;28(1):29–41. https://doi.org/10.1016/j.ymthe.2019.09.006.
15. Jubair L, Fallaha S, McMillan NAJ. Systemic delivery of CRISPR/Cas9 targeting HPV oncogenes is effective at eliminating established tumors. *Mol Ther.* 2019;27(12):2091–2099. https://doi.org/10.1016/j.ymthe.2019.08.012.
16. El Fatimy R, Subramanian S, Uhlmann EJ, Krichevsky AM. Genome editing reveals glioblastoma addiction to MicroRNA-10b. *Mol Ther.* 2017;25(2):368–378. https://doi.org/10.1016/j.ymthe.2016.11.004.
17. Phelps MP, Yang H, Patel S, Rahman MM, McFadden G, Chen E. Oncolytic virus-mediated RAS targeting in rhabdomyosarcoma. *Mol Ther Oncol.* 2018; 11:52–61. https://doi.org/10.1016/j.omto.2018.09.001.
18. Ghosh D, Venkataramani P, Nandi S, Bhattacharjee S. CRISPR-Cas9 a boon or bane: the bumpy road ahead to cancer therapeutics. *Cancer Cell Int.* 2019;19:12. https://doi.org/10.1186/s12935-019-0726-0.
19. Stadtmauer EA, Fraietta JA, Davis MM, et al. CRISPR-engineered T cells in patients with refractory cancer. *Science.* 2020;367(6481):eaba7365. https://doi.org/10.1126/science.aba7365.
20. Hamilton JR, Doudna JA. Knocking out barriers to engineered cell activity. *Science.* 2020;367(6481):976–977. https://doi.org/10.1126/science.aba9844.
21. Lu Y, Xue J, Deng T, et al. Safety and feasibility of CRISPR-edited T cells in patients with refractory non-small-cell lung cancer. *Nat Med.* 2020;26(5):732–740. https://doi.org/10.1038/s41591-020-0840-5.
22. Eyquem J, Mansilla-Soto J, Giavridis T, et al. Targeting a CAR to the TRAC locus with CRISPR/Cas9 enhances tumour rejection. *Nature.* 2017;543(7643):113–117. https://doi.org/10.1038/nature21405.
23. Hou J, Cao X, Cheng Y, Wang X. Roles of TP53 gene in the development of resistance to PI3K inhibitor resistances in CRISPR-Cas9-edited lung adenocarcinoma cells. *Cell Biol Toxicol.* 2020;36(5):481–492. https://doi.org/10.1007/s10565-020-09523-7.

24. Patel NH, Xu J, Saleh T, Wu Y, Lima S, Gewirtz DA. Influence of non-protective autophagy and the autophagic switch on sensitivity to cisplatin in non-small cell lung cancer cells. *Biochem Pharmacol.* 2020;175. https://doi.org/10.1016/j.bcp.2020.113896, 113896.
25. Xue W, Chen S, Yin H, et al. CRISPR-mediated direct mutation of cancer genes in the mouse liver. *Nature.* 2014;514(7522):380–384. https://doi.org/10.1038/nature13589.
26. Rooney N, Mason SM, McDonald L, et al. RUNX1 is a driver of renal cell carcinoma correlating with clinical outcome. *Cancer Res.* 2020;80(11):2325–2339. https://doi.org/10.1158/0008-5472.CAN-19-3870.
27. Mitra R, Adams CM, Jiang W, Greenawalt E, Eischen CM. Pan-cancer Analysis reveals cooperativity of both strands of microRNA that regulate tumorigenesis and patient survival. *Nat Commun.* 2020;11(1):968. https://doi.org/10.1038/s41467-020-14713-2.
28. Patel H, Li J, Herrero A, et al. Novel roles of PRK1 and PRK2 in cilia and cancer biology. *Sci Rep.* 2020;10(1):3902. https://doi.org/10.1038/s41598-020-60604-3.
29. Whittle JR, Vaillant F, Surgenor E, et al. Dual targeting of CDK4/6 and BCL2 pathways augments tumor response in estrogen receptor positive breast cancer. *Clin Cancer Res.* 2020;26(15):4120–4134. https://doi.org/10.1158/1078-0432.CCR-19-1872.
30. Wang H, Sun W. CRISPR-mediated targeting of HER2 inhibits cell proliferation through a dominant negative mutation. *Cancer Lett.* 2017;385:137–143. https://doi.org/10.1016/j.canlet.2016.10.033.
31. Kastrati I, Joosten SEP, Semina SE, et al. The NFκB pathway promotes tamoxifen tolerance and disease recurrence in estrogen receptor-positive breast cancers. *Mol Cancer Res.* 2020;18(7):1018–1027. https://doi.org/10.1158/1541-7786.MCR-19-1082.
32. Georgiadis C, Preece R, Nickolay L, et al. Long terminal repeat CRISPR-CAR-coupled "universal" T cells mediate potent anti-leukemic effects. *Mol Ther.* 2018;26(5):1215–1227. https://doi.org/10.1016/j.ymthe.2018.02.025.
33. Hou P, Wu C, Wang Y, et al. A genome-wide CRISPR screen identifies genes critical for resistance to FLT3 inhibitor AC220. *Cancer Res.* 2017;77(16):4402–4413. https://doi.org/10.1158/0008-5472.CAN-16-1627.
34. Williams JB, Li S, Higgs EF, et al. Tumor heterogeneity and clonal cooperation influence the immune selection of IFN-γ-signaling mutant cancer cells. *Nat Commun.* 2020;11(1):602. https://doi.org/10.1038/s41467-020-14290-4.
35. Guo L, Li J, Zeng H, et al. A combination strategy targeting enhancer plasticity exerts synergistic lethality against BETi-resistant leukemia cells. *Nat Commun.* 2020;11(1):740. https://doi.org/10.1038/s41467-020-14604-6.
36. Liu Q, Li A, Wang L, et al. Stomatin-like protein 2 promotes tumor cell survival by activating the JAK2-STAT3-PIM1 pathway, suggesting a novel therapy in CRC. *Mol Ther Oncol.* 2020;17:169–179. https://doi.org/10.1016/j.omto.2020.03.010.
37. Michels BE, Mosa MH, Streibl BI, et al. Pooled in vitro and in vivo CRISPR-Cas9 screening identifies tumor suppressors in human colon organoids. *Cell Stem Cell.* 2020;26(5):782–792.e7. https://doi.org/10.1016/j.stem.2020.04.003.
38. Hai J, Zhang H, Zhou J, et al. Generation of genetically engineered mouse lung organoid models for squamous cell lung cancers allows for the study of combinatorial immunotherapy. *Clin Cancer Res.* 2020;26(13):3431–3442. https://doi.org/10.1158/1078-0432.CCR-19-1627.
39. Conteduca V, Ku SY, Puca L, et al. SLFN11 expression in advanced prostate cancer and response to platinum-based chemotherapy. *Mol Cancer Ther.* 2020;19(5): 1157–1164. https://doi.org/10.1158/1535-7163.MCT-19-0926.
40. Artegiani B, Hendriks D, Beumer J, et al. Fast and efficient generation of knock-in human organoids using homology-independent CRISPR-Cas9 precision genome editing. *Nat Cell Biol.* 2020;22(3):321–331. https://doi.org/10.1038/s41556-020-0472-5.

41. Seino T, Kawasaki S, Shimokawa M, et al. Human pancreatic tumor organoids reveal loss of stem cell niche factor dependence during disease progression. *Cell Stem Cell*. 2018;22(3):454–467.e6. https://doi.org/10.1016/j.stem.2017.12.009.
42. Wu Y, Yang L, Chang T, Kandeel F, Yee JK. A small molecule-controlled Cas9 repressible system. *Mol Ther Nucleic Acids*. 2020;19:922–932. https://doi.org/10.1016/j.omtn.2019.12.026.
43. Neggers JE, Kwanten B, Dierckx T, et al. Target identification of small molecules using large-scale CRISPR-Cas mutagenesis scanning of essential genes. *Nat Commun*. 2018; 9(1):502. https://doi.org/10.1038/s41467-017-02349-8.
44. Abudayyeh OO, Gootenberg JS, Essletzbichler P, et al. RNA targeting with CRISPR-Cas13. *Nature*. 2017;550(7675):280–284. https://doi.org/10.1038/nature24049.
45. Chertow DS. Next-generation diagnostics with CRISPR. *Science*. 2018;360(6387): 381–382. https://doi.org/10.1126/science.aat4982.
46. Bradford J, Perrin D. A benchmark of computational CRISPR-Cas9 guide design methods. *PLoS Comput Biol*. 2019;15(8): e1007274. https://doi.org/10.1371/journal.pcbi.1007274.
47. Doench JG, Fusi N, Sullender M, et al. Optimized sgRNA design to maximize activity and minimize off-target effects of CRISPR-Cas9. *Nat Biotechnol*. 2016;34(2):184–191. https://doi.org/10.1038/nbt.3437.
48. Zhan T, Rindtorff N, Betge J, Ebert MP, Boutros M. CRISPR/Cas9 for cancer research and therapy. *Semin Cancer Biol*. 2019;55:106–119. https://doi.org/10.1016/j.semcancer.2018.04.001.
49. Labun K, Montague TG, Krause M, Torres Cleuren YN, Tjeldnes H, Valen E. CHOPCHOP v3: Expanding the CRISPR web toolbox beyond genome editing. *Nucleic Acids Res*. 2019;47(W1):W171–W174. https://doi.org/10.1093/nar/gkz365.
50. Bae S, Park J, Kim JS. Cas-OFFinder: A fast and versatile algorithm that searches for potential off-target sites of Cas9 RNA-guided endonucleases. *Bioinformatics*. 2014; 30(10):1473–1475. https://doi.org/10.1093/bioinformatics/btu048.
51. Moreno-Mateos MA, Vejnar CE, Beaudoin JD, et al. CRISPRscan: Designing highly efficient sgRNAs for CRISPR-Cas9 targeting in vivo. *Nat Methods*. 2015;12(10): 982–988. https://doi.org/10.1038/nmeth.3543.
52. Heigwer F, Kerr G, Boutros M. E-CRISP: Fast CRISPR target site identification. *Nat Methods*. 2014;11(2):122–123. https://doi.org/10.1038/nmeth.2812.
53. Hough SH, Kancleris K, Brody L, et al. Guide Picker is a comprehensive design tool for visualizing and selecting guides for CRISPR experiments. *BMC Bioinf*. 2017;18(1):167. https://doi.org/10.1186/s12859-017-1581-4.
54. Chari R, Mali P, Moosburner M, Church GM. Unraveling CRISPR-Cas9 genome engineering parameters via a library-on-library approach. *Nat Methods*. 2015;12(9): 823–826. https://doi.org/10.1038/nmeth.3473.
55. Concordet JP, Haeussler M. CRISPOR: intuitive guide selection for CRISPR/Cas9 genome editing experiments and screens. *Nucleic Acids Res*. 2018;46(W1):W242–W245. https://doi.org/10.1093/nar/gky354.
56. Park J, Bae S, Kim JS. Cas-designer: a web-based tool for choice of CRISPR-Cas9 target sites. *Bioinformatics*. 2015;31(24):4014–4016. https://doi.org/10.1093/bioinformatics/btv537.
57. Hiranniramol K, Chen Y, Liu W, Wang X. Generalizable sgRNA design for improved CRISPR/Cas9 editing efficiency. *Bioinformatics*. 2020;36(9):2684–2689. https://doi.org/10.1093/bioinformatics/btaa041.
58. Xie S, Shen B, Zhang C, Huang X, Zhang Y. sgRNAcas9: a software package for designing CRISPR sgRNA and evaluating potential off-target cleavage sites. *PLoS One*. 2014;9(6): e100448. https://doi.org/10.1371/journal.pone.0100448.

59. Ma J, Köster J, Qin Q, et al. CRISPR-DO for genome-wide CRISPR design and optimization. *Bioinformatics*. 2016;32(21):3336–3338. https://doi.org/10.1093/bioinformatics/btw476.
60. Perez AR, Pritykin Y, Vidigal JA, et al. GuideScan software for improved single and paired CRISPR guide RNA design. *Nat Biotechnol*. 2017;35(4):347–349. https://doi.org/10.1038/nbt.3804.
61. O'Brien A, Bailey TL. GT-scan: identifying unique genomic targets. *Bioinformatics*. 2014;30(18):2673–2675. https://doi.org/10.1093/bioinformatics/btu354.
62. McKenna A, Shendure J. FlashFry: a fast and flexible tool for large-scale CRISPR target design. *BMC Biol*. 2018;16(1):74. Published 2018 Jul 5. https://doi.org/10.1186/s12915-018-0545-0.
63. Liu H, Wei Z, Dominguez A, Li Y, Wang X, Qi LS. CRISPR-ERA: a comprehensive design tool for CRISPR-mediated gene editing, repression and activation. *Bioinformatics*. 2015;31(22):3676–3678. https://doi.org/10.1093/bioinformatics/btv423.
64. Xiao A, Cheng Z, Kong L, et al. CasOT: a genome-wide Cas9/gRNA off-target searching tool. *Bioinformatics*. 2014;30(8):1180–1182. https://doi.org/10.1093/bioinformatics/btt764.
65. Prykhozhij SV, Rajan V, Gaston D, Berman JN. CRISPR multitargeter: a web tool to find common and unique CRISPR single guide RNA targets in a set of similar sequences. *PLoS One*. 2015;10(3): e0119372. https://doi.org/10.1371/journal.pone.0119372.
66. Liu H, Ding Y, Zhou Y, Jin W, Xie K, Chen LL. CRISPR-P 2.0: an improved CRISPR-Cas9 tool for genome editing in plants. *Mol Plant*. 2017;10(3):530–532. https://doi.org/10.1016/j.molp.2017.01.003.
67. Naito Y, Hino K, Bono H, Ui-Tei K. CRISPRdirect: software for designing CRISPR/Cas guide RNA with reduced off-target sites. *Bioinformatics*. 2015;31(7): 1120–1123. https://doi.org/10.1093/bioinformatics/btu743.
68. Zhu LJ, Holmes BR, Aronin N, Brodsky MH. CRISPRseek: a bioconductor package to identify target-specific guide RNAs for CRISPR-Cas9 genome-editing systems. *PLoS One*. 2014;9(9): e108424. https://doi.org/10.1371/journal.pone.0108424.
69. Sander JD, Zaback P, Joung JK, Voytas DF, Dobbs D. Zinc finger targeter (ZiFiT): an engineered zinc finger/target site design tool. *Nucleic Acids Res*. 2007;35(Web Server issue):W599–W605. https://doi.org/10.1093/nar/gkm349.
70. Stemmer M, Thumberger T, Del Sol Keyer M, Wittbrodt J, Mateo JL. CCTop: an intuitive, flexible and reliable CRISPR/Cas9 target prediction tool. *PLoS One*. 2015;10(4): e0124633. https://doi.org/10.1371/journal.pone.0124633.
71. Jeong HH, Kim SY, Rousseaux MWC, Zoghbi HY, Liu Z. CRISPRcloud: a secure cloud-based pipeline for CRISPR pooled screen deconvolution. *Bioinformatics*. 2017;33(18):2963–2965. https://doi.org/10.1093/bioinformatics/btx335.
72. Horie K, Inoue K, Suzuki S, et al. Oxytocin receptor knockout prairie voles generated by CRISPR/Cas9 editing show reduced preference for social novelty and exaggerated repetitive behaviors. *Horm Behav*. 2019;111:60–69. https://doi.org/10.1016/j.yhbeh.2018.10.011.
73. Cho SW, Kim S, Kim Y, et al. Analysis of off-target effects of CRISPR/Cas-derived RNA-guided endonucleases and nickases. *Genome Res*. 2014;24(1):132–141. https://doi.org/10.1101/gr.162339.113.
74. Li J, Hong S, Chen W, Zuo E, Yang H. Advances in detecting and reducing off-target effects generated by CRISPR-mediated genome editing. *J Genet Genomics*. 2019; 46(11):513–521. https://doi.org/10.1016/j.jgg.2019.11.002.
75. Yin H, Song CQ, Dorkin JR, et al. Therapeutic genome editing by combined viral and non-viral delivery of CRISPR system components in vivo. *Nat Biotechnol*. 2016; 34(3):328–333. https://doi.org/10.1038/nbt.3471.

76. Kim SM, Yang Y, Oh SJ, Hong Y, Seo M, Jang M. Cancer-derived exosomes as a delivery platform of CRISPR/Cas9 confer cancer cell tropism-dependent targeting. *J Control Release.* 2017;266:8–16. https://doi.org/10.1016/j.jconrel.2017.09.013.
77. Ye Y, Zhang X, Xie F, et al. An engineered exosome for delivering sgRNA:Cas9 ribonucleoprotein complex and genome editing in recipient cells. *Biomater Sci.* 2020; 8(10):2966–2976. https://doi.org/10.1039/d0bm00427h.
78. de Jong OG, Murphy DE, Mäger I, et al. A CRISPR-Cas9-based reporter system for single-cell detection of extracellular vesicle-mediated functional transfer of RNA. *Nat Commun.* 2020;11(1):1113. https://doi.org/10.1038/s41467-020-14977-8.
79. Cox DB, Platt RJ, Zhang F. Therapeutic genome editing: prospects and challenges. *Nat Med.* 2015;21(2):121–131. https://doi.org/10.1038/nm.3793.
80. Mali P, Esvelt KM, Church GM. Cas9 as a versatile tool for engineering biology. *Nat Methods.* 2013;10(10):957–963. https://doi.org/10.1038/nmeth.2649.
81. Crudele JM, Chamberlain JS. Cas9 immunity creates challenges for CRISPR gene editing therapies. *Nat Commun.* 2018;9(1):3497. Published 2018 Aug 29. https://doi.org/10.1038/s41467-018-05843-9.

CHAPTER SEVEN

Mitochondrial DNA modification by CRISPR/Cas system: Challenges and future direction

Rajalakshmi Prakash[a,b] and Anbarasu Kannan[a,b,*]
[a]Department of Protein Chemistry and Technology, CSIR-Central Food Technological Research Institute, Mysuru, India
[b]Academy of Scientific and Innovative Research (AcSIR), CSIR-Central Food Technological Research Institute (CSIR-CFTRI) Campus, Mysuru, India
*Corresponding author: e-mail address: anbarasu@cftri.res.in

Contents

Abstract

CRISPR (Clustered Regularly Interspaced Short Palindromic Repeats)/Cas (CRISPR associated endonuclease), a hotshot genome editing tool which is originally known to be the form of prokaryotic adaptive immune system against viral infections has gained all the attention of scientific community as a promising genome editing platform. This review encompasses a brief description of mitochondrial disease conditions associated with the alteration in mitochondrial genome (mtDNA) and highlights the key role of the CRISPR/Cas system pertaining to its working mechanism and its involvement in gene-based therapeutics in treating the foresaid mitochondrial diseases. Here, we also extend the perception related to the detailed mechanism of CRISPR/Cas system in mtDNA modification.

Progress in Molecular Biology and Translational Science, Volume 178
ISSN 1877-1173
https://doi.org/10.1016/bs.pmbts.2020.12.009

1. Introduction

1.1 Mitochondria

Mitochondria, the power house of the cell responsible for the production energy currencies, the ATPs. Mitochondria are emerged as the significant cellular organelle around ~2.5 billion years ago inside the eukaryotic cell as per the theory of endosymbiosis between the ancestor of the mitochondria and the progenitor of eukaryotic cells.[1] Each cell contains hundreds and thousands of mitochondria in the cytoplasm depending on the cell type, tissue and organism. Whereas, cells devoid of mitochondria also do exists such as, erythrocytes (RBC). As far as the structural aspects of the mitochondria are concerned, the size of this double membraned organelle are approximately 0.75–3 μm^2 in area and tend to vary in both size and structure.[2]

The structural components of mitochondria involves outer mitochondrial membrane, intermembrane space, inner mitochondrial membrane, cristae, matrix, matrix granules, mitochondrial genome and mitochondrial ribosomes. Amongst the two mitochondrial membranes, the outer one is composed of voltage dependent anion—selective channels (VDAC) that confers the permeability of most of the molecules below 5 kDa (ATP, ADP, pyruvate, Malate and other Metabolites) between the cytosol and intermembrane space called Porins.[3] Such transportation is termed as passive diffusion that does not require any extra energy for entering or crossing the membrane. On the other hand, larger proteins gets actively translocated via large multiprotein translocase complex. Unlike the outer membrane, the inner mitochondrial membranes are characterized with the presence of several invaginations (folding) termed as cristae, which are known to provide an increased surface area (almost 5 times than that of outer membrane) for hosting the cellular components involved in electron transport chain and ATP synthesis.[4]

Apart from being required for the energy homeostasis, this cellular organelle is also pivotal for many cellular processes viz… protein synthesis using its own machinery of DNA, tRNA and ribosomes.[5] Mitochondria are also involved in thermogenesis i.e., in maintaining the body temperature in case of newborns and in some warm blooded hibernating animals. Intracellular regulation of Ca^{2+}, which in turn are responsible for the regulation of respiration is also one such important function of mitochondrial VDAC, and it is also involved in control of the process of apoptosis by releasing two of its intermembrane space proteins cytochrome *c* and the apoptosis inducing factor (AIF), there by triggering the cascade of apoptosis.

Together with the previously mentioned functions, the mitochondria are also considered as the main generation site of ROS (reactive oxygen species), which further contribute to necrotic cell death. In normal physiological conditions, these free radicals generated are efficiently removed or detoxified by the mitochondrial antioxidant system, which prevents any permanent damage to the mitochondria[6] whereas, in pathological conditions and with age, the detoxifying efficiency of the mitochondria diminishes and ROS starts getting accumulated in the mitochondria causing cell death. Apart from the fore mentioned functions, mitochondria are also involved in the biosynthesis of amino acids and synthesis of phospholipids for the membrane biogenesis.[7]

1.2 Mitochondrial genome

Along with the genomic DNA present inside the nucleus, cells do contain a circular double stranded mitochondrial DNA (mtDNA) with limited number of genes. The size of the mtDNA is around 16.5 kB length encoding 37 genes out of which 13 encodes for proteins, which are all the subunits of the enzyme complex involved in oxidative phosphorylation, 22 genes encodes for tRNAs and 2 genes encodes for rRNAs. These are present as multiples copies of a single mitochondrial chromosome in the matrix of the mitochondria. The mammalian mitochondrial genome is known to be sternly maternal since; it is transmitted only from oocyte during the process of fertilization. Whereas, a patient of mitochondrial myopathy was reported to have inherited the pathogenic mtDNA from paternal origin.[8] Unlike the nuclear genome, the mitochondrial DNA utilizes formylmethionyl tRNA for the initiation of translation similar to that of in the prokaryotes, which serves as one of the evidence in support of the theory of endosymbiosis.[9] The asymmetric distribution of G and C in the mtDNA enables its differentiation as cytosine rich light strand (L), encoded with the genes of 8 tRNAs and one polypeptide *MT-ND6* and guanine rich heavy strand (H) with maximum information encoded. The H strand encodes for around 12 polypeptides (*MT-ND1* to *MT-ND5, MT-CYB, MT-CO1* to *MT-CO3, MT-ATP6* and *MT-ATP8*), 2 rRNAs (*MT-RNR1* and *MT-RNR2*) and 14 tRNAs.[4,9,10] The process of transcription in case of mitochondrial DNA is said to be bidirectional and are proceeded by one predominant promotor of both light and heavy strand viz… PL for the light strand and PH1 for the heavy strand. It is reported that the mitochondrial genome contains only 3% of introns i.e., noncoding region when compared to that of nuclear genome, which consists of around 98.8% of the noncoding DNA. However, amongst the coding region the nucleotide bases shows overlapping between two consecutive genes, with this it is revealed that the mitochondrial genome do not follow the universal rule of genetic code.[4]

Out of the 13 genes, involved in polypeptide synthesis, 7 genes (*MT-ND1 to MT-ND6*) are involved in the synthesis of polypeptides associated with complex I of the oxidative phosphorylation such as, NADH dehydrogenase subunits. 1 gene viz... *MT-CYB* encodes polypeptide associated with Complex III (cytochrome *b*). 3 genes out of 13 (*MT-CO1 to MT-CO3*) encodes for polypeptides associated with complex IV (cytochrome *c* oxidase subunits) and 2 of the genes (*MT-ATP6 and MT-ATP8*) encodes for the proteins associated with complex V (ATP synthase). In the year 2001, one more protein got added up to the list of 13 mitochondrial encoded protein called Humanin, which is known to be encoded by the same gene that encodes for one of the two rRNAs, *MT-RNR2*.

1.3 CRISPR/Cas in and out

In recent years, plethora of researches pertaining to the use of CRISPR/Cas system as a pivotal tool in genome editing technology are flooding the field for modern gene therapy as a promising genome manipulation platform. What is CRISPR/Cas? The CRISPR (Clustered Regularly Interspaced Short Palindromic Repeats)/Cas (CRISPR associated endonuclease) system are originally the form of prokaryotic (bacterial and archaeal) adaptive immune system against viral infections, which utilizes the fragment of previously invaded viral genome as a means to remember and defend any further invasion from the cognate viral entity.

Here this viral DNA fragment, termed as spacers after getting transcribed and incorporated along with short repetitive sequence into CRISPR cassettes, play a key role as a guide CRISPR RNA (guide crRNA) targeting the invading viral DNA and destroying it in association with Cas protein. This CRISPR/Cas array is mainly composed of an AT rich leader sequence which is bordered by a set of Cas genes encoding the Cas proteins. The Cas component here, is an endonuclease enzyme that acts as a molecular scissor to cut the double stranded DNA at a specific location on the target genome either to insert or to delete DNA fragments. Depending on the involvement of the type of Cas protein, which confers immunity via cleaving the viral genome or the genome of interest i.e., depending on the complexity of the components in CRISPR/Cas system it is divided mainly into 2 classifications, and are classified further into six types which are in turn divided into 19 sub types.[11] Briefly, in the class 1 type of division comprising of type I, III and IV multiple Cas proteins are involved whereas, on the other hand in type 2 classification comprising of types II, V and VI only one effector Cas protein will be associated alongside the CRISPR array of processed guide RNA which are transcribed complimentary to the target genome.

Amongst the four effective, customizable genome editing tools, such as, genome editing via meganucleases, zinc finger nucleases (ZFNs), transcription activator like effectors (TALEs) and the more recent CRISPR/Cas, the latter has bagged all the excitements in the scientific community for being economical, precise, faster and comparatively more efficient than the former mentioned genome editing tools.[12,13] These genome editing tools acts through introduction of site—specific DNA double strand breaks. However, the prior mentioned genome editing tools acts through generating the site-specific DSBs, they all vary in degrees of specificity and efficiency, which depends mainly on the type of nuclease involved, target site in the genome to be edited, and amount of RNA injected.

Mitochondrially targeted zinc finger nuclease (mito ZFN/mtZFN) is one of the widely used potential genome editing tool in modifying mitochondrial DNA. ZFN is a synthetic nuclease constructed using zinc finger DNA binding domain and a DNA cleavage domain of FoK1, which is a restriction enzyme. mtZFN is known to contain both mitochondrial targeting sequence (MTS) as well as nuclear export sequence that efficiently localizes mtZFN into mitochondria. Recent work by gammage et al. Highlights the importance of mtZFN in mitochondrial genome editing by targeting and cleaving mtDNA harboring the m.8993T>G point mutation associated with NARP, a 4977-bp repeat-flanked deletion associated with adult onset chronic progressive external ophthalmoplegia and, less frequently, Kearns-Sayre and Pearson's marrow pancreas syndromes.

TALENs (transcription activator like effector nuclease) are yet another widely used genome editing tool which is again synthetically engineered nucleases created by the addition of the TALE DNA binding domain from Xanthomonas which is fused with FoK1 type-II nuclease domain. These are imported into mitochondria by adding an MTS and UTR sequence from ATP5B and SOD2. The mitochondrially targeted TALENs are known to cleave the damaged or the mutated mtDNA in mitochondrial diseases such as, LHON, MELAS and Leigh's syndrome etc. the construction of TALEN is considered to be a lot easier compared to that of ZFN but are known to show off target effects compared to ZFN.

The most beneficial facet of CRISPR/Cas over the other genome editing tools is that, the ease of its use and the property of multiplicity. The former can be justified by the fact where, only 20 nucleotides in the guide RNA needs to be changed in order to retarget the Cas in CRISPR/Cas. This is in contrast with the other 2 genome editing tools, ZFNs and TALENs (mitoZFN and mitoTALEN when used in association with mitochondrial genome editing), which needs recoding of proteins with a large DNA segments of around 500–1500 base pairs for each new target site.[13] This advantage of CRISPR over others have enabled it to develop a genome wide CRISPR/Cas knockout library of mouse and embryonic stem cells using lentiviral vector to express the guide RNA.[14] The latter potential advantage of multiplicity of CRISPR/Cas makes it more upfront for mutating multiple genes the same time where, it uses multiple guide RNAs in parallel to target multiple sites simultaneously in the same cell.[13]

2. Mitochondrial DNA modification in disease condition

The variations in the mtDNA are associated with several disease conditions. The mitochondrial genome is known to have 10–17 times higher mutation rate than that of the nuclear genome, which makes it heteroplasmic due to the presence of heterogenous population of mitochondrial DNA. Till date, several clinical manifestations have been reported in accordance with the mitochondrial diseases caused by mutation in mtDNA, out of which a few prevalent such conditions are briefed in this book chapter (Fig. 1).

2.1 Leber's hereditary optic neuropathy (LHON)

Amongst all mitochondrial diseases, Leber's hereditary optic neuropathy is the first mitochondrial disease to be described (by the German ophthalmologist Theodore Leber in the year 1871) and is the first one to be reported in association with mtDNA mutation.[15] LHON is a type of maternally inherited mitochondrial disease, characterized by the cellular degeneration in retinal ganglion. This leads to acute or subacute, bilateral, painless central vision loss and eventually optic atrophy in either one eye followed by the other with an interval of around 8 weeks (75%) or it can simultaneously occur in both eyes (25%).[16] In almost 95% of cases Leber's 3 hereditary optic neuropathy is an outcome of one of the three pathogenic point mutation in

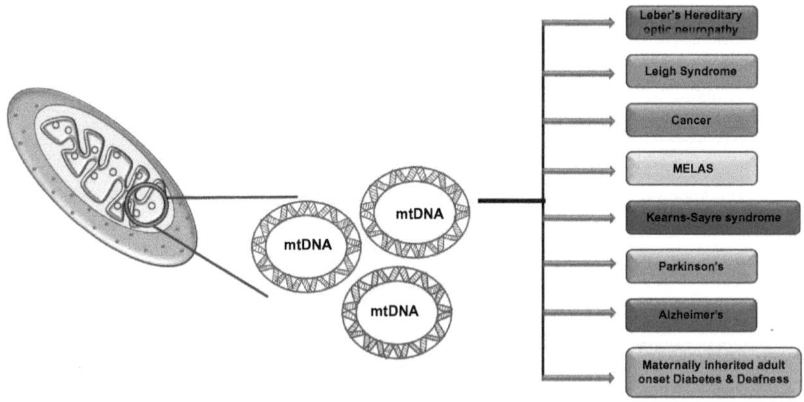

Fig. 1 Mitochondrial disease associated with mutation in mitochondrial genome (mtDNA).

the genes encoding for the protein NADH dehydrogenase (Complex-I) of electron transport chain. The above three point mutation takes place in the nucleotide positions 11,778 (G to A causing conversion arginine to histidine and is about 50–70%), 14,484 (T to C which accounts for around 10–15%), and 3460 (G to A which accounts for 8–25%) affecting the mitochondrial subunit genes, *MT-ND4, MT-ND6* and *MT-ND1* respectively.[17] Earlier LHON was considered to be the disease of younger male with a preponderance of around 86% whereas, another report states that, both male and female of all ages are affected by the condition with the ratio 3:1 respectively which is contrast with the traditional 5:1 ratio.[18,19]

The diagnosis of the disease condition includes neuro ophthalmic evaluation, neuroimaging (MRI) and biochemical studies for the determination of respiratory chain dysfunction. Apart from the prior mentioned techniques targeted diagnosis, multigene panel or complete mtDNA sequencing are the molecular testing approaches employed for the diagnosis of LHON.[20] A recent study reports as a rapid, cost effective, user friendly and high throughput approach of TaqMan-MGB probe based fluorescence qPCR assay in order to perform both qualitative and quantitative analysis of the fore mentioned three primary mutations in LHON mtDNA, which offers to be a promising approach for genetic screening and testing of these mutations.[21]

The current therapeutics involves the use of nutritional supplements (several combinations of vitamins—B2, B3, B12, C, E and folic acid, coenzyme Q10 and other supplements), activators of mitochondrial biogenesis via activation of PGC1α (fibrates, rosiglitazone, metformin and 5-aminoimidazole-4-carboxamide ribonucleoside (AICAR)), brimonidine, phytoestrogens, cyclosporine A, etc. The pharmaceutical compounds such as idebenone, which are a short chain derivative of coenzyme Q10, used as CoQ replacement and this restore electron flow or increase antioxidant defenses. Apart from these, the clinical trials for the successful use of gene therapy and stem cells are under progress as a promising approach towards treating LHON.[18,22]

2.2 Leigh's syndrome (LS)

Leigh's Syndrome, in other words also referred to as, subacute necrotic encephalomyelopathy, is a heterogenous neurologic disorder again caused due to the damage in mitochondrial energy production. The disease condition was described for the first time in the 1951 by a British neuropsychiatrist Archibald Denis Leigh.[23] The possible mutations involves both mitochondrial as well as nuclear genes. Amongst all, the mutation seen in *MT-ATP6*, the gene that codes for ATP synthase, of electron transport chain, accounts for around 10–20% of Leigh's disease, other mtDNA mutations that may involve are *MT-ND2, MT-ND3, MT-ND5, MT-ND6, and MT-CO1*. Apart from the fore mentioned mitochondrial DNA mutation, Leigh's syndrome is also an outcome of several mutations in the nuclear genome, amongst which the most commonly involved nuclear gene is SURF1. The clinical manifestations of the disease includes bilateral, focal lesions in brainstem causing inability to communicate with other neurons causing hypotonia, ataxia, diplagia, oculomotor dysfunction etc. At present, the available medications and nutritional supplementations in order to control and manage the symptoms caused due to Leigh's disease are vit B1, sodium bicarbonate, sodium citrate, Dichloroacetate and CoQ_{10}. The use of EPI-743, a CoQ_{10} analogue is under Clinical trial. There are reports stating the use of CRISPR/Cas9 for mutations in nuclear SURF1, further the use of successful gene therapy for treating LS requires many clinical trials.[23]

Several other mtDNA associated diseases both occurring often and rare have been enlisted in Table 1. Till date many approaches have been employed targeting mitochondrial disease yet there are no FDA approved drug for the treatment of these disease conditions. The available non-FDA approved medications can only help alleviate the symptoms of the fore mentioned mitochondrial diseases. This trigger the researchers to aim at an alternative approaches such as genome editing and other molecular biology based techniques.

Table 1 Mitochondrial diseases associated with mtDNA mutation.

SL. No.	Name of the mitochondrial disease	Year of discovery	Mutation site	Clinical manifestation	Possible management	Reference
1	Leber's hereditary optic neuropathy	1871	MT-ND4, MT-ND6 and MT-ND1	Acute/subacute, bilateral, painless central vision loss finally causing optic atrophy	Vitamins—B2, B3, B12, C, E and folic acid, coenzyme Q10, rosiglitazone, metformin, brimonidine, phytoestrogens, cyclosporine A, gene therapy, mitochondrial replacement therapy	15–22,24
2	Leigh syndrome	1951	MT-ATP6, MT-ND2, MT-ND3, MT-ND5, MT-ND6, and MT-CO1 and mutation in nuclear gene SURF1	Bilateral, focal lesions in brainstem causing inability to communicate with other neurons causing hypotonia, ataxia, diphlagia, oculomotor dysfunction, etc.	Vit B1, sodium bicarbonate, sodium citrate, dichloro acetate and CoQ10, EPI-743 (Clinical trial), CRISPR/Cas9 for SURF1, gene therapy	23
3	Neurogenic muscle weakness, ataxia and retinitis pigmentosa (NARP)	1990	MT-ATP6	Sensory neuropathy, ataxia, dementia, vision loss due to Retinitis pigmentosa	Supplementation with antioxidants	25–27
4	Familial bilateral striatal necrosis (FBSN)	1995	MT-ATP6	Oculomotor dyspraxia, opthalmoparesis	CoQ10, ubiquinone	28,29
5	Cancer	1931	MT-ATP6	Manifestation associated with cancer	Current therapeutics for Cancer (most of the drugs and CRISP/Cas-based genome editing are under clinical trial)	30,31

Continued

Table 1 Mitochondrial diseases associated with mtDNA mutation.—cont'd

SL. No.	Name of the mitochondrial disease	Year of discovery	Mutation site	Clinical manifestation	Possible management	Reference
6	Mitochondrial encephalopathy, lactic acidosis and stroke like episodes (MELAS)	1984	MT-ND1, MT-ND5, MT-TH, MT-TL1, MT-TV	Stroke like episodes, dementia, epilepsy, lactic acidemea, myopathy, hearing impairment, diabetes	Supplementation with antioxidants and cofactors, L-arginine therapy, citrulline supplementation, CoQ10, creatinine monohydrate, L-carnitine etc. (genome editing under clinical consideration)	32
7	Kearns-Sayre syndrome (KSS)	1958	Deletion in mtDNA	Pigmentory retinopathy, modest night blindness, ptosis, opthalmoplegia, hearing loss, cerebral ataxia, small stature	CoQ10 supplementation, implantation of pacemakers is suggested	33,34
8	Parkinson's	1991	Somatic mtDNA point Mutation, ATP13A2, POLG	Tremor, bradykinesia, rigidity, postural instability	Medications (Levodopa, dopamine agonists, monoamine oxidase—B inhibitors such as selegiline and rasgiline and catechol—o—methyl transferase inhibitors) and surgical therapy	35–38
9	Alzheimer's	2001	Somatic mtDNA mutations	Dementia, memory loss, confusion, loss of cognition	Donepezil (acetylcholinesterase inhibitors), mementine	38,39
10	Maternally inherited adult onset diabetes and deafness	1992	MT-TL1, MT-TK, and MT-TE	Diabetes and sensorineural hearing loss, dystrophy, focal segmental glomerulosclerosis, stroke, atrophy, myopathy	Insulin therapy, dietary changes (inclusion of hypoglycemic agents)	40,41

Note: There are no FDA approved medications as of now for the treatment of mitochondrial mtDNA mediated diseases. The fore mentioned are for the alleviating the symptoms caused by these disease conditions.

The year of discovery pertaining to Cancer, Parkinson's disease as well as Alzheimer's disease are with reference to its association with mtDNA mutation.

3. CRISPR/Cas mechanism

As mentioned in the introduction CRISPR/Cas is a RNA guided DNA targeting endonuclease system, which works in a sequence specific manner, the mechanism of which employs three main stages (I) DNA acquisition from the invading phage particle. (II) Biogenesis of CrRNA or CRISPR/Cas assembly formation and finally the (III) target DNA destruction or in other words, interference of the target DNA and Insertion of desired gene sequence (knock in). Here in this a precise mechanism pertaining to CRISPR/Cas9 i.e., the type II CRISPR/Cas system has been put in a limelight (Fig. 2).

(I) DNA acquisition from the invading phage particle:

As mentioned in the introduction part the CRISPR/Cas is said to be the dual RNA-guided DNA endonuclease system that requires a short protospacer from the invading viral entity, a noncoding trans-activating crRNA (tracrRNA), a protospacer adjacent motifs (PAMs) and a Cas endonuclease protein (Cas9 in case of type II CRISPR). The PAM here is characterized

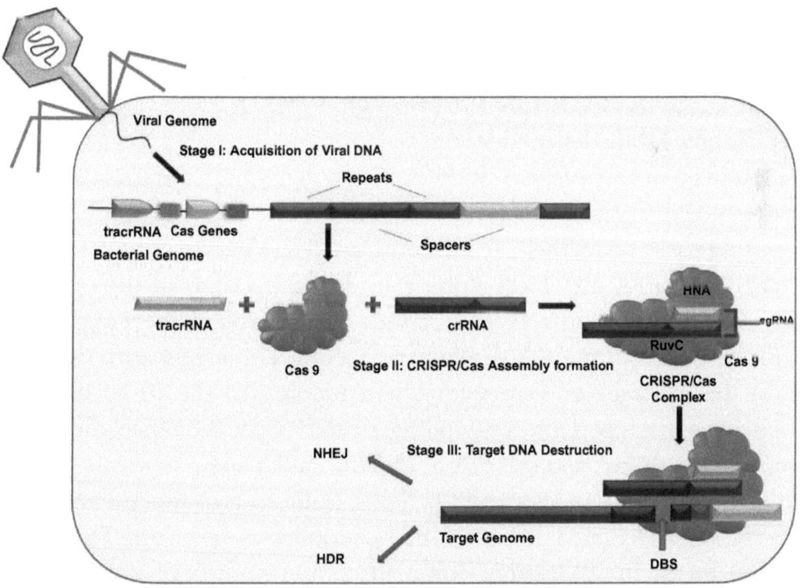

Fig. 2 General mechanism involved in genome editing mediated by the CRISPR/Cas system.

by the presence of three nucleotides more often NGG which is present flanking right next to the protospacer in the target DNA to which the CRISPR cassette binds.

Briefly, the mechanism of CRISPR based antiviral adaptive immunity shows the following step where the fragment of viral entity during the time of viral infection gets retained inside the bacterial cell called as a protospacer which facilitates the host organism to remember the pathogenic genome in case of any further invasion. The two vital Cas protein viz…Cas1 and Cas2, the presence of which is highly conserved, are instrumental in the binding and acquisition of the protospacer into the CRISPR array are known to present in almost all the CRISPR/Cas system. The selection of this short nuclear sequence, the protospacer is based on the presence of PAM, which is recognized by PAM recognition domain on Cas9.

(II) Biogenesis of CrRNA or CRISPR/Cas assembly formation:

Followed by the insertion of protospacer into the CRISPR array the latter is transcribed into o precursor crRNA, which is in turn processed into a mature guide crRNA. Unlike type I and type III systems the type-II system involving the presence of Cas9 requires a transactivating CRISPR RNA as mentioned prior, for the further processing of the pre-crRNA. This tracrRNA is known to hybridize with the pre-crRNA in order to stabilize Cas9. The complex of tracrRNA and pre-crRNA will be further recognized and processed via Rnase III and finally ending up in the maturation of crRNA know termed as sgRNA (small guide RNA).

(III) Target DNA destruction:

The last stage in case of any CRISPR/Cas knock down system is the interference of the targeted DNA molecule in which the mature sgRNA after processing now guides the Cas endonuclease (Cas9 in case of type II system) towards the target site. Here upon the second invasion of the same viral entity the mature CRISPR/Cas cassette scans through the viral genome for the presence of the PAM sequence present either upstream or downstream the protospacer sequence. Upon binding to the PAM sequence, which further triggers the unwinding of the target sequence the mature combo of tracrRNA and crRNA guides the cas9 protein to create a double stranded nick/break in the target genome using its two functional domain viz… HNH and RuvC/RNaseH-like endonuclease domain. This double stranded break (DSB) further stimulate a non-homologous end joining (NHEJ) or homology directed repair (HDR) mediated genome editing. The repair mechanism involving NHEJ pathway is said to be error-prone where, the DSB are processed by the action of the endogenous DNA repair machinery of rejoining. This results in unpredictable insertion or deletion causing frame shift in the open reading frame which finally leads to the destruction of that genome or in a single word "knockout." On the other hand, the DNA repair machinery involving the activity of HDR in the presence of a repair template in the form of plasmid allows the high fidelity and precise editing of DBS where the exogenous template DNA will be inserted into the DBS created leading to gene knock out, knock in and gene replacement. Unlike the NHEJ mediated DNA repair the HDR are less common but are considered to be the proper repair mechanism for the targeted gene editing and hence are limited to in vivo approaches.[11,12,42–44]

4. Implementation of CRISPR/Cas in mitochondrial DNA modification

The first ever, successful implementation of CRISPR/Cas9 was demonstrated/performed in the year 2015, by Jo and coworkers, which has set a breakthrough in the field of genome editing.[45] Where, they have reported mtDNA alteration by targeting the two mitochondrial protein cox1 and cox3 that disrupted the normal functioning of mitochondria since, these proteins are involved in oxidative phosphorylation. Unlike the normal Cas9, targeting towards nuclear genome using nuclear localization signals (NLS), here they had used a specialized Cas9 termed mitoCas9 by replacing the FLAG and NLS from the lentiviral construct with mitochondrial targeting sequence (MTS) of mitochondrial matrix protein cytochrome *c* and HA tag, in order to reduce the possibility of off targeting towards the nuclear genome. The use of lentiviral delivery system had eased the process of successful transfection of CRISPR/Cas into the model system. With this, they targeted the fore mentioned mitochondrial proteins utilizing the lentiviral-CRISPR/Cas system constructed with mitocas9 and sgRNA targeting cox1 and cox3. As a result, they successfully ended up with a target specific CRISPR/Cas9 system with restricted target access only towards mtDNA in order to create a mutant mtDNA, has provided an opportunity for many researchers in studying mitochondrial disease models since, the use of conventional method of studying these mutant mtDNA is quite challenging. The use of mitoCas9 and selected cell line with desired mutation enables the synthesis/designing of cellular model for studying the activity and effect of a particular mtDNA mutation related to human mitochondrial diseases.[46]

Despite the fore mentioned report, there are controversial reports stating the challenges or impossibility of the use of CRISPR/Cas9 in mtDNA editing since, there is a lack of reports or similar kind of studies and the most important consideration that there is no clear-cut accepted mechanism to employ the transport of the RNA into the mammalian mitochondria. While several other studies reports that the RNA molecule imported into the mitochondria may perform the function such as mitochondrial RNA processing.[47]

It is said that the mitochondrial genome editing or modification via the use of gene editing tool such as CRISPR-Cas9 system is highly challenging mainly due to fact that the delivery of guide RNA and Cas9 enzyme complexes into mitochondria partially competent i.e., the CRISPR/Cas9 mechanism requires the import of all three components, such as Cas9 nuclease, guide RNA, and the desired insertion sequence in case of CRISPR/Cas based knock in into the integral mtDNA present inside the mitochondria, which is so far considered challenging.[48] The lack of effective delivery methods to deliver gRNA through mitochondrial membrane has proven to be the major obstacle in the development of CRISPR-Cas9 mediated gene editing of mitochondrial genome as the import of mRNAs into the mitochondria is largely dependent on activity of tRNA. On the other hand, there are indeed many reports/studies inferring mitochondrial import of exogenous RNA and miRNA.[49]

Recent studies in the field of mitochondrial genome editing have emphasized the use of RGEN/Cas9 (RNA-guided endonuclease), which is a CRISPR/Cas system derived tool for a successive manipulation of mitochondrial genome. RGEN/Cas9 system is composed of Cas9, CRISPR RNA (crRNA) and a trans activating crRNA (tracrRNA). The RGEN sometimes do contain Cas9 and a single chain guide RNA. The use of specialized RGEN system such as MitoRGEN/SpCas9 might prove to be instrumental in overcoming the obstacles in the use of classical CRISPR/Cas9 system where the modification with MTS can effectively import SpCas9 (mitoCas9) into the matrix of the mitochondria.[50]

5. Future perspective

The recent developments in the field of genome editing has created an hope in the scientific community around the globe in combating the mitochondrial diseases caused precisely due to the alterations or mutation caused in the specific location in the mitochondrial DNA. Till now there are no available treatment or FDA approved medications for curing any of the mitochondrial diseases. The available medications are suggested or used to decrease or alleviate the clinical manifestations caused by the disease condition. With this context the researcher around the globe are looking for an effective alternative approach for combating the fore said mitochondrial diseases. This is where the gene based therapy comes into play which involves several approaches such as mitochondrial replacement therapy, use of meganucleases, Zinc finger nucleases (ZFNs), Transcription activator—like effectors (TALEs) and the more recent CRISPR/Cas. Each of the fore mentioned gene therapy approaches have their own pros and cons and are now flooding into the clinical trials.

The use of CRISPR/Cas has proven to be a fundamental tool in the field of genome editing for its customizable nature, cost effectiveness, high precision rate and the property of multiplexing when compared to the other conventional gene editing tools. Accounting for the fact that, there are only limited number of studies pertaining to the use of CRISPR mediated gene editing in human disease has created a space for more groundwork to be performed. Despite the lack in number of clinical studies, the use of CRISPR may prove to be the greatest resolution targeting the genome based diseases, accounting for the fact that CRISPR provides an opportunity in studying the genetic disorder by creating knock in and knock out models.

Although significant number of progressions have been made in order to increase the efficiency and target specificity, several improvisations are need to be made when using CRISPR/Cas for the successful mitochondrial genome editing in order to overcome the fore mentioned obstacles. The improvisations such as, in the import of nucleic acids inside the mitochondria can significantly advance the functionality of this influential genome manipulating tool. Recent studies included the use of newly emerging CRISPR/Cas systems such as spCas9-NG, base editing, xCas9, Cpf1, Cas13, Cas14 and RGEN/Cas9 system.

The internalization of CRISPR/Cas system components or the RNA moieties inside the target mitochondria may be improvised by the use of mitochondria targeting liposome called "MITO porter." By this, the problem of nucleic acid internalization could be resolved along with yet another major consideration "the size" using MITO porter delivery of wide variety of encapsulated molecules into the mitochondrial matrix can be achieved regardless of the size of the molecule. With this advantage Cas9 protein/gRNA can easily be incorporated inside the mitochondria.

It is also that exosome, which are 30–150 nm sized extracellular vesicles can be used as a nano carrier for the effective delivery of the CRISPR/Cas9 system targeting towards the mitochondria. This can be achieved by tagging the CRISPR/Cas9 loaded exosomes with the available mitochondrial targeting sequence (MTS). This engineered exosome being target specific, are proposed to deliver the cargo to the mitochondrial matrix as they fuse with the mitochondrial membrane. The future prospect can be aimed at this direction in order to overcome all the obstacles hindering the wide range application of CRISPR/Cas9 as an efficient genome editing tool especially for editing mitochondrial genome.

References

1. Shtolz N, Mishmar D. The mitochondrial genome—on selective constraints and signatures at the organism, cell, and single mitochondrion levels. *Front Ecol Evol.* 2019;7:342. https://doi.org/10.3389/fevo.2019.00342.
2. Miyazono Y, Hirashima S, Ishihara N, et al. Uncoupled mitochondria quickly shorten along their long axis to form indented spheroids, instead of rings, in a fission-independent manner. *Sci Rep.* 2018;8:350. https://doi.org/10.1038/s41598-017-18582-6.
3. Kuhlbrandt W. Structure and function of mitochondrial membrane protein complexes. *BMC Biol.* 2015;13:89. https://doi.org/10.1186/s12915-015-0201-x.
4. Chial H, Craig J. mtDNA and mitochondrial diseases. *Nat Educ.* 2008;1(1):217.
5. Kuzmenko AV, Levitskii SA, Vinogradova EN, et al. Protein biosynthesis in mitochondria. *Biochem Mosc.* 2013;78:855–866. https://doi.org/10.1134/S0006297913080014.
6. Kowaltowski AJ. Alternative mitochondrial functions in cell physiopatology: beyond ATP production. *Braz J Med Biol Res.* 2000;33(2):241–250. https://doi.org/10.1590/s0100-879x2000000200014.
7. Gohil VM, Greenberg ML. Mitochondrial membrane biogenesis: phospholipids and proteins go hand in hand. *J Cell Biol.* 2009;184(4):469–472. https://doi.org/10.1083/jcb.200901127.
8. Schwartz M, Vissing J. Paternal inheritance of mitochondrial DNA. *N Engl J Med.* 2002;347:576–580. https://doi.org/10.1056/NEJMoa020350.
9. Taanman J-W. The mitochondrial genome: structure, transcription, translation and replication. *Biochim Biophys Acta (BBA).* 1999;1410(2):103–123. https://doi.org/10.1016/s0005-2728(98)00161-3.
10. Barchiesi A, Vascotto C. Transcription, processing, and decay of mitochondrial RNA in health and disease. *Int J Mol Sci.* 2019;20(9):2221. https://doi.org/10.3390/ijms20092221.
11. Hille F, Charpentier E. CRISPR-Cas: biology, mechanisms and relevance. *Philos Trans R Soc B.* 2016;371:20150496. https://doi.org/10.1098/rstb.2015.0496.
12. Hsu PD, Lander ES, Zhang F. Development and applications of CRISPR-Cas9 for genome engineering. *Cell.* 2014;157(6):1262–1278. https://doi.org/10.1016/j.cell.2014.05.010.
13. Gupta RM, Musunuru K. Expanding the genetic editing tool kit: ZFNs, TALENs, and CRISPR-Cas9. Review. *J Clin Invest.* 2014;124(10):4154–4161. https://doi.org/10.1172/JCI72992.
14. Koike-Yusa H, Li Y, Tan E-P, Velasco-Herrera MDC, Yusa K. Genome-wide recessive genetic screening in mammalian cells with a lentiviral CRISPR-guide RNA library. *Nat Biotechnol.* 2014;32(3):267–273. https://doi.org/10.1038/nbt.2800.

15. Wallace DC, Singh G, Lotr MT, et al. Mitochondrial DNA mutation associated with Leber's hereditary optic neuropathy. *Science*. 1988;242(4884):1427–1430. https://doi.org/10.1126/science.3201231.
16. abu-amero k. Leber's hereditary optic neuropathy: the mitochondrial connection revisited. *Middle East Afr J Ophthalmol*. 2011;18(1):17–23. https://doi.org/10.4103/0974-9233.75880.
17. Catarino CB, Ahting U, Gusic M, et al. Characterization of a Leber's hereditary optic neuropathy (LHON) family harboring two primary LHON mutations m.11778G>A and m.14484T>C of the mitochondrial DNA. *Mitochondrion*. 2017;36:15–20. https://doi.org/10.1016/j.mito.2016.10.002.
18. Koilkonda RD, Guy J. Leber's hereditary optic neuropathy-gene therapy: from benchtop to bedside. *J Ophthalmol*. 2011;2011. https://doi.org/10.1155/2011/179412, 179412.
19. Poincenot L, Pearson AL, Karanjia R. Demographics of a large international population of patients affected by Leber's hereditary optic neuropathy. *Ophthalmology*. 2020;127:679–688. https://doi.org/10.1016/j.ophtha.2019.11.014.
20. Yu-Wai-Man P, Turnbull DM, Chinnery PF. Leber hereditary optic neuropathy. *J Med Genet*. 2002;39(3):162–169. https://doi.org/10.1136/jmg.39.3.162.
21. Xue B, Yang L, Wang X, et al. TaqMan-MGB probe quantitative PCR assays to genotype and quantify three mtDNA mutations of Leber hereditary optic neuropathy. *Sci Rep*. 2020;10:12264. https://doi.org/10.1038/s41598-020-69220-7.
22. Theodorou-Kanakari A, Karampitianis S, Karageorgou V, et al. Current and emerging treatment modalities for Leber's hereditary optic neuropathy: a review of the literature. *Adv Ther*. 2018;35(10):1510–1518. https://doi.org/10.1007/s12325-018-0776-z.
23. Schubert Baldo M, Vilarinho L. Molecular basis of Leigh syndrome: a current look. *Orphanet J Rare Dis*. 2020;15:31. https://doi.org/10.1016/j.ymgme.2015.06.004.
24. Berardo A, Emmanuele V, Vargas W, Tanji K, Naini A, Hirano M. Leber hereditary optic neuropathy plus dystonia, and transverse myelitis due to double mutations in MT-ND4 and MT-ND6. *J Neurol*. 2020;267:823–829. https://doi.org/10.1007/s00415-019-09619-z.
25. Holt IJ, Harding AE, Petty RK, Morgan-Hughes JA. A new mitochondrial disease associated with mitochondrial DNA heteroplasmy. *Am J Hum Genet*. 1990;46(3):428–433.
26. Kerrison JB, Biousse V, Newman NJ. Retinopathy of NARP syndrome. *Arch Ophthalmol*. 2000;118(2):298. https://doi.org/10.1001/archopht.118.2.298.
27. Parfait B, de Lonlay P, von Kleist-Retzow J, et al. The neurogenic weakness, ataxia and retinitis pigmentosa (NARP) syndrome mtDNA mutation (T8993G) triggers muscle ATPase deficiency and hypocitrullinaemia. *Eur J Pediatr*. 1999;158:55–58. https://doi.org/10.1007/s004310051009.
28. Thyagarajan D, Shanske S, Vazquez-Memije M, Devivo D, Dimauro S. A novel mitochondrial ATPase 6 point mutation in familial bilateral striatal necrosis. *Ann Neurol* 1995;38:468–472. https://doi.org/10.1002/ana.410380321.
29. Schon EA, Santra S, Pallotti F, Girvin ME. Pathogenesis of primary defects in mitochondrial ATP synthesis. *Semin Cell Dev Biol*. 2001;12(6):441–448. https://doi.org/10.1006/scdb.2001.0281.
30. Scheid AD, Beadnell TC, Welch DR. The second genome: effects of the mitochondrial genome on cancer progression. *Adv Cancer Res*. 2019;142:63–105. https://doi.org/10.1016/bs.acr.2019.01.001.
31. Wallace DC. Mitochondria and cancer: Warburg addressed. *Cold Spring Harb Symp Quant Biol*. 2005;70:363–374. https://doi.org/10.1101/sqb.2005.70.035.
32. El-Hattab AW, Adesina AM, Jones J, Scaglia F. MELAS syndrome: clinical manifestations, pathogenesis, and treatment options. *Mol Genet Metab*. 2015;116(1–2):4–12. https://doi.org/10.1016/j.ymgme.2015.06.004.

33. Yu M, Yu L, Wang ZX. Diagnosis and management of Kearns-Sayre syndrome rely on comprehensive clinical evaluation. *Chin Med J (Engl)*. 2016;129(20):2519–2520. https://doi.org/10.4103/0366-6999.191855.
34. Finsterer J, Winklehner M, Stöllberger C, Hummel T. Unusual phenotype and disease trajectory in Kearns-Sayre syndrome. *Case Rep Neurol Med*. 2020;2020:7368527. https://doi.org/10.1155/2020/7368527.
35. Moore DJ, West AB, Dawson VL, Dawson TM. Molecular pathophysiology of Parkinson's disease. *Annu Rev Neurosci*. 2005;28:57–87. https://doi.org/10.1146/annurev.neuro.28.061604.135718.
36. Deas E, Wood NW, Plun-Favreau H. Mitophagy and Parkinson's disease: the PINK1-parkin link. *Biochim Biophys Acta*. 2011;1813(4):623–633. https://doi.org/10.1016/j.bbamcr.2010.08.007.
37. Martinez TN, Greenamyre JT. Toxin models of mitochondrial dysfunction in Parkinson's disease. *Antioxid Redox Signal*. 2012;16(9):920–934. https://doi.org/10.1089/ars.2011.4033.
38. Johri A, Beal MF. Mitochondrial dysfunction in neurodegenerative diseases. *J Pharmacol Exp Ther*. 2012;342(3):619–630. https://doi.org/10.1124/jpet.112.192138.
39. Ridge PG, Kauwe JSK. Mitochondria and Alzheimer's disease: the role of mitochondrial genetic variation. *Curr Genet Med Rep*. 2018;6(1):1–10. https://doi.org/10.1007/s40142-018-0132-2.
40. Sugai K, Ueda H, Morimoto K, et al. Maternally inherited diabetes and deafness complicated by mesangial galactose-deficient IgA1 deposits: a case report. *BMC Nephrol*. 2018;19:350. https://doi.org/10.1186/s12882-018-1152-6.
41. Ouweland J, Lemkes H, Trembath R, et al. Maternally inherited diabetes and deafness is a distinct subtype of diabetes and associates with a single point mutation in the mitochondrial $tRNA^{Leu(UUR)}$ gene. *Diabetes*. 1994;43(6):746–751. https://doi.org/10.2337/diab.43.6.746.
42. Jiang F, Doudna JA. CRISPR-Cas9 structure and mechanism. *Annu Rev Biophys*. 2017;46:505–529. https://doi.org/10.1146/annurev-biophys-062215-010822.
43. Ran F, Hsu P, Wright J, Agarwala V, Scott D, Zhang F. Genome engineering using the CRISPR-Cas9 system. *Nat Protoc*. 2013;8:2281–2308. https://doi.org/10.1038/nprot.2013.143.
44. Verma N, Zhu Z, Huangfu D. CRISPR/Cas-mediated Knockin in human pluripotent stem cells. *Methods Mol Biol*. 2017;1513:119–140. https://doi.org/10.1007/978-1-4939-6539-7_9.
45. Gammage PA, Moraes CT, Minczuk M. Mitochondrial genome engineering: the revolution may not be CRISPR-Ized. *Trends Genet*. 2018;34(2):101–110. https://doi.org/10.1016/j.tig.2017.11.001.
46. Jo A, Ham S, Lee GH, et al. Efficient mitochondrial genome editing by CRISPR/Cas9. *Biomed Res Int*. 2015;2015:305716. https://doi.org/10.1155/2015/305716.
47. Cappa R, de Campos C, Maxwell AP, McKnight AJ. "Mitochondrial toolbox"—a review of online resources to explore mitochondrial genomics. *Front Genet*. 2020;11:439. https://doi.org/10.3389/fgene.2020.00439.
48. Slone J, Huang T. The special considerations of gene therapy for mitochondrial diseases. *NPJ Genom Med*. 2020;5:7. https://doi.org/10.1038/s41525-020-0116-5.
49. Rai PK, Craven L, Hoogewijs K, Russell OM, Lightowlers RN. Advances in methods for reducing mitochondrial DNA disease by replacing or manipulating the mitochondrial genome. *Essays Biochem*. 2018;62(3):455–465. https://doi.org/10.1042/EBC20170113.
50. Verechshagina N, Nikitchina N, Yamada Y, et al. Future of human mitochondrial DNA editing technologies. *Mitochondrial DNA A DNA Mapp Seq Anal*. 2019;30(2):214–221. https://doi.org/10.1080/24701394.2018.1472773.

Further reading

51. Makarova K, Wolf Y, Alkhnbashi O, et al. An updated evolutionary classification of CRISPR–Cas systems. *Nat Rev Microbiol.* 2015;13:722–736. https://doi.org/10.1038/nrmicro3569.
52. Makarova KS, Haft DH, Barrangou R, et al. Evolution and classification of the CRISPR-Cas systems. *Nat Rev Microbiol.* 2011;9(6):467–477. https://doi.org/10.1038/nrmicro2577.
53. Jinek M, Chylinski K, Fonfara I, Hauer M, Doudna JA, Charpentier E. A programmable dual-RNA-guided DNA endonuclease in adaptive bacterial immunity. *Science.* 2012; 337(6096):816–821. https://doi.org/10.1126/science.1225829.

CHAPTER EIGHT

Immune responses to CRISPR-Cas protein

Sobhan Roy*

Departments of Biochemistry and Molecular Biology, University of Chicago, Chicago, IL, United States
*Corresponding author: e-mail address: sobhan.roy@gmail.com

Contents

Abstract

The clustered regularly interspaced short palindromic repeats (CRISPR), and CRISPR-associated (Cas) protein technologies have evolved as promising, cost-effective, and efficient methods for editing genomes. Editing genomes with high specificity and precision is a daunting task, where errors can lead to undesirable outcomes. Many elegant studies have successfully shown that the CRISPR-Cas9 system can modify, disrupt, and add new DNA sequences directly into the genomes of the cells or animals being studied. As such, the CRISPR-Cas9 technology holds immense potential for biomedical research as well as agricultural and therapeutic applications, further emphasized by its unprecedented movement into the clinical setting.

Throughout every stage of life, missense mutations can lead to highly unfavorable outcomes, syndromes, and diseases. Many of these mutations are transferred directly through the fertilization process and, thereby, acquired at birth and propagated to the next generation. As such, it has been of great interest to develop techniques to

Progress in Molecular Biology and Translational Science, Volume 178
ISSN 1877-1173
https://doi.org/10.1016/bs.pmbts.2020.12.003

repair these mutations using genetic manipulation, prior to or following birth. CRISPR-Cas9 has many advantages in this regard over numerous other existing technologies. Regardless, editing bases within a genome can be associated with numerous challenges that were previously unrecognized and lead to unforeseen consequences. While the CRISPR-Cas9 method is perfectly suitable for editing cells outside the body with limited risk to the normal functioning of the cell, recent publications have illustrated a number of challenging conditions resulting from its use. One of them is directed to the host immune response toward CRISPR-Cas9. With this in mind, this review will discuss recent observations on the host immune response to CRISPR-Cas9 and the associated challenges that arise as a result.

1. Introduction

Every living organism has an intracellular genome to store all the genetic information in the form of deoxyribonucleic acid (DNA). DNA encodes various genes that are transcribed directly into mRNAs and subsequently translated into proteins to carry out specific functions following the central dogma. As such, DNA at the genomic level is upstream of all cellular and molecular processes taking place within the host, both at the cellular and tissue level. As such, missense genetic mutations can directly affect the downstream associated protein functions and traits. To overcome these errors, every host and cell needs to modify or revert these alterations back into the wild-type genome sequence. This is a seemingly impossible task that is limited by fidelity within the cell and potentially associated with off-target side effects as a result of the repair process. However, over the past few decades, with the discovery of gene-editing techniques like CRISPR-Cas9, there is hope to revert these alterations back to their wild-type state with a reduced risk of mistake and increased fidelity.

To date, many publications have successfully implemented the CRISPR-Cas9 technology within the laboratory setting, serving as a basis for its potential use in therapeutic applications. As a result, it is becoming far more popular than earlier genome editing methods, such as the Zinc-finger nuclease (ZFNs)[1] and transcription activator-like effector nuclease (TALEN) methods.[2] There are currently ongoing clinical trials based upon CRISPR-Cas9 that are focused on treating diseases like β thalassemia, sickle cell anemia, and HIV-1-infection, where the list of applications and therapeutic strategies is rapidly expanding.[3] For most of the treatments, the genomes of specific cells are edited outside of the host body, or in *ex vivo* conditions, and then introduced back into the host.[4] However, while a number of diseases can potentially be treated through *ex vivo* applications, which has been promising in terms of safety, a wide number of diseases and ailments may require direct genome editing inside the body of the host, which presents a wide array of new challenges for the use of CRISPR-Cas9 as a direct therapeutic.

2. The overview of CRISPR-Cas9 technology

The CRISPR-Cas systems are widely present in microbial genomes and show significant variation across species.[5] Fundamentally, the CRISPR-Cas system has evolved to provide support in fighting against invading phages that pose a substantial risk to microbial life. As a result, CRISPR-Cas systems are often referred to as microbial adaptive immune systems as they store information about specific phages that can protect in the future.

Due to high variability, CRISPR-Cas systems are categorized in many ways. However, three basic categories have been reported. A CRISPR-Cas system consists of a Cas gene and its associated CRISPR arrays of repetitive sequences (direct repeats) interspaced by nonrepetitive DNA sequences (spacers) derived from non-self-DNA material. The CRISPR repetitive sequences are transcribed into short RNAs and termed CRISPR RNAs (crRNA). Cas proteins bind to these crRNAs to make the functional units of the system. The crRNA then directs hybridization using specific DNA sequences, classical Watson-Crick base pairing, and DNA strand digestion carried out by specific nuclease domains within the Cas protein. Type I and III CRISPR systems rely on ensembles of Cas proteins to form the holoenzyme. In comparison, the type-II system uses a single large Cas9 protein that contains all of the necessary functional domains. Further, Cas9 proteins bind their targets using two RNA molecules: crRNA and trans-activating crRNA (tracrRNA). The crRNA and tracrRNA hybridize to form a duplex and complex with Cas9 and direct cognate DNA sequence cleavage with protospacer-adjacent motifs (PAM).

3. The structure of CRISPR-Cas9

The domain architecture of Cas9 proteins across various microbes is, overall, very similar, although the sizes of each Cas9 enzymes may vary.[6,7] For example, *Staphylococcus pyogenes* Cas9 (SpCas9) is a large protein with 1368 amino acids and consists of six domains[6,8]; *Staphylococcus aureus* Cas9 (SaCas9) is relatively small with 1053 amino acids and six domains; *Actinomyces* Cas9 is 1101 amino acid long with six domains, and *Francisella*

Cas9 is the largest with 1629 amino acids and six domains.[6] A thermostable Cas9 of 1083 amino acid long from thermophilic bacterium *Geobacillus thermodenitrificans* has also been characterized.[9] SpCas9 has been adopted and engineered for CRISPR mediated genome engineering for various scientific applications. As a result, determining the structure of SpCas9 was essential to understand its mechanism of action.

SpCas9 is a crescent-shaped molecule and composed of two distinct lobe architectures: the recognition lobe (REC) and the nuclease (NUC) lobe (Fig. 1).[7,8,10] The REC lobe is subdivided into bridge helix (BH: an extended α helical region), REC1, and REC2 domains. The NUC lobe comprises the RuvC domain, HNH domain, and PAM-interacting (PI) domain (Fig. 1A and B). The REC and NUC lobe interfaces are positively charged and binds with the negatively charged RNA: target DNA heteroduplex (Fig. 1C). The RuvC domain is assembled from three separate motifs. The HNH domain cleaves the DNA strand that is complementary to guide RNA. The RuvC domain mediates the non-complementary strand's cleavage, collectively resulting in double-stranded DNA breaks (DSBs). Two mechanisms then repair the cleaved DNA; host cell endogenous non-homologous end joining (NHEJ) or by an exogenously supplied template. NHEJ often results in the insertion or deletion of one or more nucleotides, leading to frameshift mutations within the edited genome. The provided DNA template can modify the sequence to the correct desired sequence without this fallback. The crystal structures Cas9 were solved in complex with and without guide RNA and target DNA and directly helped in understanding the mechanism of the CRISPR-Cas9 system.[7,8,10]

4. Challenges associated with the implication of CRISPR-CAS9 technology

Every new technology is associated with new challenges. As such, the CRISPR-Cas system has unveiled a number of new complications that need to be addressed to move forward faster in the realm of therapeutics. The most significant difficulties in this system are the ethical considerations consequences of heritable genome editing, off-target effects as a result of its use as a therapeutic, and the human immune response against various Cas proteins. Ethical issues are critical for the application of CRISPR-Cas and are being discussed across a number of scientific communities. Additionally, CRISPR-Cas9 nuclease often causes unwanted mutations at off-target sites that match the target site.[11,12] Recently, a high-fidelity version,

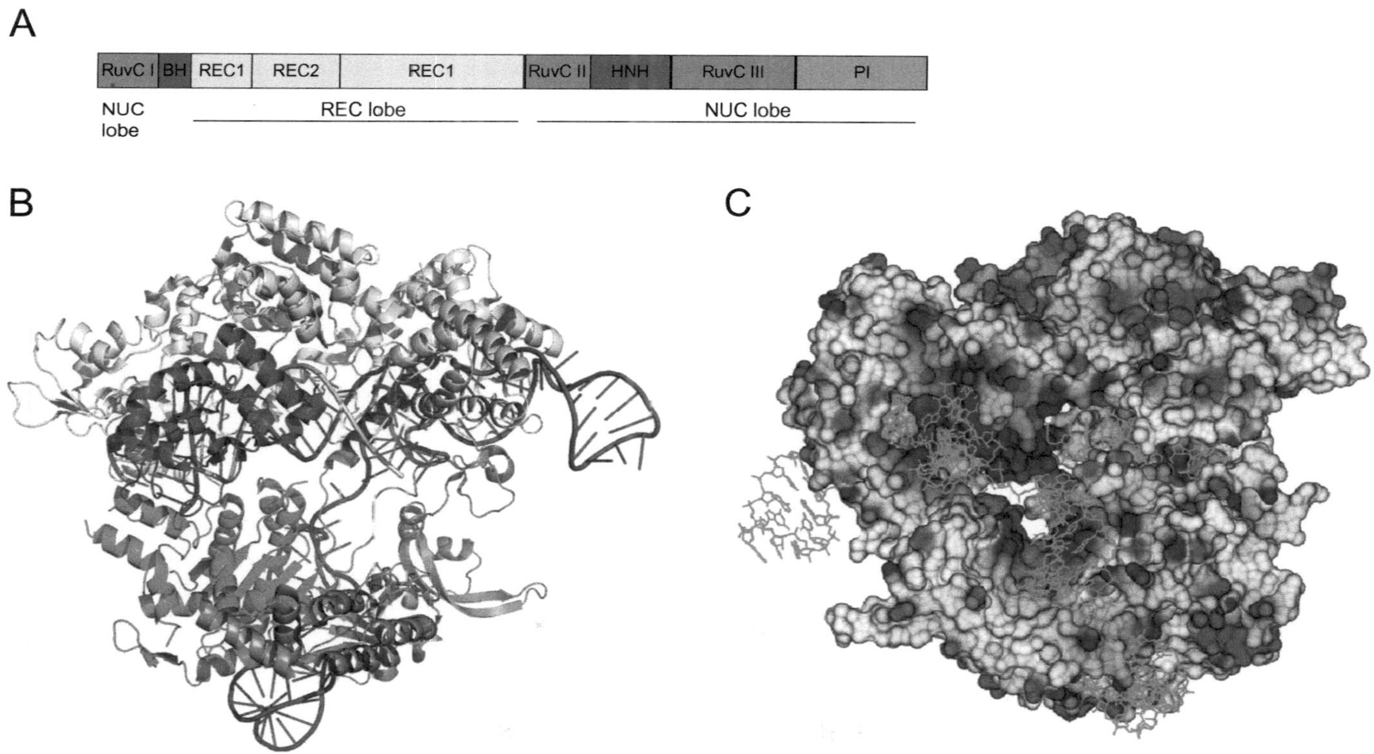

Fig. 1 The crystal structure of *Staphylococcus pyogenes* CRISPR-Cas9 in complex with guide RNA and target DNA. (A) The domain architecture of SpCas9 protein. A single polypeptide chain folds into different functional domains of Cas9. The CRISPR-Cas9 complex structure as a cartoon (B) and electrostatic surface (C) are shown. The helices are color-coded, as depicted in Fig. A. The sgRNA is colored as red and bound to target DNA colored as wheat. The positively charged pockets of Cas9 around the DNA binding grooves are shown as blue color (C). DNA and RNA are colored as green. The figure was generated in Pymol software using Protein Data Bank code: 4OO8.

SpCas9-HF1, has been developed for a safer CRISPR-Cas9 application in regard to off-target effects.[13] Finally, Cas9 is a large microbial protein and the immunogenicity of Cas9 proteins may induce an immune response directly from somatic cell editing. Therefore, understanding the human immune response will be essential to identify antigenic epitopes and cell populations that recognize Cas9 epitopes and, finally, help synthesize a better strategy for *in vivo* applications.

5. Immunogenicity of Cas9 proteins

The human immune response against Cas9 proteins may lead to unwanted outcomes and side effects during *in vivo* use of CRISPR-Cas9 therapies. However, the full extent of these potential shortcomings is not entirely clear. For example, *in vivo* studies in mice supported the idea of immune activation following treatment. Further, *ex vivo* studies demonstrated the human immune system could successfully detect Cas9 expressing cells and mount targeted immune responses. Therefore, understanding the immunogenicity of Cas9 is crucial for future therapeutic applications. This will help to design immunologically inert Cas9 and other strategies to circumvent the patients' immune reactions and any adverse effects as a result. In addition to these findings, the *Staphylococcus* species already routinely infect humans, and the role of Cas proteins during these infections remains unknown.

5.1 Human immune responses to *Staphylococcus aureus* and *Staphylococcus pyrogens*

CRISPR-Cas proteins are widely distributed in microorganisms like bacteria and archaea,[6] including large numbers that are present as human microbiota and pathogens. The orthologs of Cas9 proteins that are being studied for various applications include *Staphylococcus aureus* (SaCas9) and *Staphylococcus pyrogens* (SpCas9), *Francisella novicida* (FnCas9), *and Actinomyces naeslundii* (AnaCas9). *Francisella* species can be intracellular parasites for a variety of organisms. However, *F. novicida* is non-pathogenic and is found to infect humans[14] rarely. *Actinomyces naeslundii* is found in the oral microflora and can act as an opportunistic pathogen and cause periodontal disease and pulmonary actinomycosis.[15] *Staphylococcus* species are facultative colonizers of human skin and represent a normal part of the oral microbiota. However, they can cause infection in a sizeable human population every year around the globe.[16] *Staphylococcus* species causes diseases like scalded skin syndrome, pharyngitis, and pyoderma, and may involve severe invasive infections like meningitis, septicemia, and endocarditis. Therefore, it is plausible that *Staphylococcus* antigens are exposed to the immune system during the illness.

Cas9 proteins are large, derived from the microbial genome, and exposed in combination with other bacterial potent agonists, which may act as adjuvants for the immune system. The immune system is highly efficient and capable of recognizing foreign proteins, peptides, lipids, and vitamin metabolites as antigens.[17,18] The adaptive immune system uses recombined receptors (T cells receptor or TCR and B cell receptor or BCR) to recognize these antigens. Antibodies can bind to secreted proteins, surface-exposed proteins, and sugar molecules of *Staphylococcus* species. In contrast, T cells can recognize processed antigens generated from the whole protein, lipids, and metabolites. Therefore, during a *Staphylococcus* infection, the human immune system can recognize microbial Cas9 proteins as antigens and other *Staphylococcus* materials, which may aid in the development of immunogenic memory to Cas9 proteins in humans.

5.2 Immune responses to Cas9 in mice

A recent study has shown that both humoral and cell-mediated immunity recognize Cas9 proteins as antigens and mount corresponding immune responses in mice.[19] For example, Cas9 exposure induces cellular responses after delivery by adeno-associated virus vector (AAV) and viral vector free intramuscular DNA electroporation. T cells, B cells, and myeloid cells (monocytes, macrophages, and dendritic cells) are elevated following these delivery methods. T cells specific to Cas9 antigens increased and showed the characteristic features of memory responses. T cells use multiple TCR genes to mount immune responses. Therefore, TCRs may be capable of recognizing various peptide antigens derived from the large Cas9 protein. Beyond this, the same study found T cells localized in Cas9 expressing tissues. However, these cells only displayed naïve or immature phenotypes. This is critical as T cell-mediated cytotoxicity is carried out only by activated T cells. Mice also developed Cas9 specific antibodies to multiple Cas9 epitopes that include linear epitopes regardless of the Cas9 delivery system. Beyond this, these studies have also suggested that the Cas9 protein is immunogenic and may elicit a moderate immune response.

5.3 Immune responses to Cas9 in human

It was recently demonstrated that the human immune system is reactive to Cas9 proteins, where both B cells and T cells are activated in the presence of the Cas9 protein, as observed in mice. This indicates pre-existing immunity in many humans, which may pose obstacles for the use of CRISPR-Cas9 as an application to treat patients[20] directly. Pre-existing immunity may hinder the technique's efficacy and generate undesirable toxicity and complications from unwarranted immune responses (Fig. 2). However, to expand upon this point, there are currently two methods to introduce edited cells into a host; *ex vivo* and *in vivo*. In *ex vivo* applications, cells will be transduced with CRISPR-Cas9 outside the body.[4] Therefore, the immune response to Cas9 will be negligible after these cells are introduced back into the host body (Fig. 2). As such, *ex vivo* applications may prove to be safer for therapeutic treatment until new technologies can better address the host response to *in vivo* applications.

5.4 Human antibody responses to Cas9

The Cas9 protein is an intracellular protein and not exposed to the surface or secreted outside the cell envelope. Antibody responses are mostly directed toward exposed and secreted proteins from pathogens. Antibodies bind with high affinities to the extracellular surface of bacteria and viruses, where they

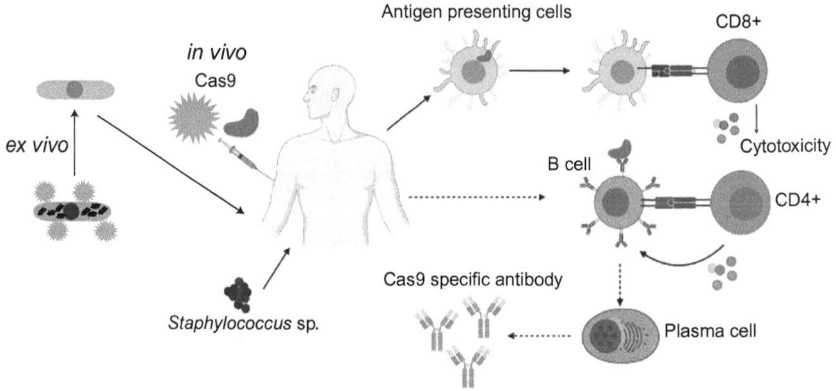

Fig. 2 Antigenicity of Cas9 and Cas9 expressing microbes in humans. Antigen-presenting cells process and present the antigens to T cells. That leads to cytotoxicity by $CD8^+$ T cells and immune modulation by $CD4^+$ T cells. $CD4^+$ T cells also help in the process of B cell maturation and antibody response. The BCR will recognize the Cas9 protein and give rise to plasma cells to secrete Cas9 specific antibodies. The figure was partially generated using the Biorender website.

block interactions with host counterparts and prevent entry into host cells. However, antibodies can be generated to intracellular proteins as well. Recently, a few studies were conducted to understand the immunogenicity of Cas9 proteins, and all of them have concluded that the majority of immune cells are activated by Cas9 proteins.[21–23] In an elegant study, Charlesworth et al.[22] reported antibodies in human blood that can bind to SpCas9 and SaCas9 proteins. The antibodies can detect Cas9 in immunoblot experiments, indicating the presence of high-affinity interactions. The authors analyzed cord and peripheral blood samples from adult donors as sources of antibodies. The cord blood samples constitute immunoglobin-G (IgG) from maternal sources and are an easy way to detect IgG reactions against Cas9. A positive IgG response signifies the presence of affinity-matured antibodies in normal adult humans, and it was found that a large fraction of donors (~58–78%) possess specific antibodies for Cas9 proteins. Another study by Ferdosi et al. reported a lower number (~5%) of donors with antibody reactivity. As such, the positive donor numbers might vary depending on the immune exposure and immune system of the specific individuals tested.[24] Despite these findings, the recognition of Cas9 by antibodies may not be detrimental for therapeutic applications. However, it indicates pre-exposure of the donors to *Staphylococcus aureus* and *Staphylococcus pyrogens,* which resulted in the development of specific and affinity matured antibodies in the blood. Furthermore, this could indicate Cas9 reactive T cells in the donor as well, which might have significant consequences for the therapeutic applications of CRISPR-Cas9. B cell epitopes for Cas9 are not characterized in humans but can easily be predicted using bioinformatic tools. B cell epitope prediction analysis shows multiple peptides that can potentially be targeted by antibodies (The Immune Epitope Database-IEDB)[25] (Fig. 3, left). Therefore, several different polyclonal antibodies may target Cas9 proteins.

5.5 Human T cell response to Cas9

T cell responses are usually generated from recognizing a foreign antigen presented by classical and nonclassical major histocompatibility (MHC) like molecules.[17,18,26–28] Classical MHC I and II are expressed on the cell surface and present short peptide antigens generated by proteasomal degradation inside the cell. Antigen-presenting cells (APCs) capture the foreign antigens and process and present the antigen in association with MHC molecules. T cells recognize these antigens using recombined T cell receptors (TCRs) expressed on the cell surface. Successful T cell recognition of a

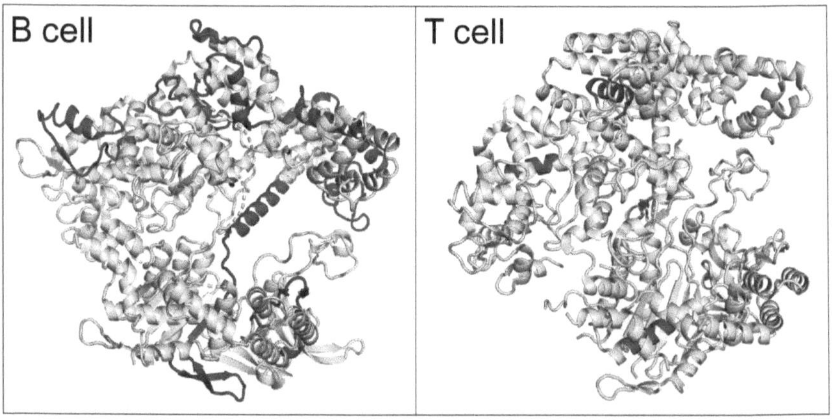

Fig. 3 Antigenic epitopes present on Cas9 protein: predicted B cell epitopes (left) and identified T cell epitopes (right). Some of the predicted B cell epitopes are showing as the color red. Four identified T cell antigens presented by HLA-A*02 shown as violet color.[24] The figures were generated using Pymol software.

particular antigen leads to activation of downstream signaling cascades, cell killing, cytokine secretion, and modulation of immune responses. $CD4^+$ T cells recognize antigens in association with MHC class II molecules and play critical functions in orchestrating the immune reaction. A unique subset of $CD4^+$ T cells, T follicular helper (T_{fh}), cells are required for the affinity maturation of antibodies in the germinal center. T regulatory cells (T_{reg}) are also crucial for the downregulation of immune responses. In any immune response, multiple immune cell populations take part together, where their functions are interconnected. Therefore, in the immune response to Cas proteins, different T cell subsets are most likely involved. Two excellent studies have delineated the reactions of different T cell subsets in humans[22,23] and reported Cas9 reactive T cells in the adult human population in high frequencies.

The Cas9 proteins can stimulate both $CD4^+$ and $CD8^+$ T cells when added directly to peripheral blood nuclear monocytes (PBMCs). The antigen-presenting cells, such as dendritic cells, macrophages, and B cells, take up protein and process it into shorter peptide fragments that are presented in association with MHC molecules on the cell surface. This technique has advantages over other activation assays as no prior information about MHC molecules and typing is required, and the entire Cas9 protein can be assayed and utilized. As a result of this whole protein method, MHC complexes will only present naturally processed antigens as the assay uses the natural cellular machinery to process the Cas9 protein. Therefore, peptide antigens generated during this process are presented and processed in a nearly physiological manner and are immunologically significant. This method also allows for the identification of different T cell markers for activation (*e.g.*, CD137 and CD154) and cytokine secretion, such as interferon-γ (IFN-γ), tumor necrosis factor-α (TNF-α), and interleukin 2 (IL-2). Further, in these studies, a high percentage (~96%) of donors showed effector T (T_{eff}) cell phenotypes and expressed one or multiple cytokines of activation as a result. The $CD4^+$ $CD137^+$ and $CD8^+$ $CD137^+$ T cell subsets secreted IFN-γ and TNF-α separately or both together, although in lower numbers for the $CD8^+$ T cells.

Based on the expression of CD45RO and chemokine CCR7, past exposure of T cells to antigens can be assessed and phenotypically classified into naïve ($CCR7^+$ $CD45RO^-$), central memory (T_{CM}, $CCR7^+$ $CD45RO^+$), effector memory (T_{EM}, $CCR7^-$ $CD45RO^+$), and terminally differentiated effector T cells (T_{EMRA}, $CCR7^-$ $CD45RO^-$). Effector-memory (T_{EM}) cells mostly present in the activated $CD4^+$ T cells, whereas T_{EM} and T_{EMRA} cells are significantly present in the activated $CD8^+$ T cells (Fig. 4). This may imply that the repetitive exposure with *Staphylococcus* sp. and Cas9 proteins to the donors and past induction of immune responses leads to memory

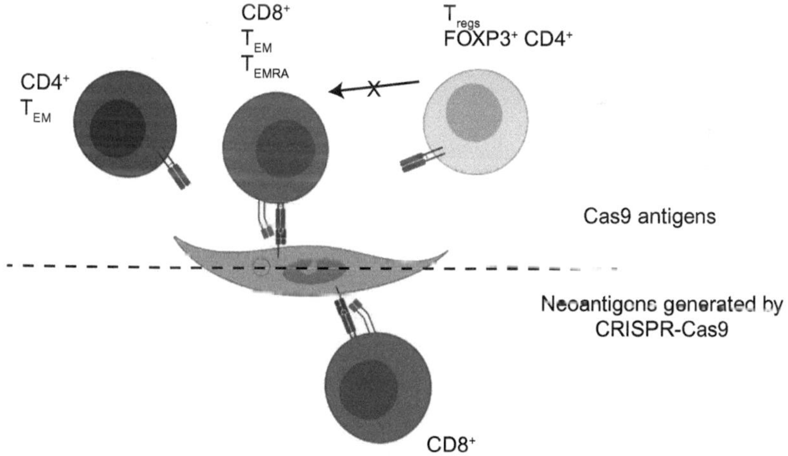

Fig. 4 T cell population in the immune reaction of Cas9 antigen and neoantigen generated by CRISPR-Cas9 editing. $CD4^+$ T_{EM} cells, and $CD8^+$ T_{EM} and T_{EMRA} cells, and $CD4^+$ $FOXP3^+$ T_{reg} cells take part in the immune reactions. T_{reg} cells downregulate the immune responses or killing of $CD8^+$ T cells (upper). The lower part indicates a possible scenario where neoantigen is generated by CRISPR-Cas9 genome editing and its recognition by T cells. The figure was partially generated using the Biorender website.

T cells. This is critical because memory T cells can be activated more easily by the Cas9 antigens during therapeutic intervention. Cytotoxic $CD8^+$ T cells can detect the Cas9 peptide presenting cells and destroy the tissue, thereby hindering any positive outcome intended with the gene therapy.

T_{reg} cells are also activated by Cas9 antigens.[23] It was demonstrated that *Staphylococcus* antigenic mixes could induce T_{reg} cell activation in humans.[29] This is an interesting finding as T_{reg} cells are developed from the thymus but suggest that they may be differentiated in the periphery also. This is surprising because T_{reg} cells mostly recognize host self-peptide antigens presented by MHC-II molecules and function to protect host tissues from unwanted immune activation. However, it has been documented that T_{reg} cells can be developed by repetitive exposure of antigens to control the inflammatory responses and play roles in containing microorganisms.[30]

Beyond this, it has been shown that T_{reg} cells can be expanded and recognized by foreign pathogenic peptides,[31] where T_{reg} cells play essential roles in the development of homeostatic mechanisms with human microbiota.[30] Thereby, frequent exposure to Staphylococcus sp. may contribute to the expansion of Cas9 recognizing T_{reg} cells. Another possible explanation would be that these T_{reg} cells are cross-reactive to Cas9 antigens and can recognize more than one antigen. Wagner et al.[23] found a significant number of Cas9 reactive $CD4^+$ cells are T_{reg} ($CD25^+$ $CD137^+$ $Foxp3^+$) (Fig. 4). However, these cells do not contribute to the cytokine production with the T effector cells. Moreover, they can inhibit the activation and cytokine production of $CD4^+$ effector cells, but only to a smaller extent in $CD8^+$ effector cells.

Ferdosi et al.[24] identified Cas9 peptides presented by HLA-A*02 molecule in humans. HLA-A*02 is one of the most common classical MHC class I molecules and alleles in humans and is recognized by $CD8^+$ T cells. Four Cas9 peptides were identified to be presented by the HLA-A*02 molecule (Figs. 3 and 5). Two Cas9 peptides show potent T cell activation over the other two. MHC molecules have an antigen-binding domain with pockets that can bind to specific amino acid motifs. The peptide's affinity to MHC depends on the anchoring residues present on the peptide (Fig. 5). The anchoring residues can be mutated to disrupt the binding with MHC to engineer an immunologically silent Cas9. However, there are many MHC molecules in humans, and it may not be easy to mutate all the epitopes necessary to ensure there is no antigenic priming for another MHC complex. Some of the Cas9 residues which are antigenic may not replaceable as they might serve essential roles in Cas9 DNA recognition and cleavage. However, some of the most common epitopes can be altered, as reported.

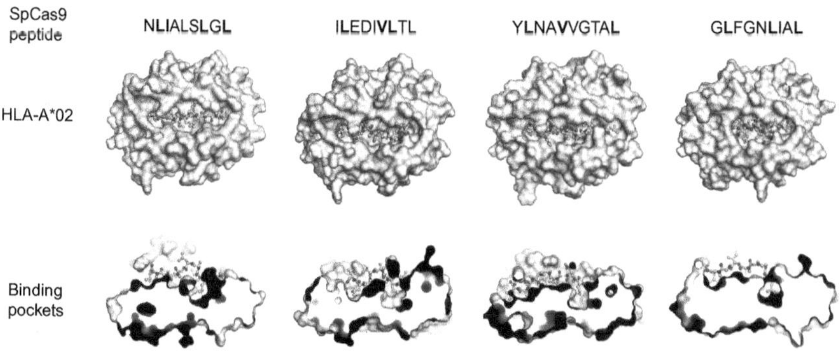

Fig. 5 Cas9 peptide antigens and their presentation by HLA-A*02 molecule. The four antigenic SpCa9 peptides are shown with the anchoring residues (bold). The top-down view of the complex model of HLA-A*02 with the peptides (middle). The models are generated by docking the respective peptide antigen into HLA-A*02 using the GalaxyPepDock server.[32] HLA-A*02 is shown as white surface, and peptides are shown as ball and stick in yellow, red, and blue colors. The anchoring residues of each peptide are shown in the slice structure. The figures were generated using Pymol software.

5.6 Neoantigen generated by CRISPR-Cas9 editing

The generation of neoantigens due to genome editing is a rare but very real possibility. However, it may still be useful to select the amino acid and analyze the corrected peptide for antigen presentation and T cell recognition. T cells undergo positive and negative selections in the thymus and are educated to mature form. T cells are exposed to all the possible self-antigens in the development process, and negative selection deletes all strongly self-reacting T cells. Otherwise, self-recognition can lead to a variety of autoimmune diseases. Therefore, in normal circumstances, T cells do not recognize host cell peptides presented by the numerous MHC molecules. However, mutations in genes can alter the peptide repertoire presented by MHC molecules. In cancerous growth, mutations generate modified self-peptides (termed neoantigens) and typically result from missense mutations, insertions, deletions, and frameshift mutations[33–35] (Fig. 6). T cells can recognize these mutations and elicit anti-cancer immune responses. Therefore, neoantigens are being targeted for cancer therapeutics.[36,37]

Similarly, somatic CRISPR-Cas9 editing can and likely generate mutant peptides that could potentially act as neoantigens for T cells. T cells may eliminate those cells expressing the modified gene from the body. However, mutation to edit the genome in diseases like sickle cell anemia (glutamate to valine substitution) in red blood cells (RBC) may not be immunogenic

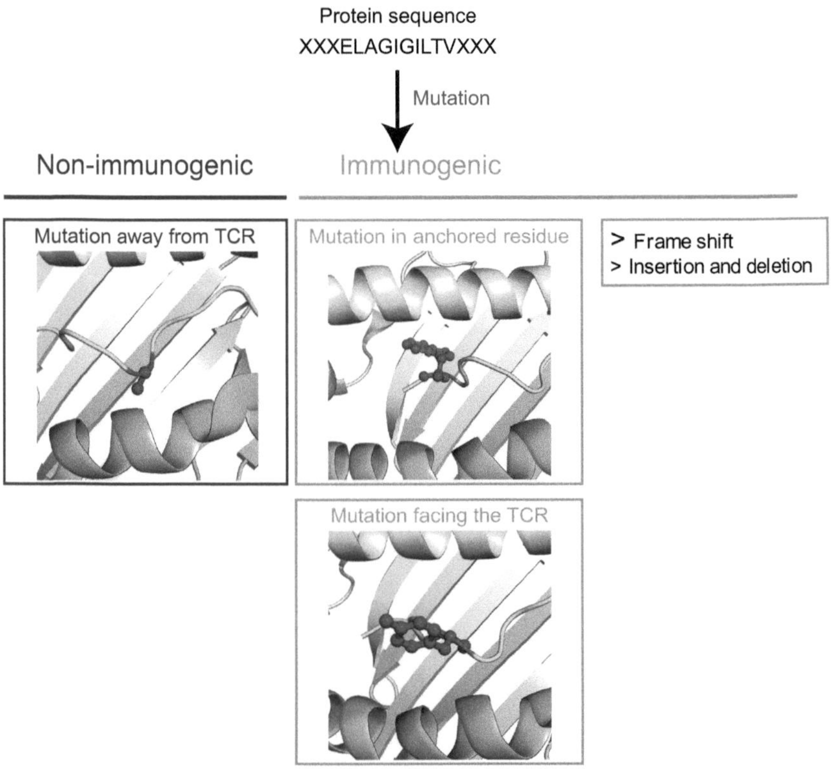

Fig. 6 Generation of neoantigens and their presentation by MHC molecules. Neoantigens can be non-immunogenic (left) or immunogenic (middle and right) based on the mutated amino acid residues. TCRs can recognize any conformational changes in the peptide arising from a mutation in the anchoring or exposed residues. The frameshift, insertion, and deletion generate new epitopes, thereby modulating immune recognition.

as matured RBCs do not express MHCs. However, it may be immunogenic in other cells expressing MHC molecules and potentially lead to unintended side effects.

6. Future direction and strategies to deal with immunogenicity

The effect of the immune response to CRISPR-Cas proteins is understood *in vitro* conditions in humans. However, based on mice and human *in vitro* studies, Cas9 protein may show immune activations. Both T cells and B cells will recognize Cas9 antigens and activate $CD4^+$, $CD8^+$, and

T_{reg} cells. However, antibody response may not affect the therapeutic efficacy of the CRISPR-Cas technology. On the contrary, T cell response may influence the outcome of the effectiveness. The $CD8^+$ T cells can eliminate the Cas9 expressing cells. Neoantigens generated by CRISPR-Cas9 genome editing can be recognized by T cells and induce immune activation as well. There are a few strategies that can be employed to tackle the immune activations and safe treatment. Anti-inflammatory or immunomodulatory drugs can be useful to downregulate the immune responses toward CRISPR-Cas. The T_{reg} cells inhibit T cell activation; therefore, T_{reg} cells activating peptides may help control the overall immune reaction. T_{reg} peptides can be administered to increase the Cas9 specific T_{reg} in the host. However, more studies will be required to understand the safety and effectiveness of the procedures. Immunodominant peptide epitopes and antigens present in Cas9 proteins can be mutated without altering the furcation to engineer an optimized CRISPR-Cas system.

7. Conclusion

CRISPR-Cas technology holds immense potential to treat genetic disorders in humans, which are not possible using available technologies. However, genome editing is associated with significant off-target effects and maybe give rise to unforeseen side effects. The human immune response can play vital roles in successfully implementing the *in vivo* CRISPR-Cas gene editing. Staphylococcus sp. are part of human microbiota and opportunistic pathogen to humans, and many people get infected every year around the world. Significant human populations showed pre-existing immunity to Cas9 proteins. *In vitro* studies showed that antibodies and T cells could recognize Cas9 antigens and start immune responses. However, it may need more research to understand the immune responses *in vivo* conditions.

Acknowledgment

I am grateful to Prof. Erin J. Adams for helpful discussions and guidance. I am thankful to Dr. Kristof Nolan for his comments and feedback on the manuscript. I am also thankful to Adams lab members for discussion. Finally, the literature available on CRISPR which are not cited.

References

1. Urnov FD, Rebar EJ, Holmes MC, Zhang HS, Gregory PD. Genome editing with engineered zinc finger nucleases. *Nat Rev Genet.* 2010;11:636–646.
2. Joung JK, Sander JD. TALENs: a widely applicable technology for targeted genome editing. *Nat Rev Mol Cell Biol.* 2013;14:49–55.

3. Mehta A, Merkel OM. Immunogenicity of Cas9 protein. *J Pharm Sci*. 2020;109:62–67.
4. Stadtmauer EA, Fraietta JA, Davis MM, et al. CRISPR-engineered T cells in patients with refractory cancer. *Science*. 2020;367(6481):eaba7365. https://doi.org/10.1126/science.aba7365. Epub 2020 Feb 6. PMID: 32029687.
5. Doudna JA, Charpentier E. Genome editing. The new frontier of genome engineering with CRISPR-Cas9. *Science*. 2014;346:1258096.
6. Jiang F, Doudna JA. CRISPR-Cas9 structures and mechanisms. *Annu Rev Biophys*. 2017;46:505–529.
7. Jinek M, Jiang F, Taylor DW, et al. Structures of Cas9 endonucleases reveal RNA-mediated conformational activation. *Science*. 2014;343:1247997.
8. Nishimasu H, Ran FA, Hsu PD, et al. Crystal structure of Cas9 in complex with guide RNA and target DNA. *Cell*. 2014;156:935–949.
9. Mougiakos I, Mohanraju P, Bosma EF, et al. Characterizing a thermostable Cas9 for bacterial genome editing and silencing. *Nat Commun*. 2017;8:1647.
10. Nishimasu H, Cong L, Yan WX, et al. Crystal structure of Staphylococcus aureus Cas9. *Cell*. 2015;162:1113–1126.
11. Fu Y, Foden JA, Khayter C, et al. High-frequency off-target mutagenesis induced by CRISPR-Cas nucleases in human cells. *Nat Biotechnol*. 2013;31:822–826.
12. Pattanayak V, Lin S, Guilinger JP, Ma E, Doudna JA, Liu DR. High-throughput profiling of off-target DNA cleavage reveals RNA-programmed Cas9 nuclease specificity. *Nat Biotechnol*. 2013;31:839–843.
13. Kleinstiver BP, Pattanayak V, Prew MS, et al. High-fidelity CRISPR-Cas9 nucleases with no detectable genome-wide off-target effects. *Nature*. 2016;529:490–495.
14. Kingry LC, Petersen JM. Comparative review of Francisella tularensis and Francisella novicida. *Front Cell Infect Microbiol*. 2014;4:35.
15. Supriya BG, Harisree S, Savio J, Ramachandran P. Actinomyces naeslundii causing pulmonary endobronchial Actinomycosis—a case report. *Indian J Pathol Microbiol*. 2019;62: 326–328.
16. Lowy FD. Staphylococcus aureus infections. *N Engl J Med*. 1998;339:520–532.
17. Roy S, Ly D, Li NS, et al. Molecular basis of mycobacterial lipid antigen presentation by CD1c and its recognition by alphabeta T cells. *Proc Natl Acad Sci U S A*. 2014;111: E4648–E4657.
18. Greene JM, Dash P, Roy S, et al. MR1-restricted mucosal-associated invariant T (MAIT) cells respond to mycobacterial vaccination and infection in nonhuman primates. *Mucosal Immunol*. 2017;10:802–813.
19. Chew WL, Tabebordbar M, Cheng JK, et al. A multifunctional AAV-CRISPR-Cas9 and its host response. *Nat Methods*. 2016;13:868–874.
20. Crudele JM, Chamberlain JS. Cas9 immunity creates challenges for CRISPR gene editing therapies. *Nat Commun*. 2018;9:3497.
21. Simhadri VL, McGill J, McMahon S, Wang J, Jiang H, Sauna ZE. Prevalence of pre-existing antibodies to CRISPR-associated nuclease Cas9 in the USA population. *Mol Ther Methods Clin Dev*. 2018;10:105–112.
22. Charlesworth CT, Deshpande PS, Dever DP, et al. Identification of preexisting adaptive immunity to Cas9 proteins in humans. *Nat Med*. 2019;25:249–254.
23. Wagner DL, Amini L, Wendering DJ, et al. High prevalence of Streptococcus pyogenes Cas9-reactive T cells within the adult human population. *Nat Med*. 2019;25;242–248.
24. Ferdosi SR, Ewaisha R, Moghadam F, et al. Multifunctional CRISPR-Cas9 with engineered immunosilenced human T cell epitopes. *Nat Commun*. 2019;10:1842.
25. Vita R, Mahajan S, Overton JA, et al. The immune epitope database (IEDB): 2018 update. *Nucleic Acids Res*. 2019;47:D339–D343.
26. Roy S, Ly D, Castro CD, et al. Molecular analysis of lipid-reactive Vdelta1 gammadelta T cells identified by CD1c tetramers. *J Immunol*. 2016;196:1933–1942.

27. Benveniste PM, Roy S, Nakatsugawa M, et al. Generation and molecular recognition of melanoma-associated antigen-specific human γδ T cells. *Sci Immunol.* 2018;3(30): eaav4036. https://doi.org/10.1126/sciimmunol.aav4036. PMID: 30552102.
28. Adams EJ, Luoma AM. The adaptable major histocompatibility complex (MHC) fold: structure and function of nonclassical and MHC class I-like molecules. *Annu Rev Immunol.* 2013;31:529–561.
29. Bacher P, Heinrich F, Stervbo U, et al. Regulatory T cell specificity directs tolerance versus allergy against aeroantigens in humans. *Cell.* 2016;167:1067–78 e16.
30. Belkaid Y, Tarbell K. Regulatory T cells in the control of host-microorganism interactions (*). *Annu Rev Immunol.* 2009;27:551–589.
31. Shafiani S, Dinh C, Ertelt JM, et al. Pathogen-specific Treg cells expand early during mycobacterium tuberculosis infection but are later eliminated in response to Interleukin-12. *Immunity.* 2013;38:1261–1270.
32. Lee H, Heo L, Lee MS, Seok C. GalaxyPepDock: a protein-peptide docking tool based on interaction similarity and energy optimization. *Nucleic Acids Res.* 2015;43: W431–W435.
33. Schumacher TN, Schreiber RD. Neoantigens in cancer immunotherapy. *Science.* 2015;348:69–74.
34. Schumacher TN, Scheper W, Kvistborg P. Cancer neoantigens. *Annu Rev Immunol.* 2019;37:173–200.
35. Gros A, Parkhurst MR, Tran E, et al. Prospective identification of neoantigen-specific lymphocytes in the peripheral blood of melanoma patients. *Nat Med.* 2016;22:433–438.
36. Ott PA, Hu Z, Keskin DB, et al. An immunogenic personal neoantigen vaccine for patients with melanoma. *Nature.* 2017;547:217–221.
37. Sahin U, Derhovanessian E, Miller M, et al. Personalized RNA mutanome vaccines mobilize poly-specific therapeutic immunity against cancer. *Nature.* 2017;547:222–226.

CHAPTER NINE

Application of CRISPR-Cas systems in neuroscience

Deepro Bonnerjee* and Sangram Bagh
Biophysics and Structural Genomics Division, Saha Institute of Nuclear Physics, Homi Bhabha National Institute (HBNI), Kolkata, India
*Corresponding author: e-mail address: deepro.bonnerjee@saha.ac.in

Contents

Abstract

CRISPR-Cas systems have, over the years, emerged as indispensable tools for Genetic interrogation in contexts of clinical interventions, elucidation of genetic pathways and metabolic engineering and have pervaded almost every aspect of modern biology. Within this repertoire, the nervous system comes with its own set of perplexities and mysteries. Scientists have, over the years, tried to draw up a clearer genetic picture of the neuron and how it functions in a network, mainly in an endeavor to mitigate diseases of the human nervous system like Alzheimer's, Parkinson's, Huntington's, Autism Spectrum Disorder (ASD), etc. With most being progressive in nature, these diseases have plagued mankind for centuries. In spite of our immense progress in modern biology, we are yet to get a grasp over these diseases and unraveling their mechanisms is of utmost importance. Before CRISPR-Cas systems came along, the elucidation of the complex interactome of the mammalian nervous system was attempted with erstwhile existing electrophysiological, histological and pharmacological techniques coupled

Progress in Molecular Biology and Translational Science, Volume 178
ISSN 1877-1173
https://doi.org/10.1016/bs.pmbts.2020.12.010

with Next Generation Sequencing and cell-specific targeting technologies. Zinc finger nucleases (ZFNs) and transcription activator-like effector nucleases (TALENs), imparted excellent sequence specific DNA targeting capabilities but came with their huge baggage of extensive protein engineering requirements, which practically rendered them unsuitable for high throughput exercises. With the discovery of Clustered Regularly Interspaced Palindromic Repeats (CRISPR) and CRISPR Associated proteins(CAS) systems by Ishino (1987)[1], the era of extensive custom made endonuclease targeting was ushered in. For the first time in 2012, Jinek et al. (2012)[2] repurposed the CRISPR-Cas mediated bacterial immune system for customizable mammalian gene editing. The CRISPR-Cas technology made it possible to easily customize Cas9 endonucleases to cleave near specifically targeted sequences, thereby facilitating knock-ins or knock-outs, silencing or activating or editing any gene, at any locus of the genome, both at the base-pair level or at the epigenetic level. With this enhanced degree of freedom, decrypting the nervous system and therapeutic interventions for neuropathies became significantly less cumbersome an exercise. Here we take a brisk walk through the several endeavors of research that show how the humble bacteria's CRISPR-Cas system gave us the "nerves" to "talk" to our nerves with ease.

1. A basic overview of the CRISPR-Cas system

Following the first observation of the CRISPR-CAS system by Ishino in 1987[1] several variations and types of the system have been discovered which varied in their Nuclease composition and mechanism of rendering anti-viral immunity to the bacterium. The systems are broadly classified into two classes namely, Class I—where the Nuclease is composed of multiple sub-units and Class II that are made of a single unit. Class II Nucleases are further categorized into Type II, V and VI. Types II and V respectively represent DNA targeting Cas9 and Cas12a while Type VI is a RNA targeting Cas13[3]. Jinek and group[2] were the first, in 2012, to repurpose the CRISPR-Cas system from *Streptococcus pyogenes* toward mammalian gene editing. spCas9, as it was termed (based on its origins in *Streptococcus pyogenes*), is by far the most commonly employed Cas variant. The CRISPR-Cas9 systems works in an RNA directed manner where a CRISPR RNA (crRNA) functions as a guide to direct the Cas9 protein toward a target sequence through base complementarity. A second RNA component, the trans-activating RNA (tracr RNA) functions in stimulating endonuclease activity besides a role in pre-crRNA processing and maturation. When a bacterium gets infected by a virus, the fragments of the genetic material of the virus gets incorporated in between certain repetitive sequences in the bacterial or archaeal genome that form the array of Clustered Regularly Interspaced short Palindromic repeats. These regions between the repeat sequences are called spacers and within each spacer region in between the repeat sequences lies a unique DNA marker from a previous DNA viral infection in the cell. This array forms a database for all previous DNA viral infections that the host has been challenged with. Upon a repeat infection from a virus, this entire CRISPR array gets transcribed along with the repeat sequences to form the pre-crRNA. Cas9 protein, encoded upstream of the CRISPR region conjugates with a tracrRNA that goes on to bind to the Repeat sequence regions and cleaves at the 3′ prime end of the same. This cleavage results in multiple fragments each carrying a Viral marker sequence toward the 5′ prime end with the 3′ segment bound to the tracrRNA and the Cas9 protein. The free 5′ end seeks out complementary DNA sequences within the cell which are presumably from a viral infection. The endonuclease activity of the Cas9 protein then generates a double strand blunt end cut to the complementary DNA strand about 3 bases upstream of the cognate sequence thereby disabling it.

Here the question that arises is, how the Cas9-tracrRNA-crRNA complex differentiates between the cell's own CRISPR library and that of foreign DNA material? This is facilitated by small (usually 2–6 nucleotides long) Protospacer Associated Motifs (PAMs) which flanks the complementary DNA signatures in a virus. The host cell chooses the signature regions of a virus from an approximately 20 bp region flanked by specific PAM sequences which are not incorporated within the CRISPR motif. The presence of PAMs signals a foreign strand while their absence signals "self." This entire process of viral DNA fragment integration to Cas9 mediated disabling is orchestrated by a group of CAS proteins situated at the upstream of the CRISPR array. Cas9 is one of them. Cas9 orthologs from different species differ in their specific Pam sequences. For example, for spCas9 the PAM sequence is NGG[2]. SaCas9 (from *Staphylococcus aureus*) recognizes the PAM sequence 5′-NNGRR(N)-3′[4]. Nme-Cas9 from Neisseria meningitides recognizes 5′-NNNNGATT-3′ as PAM[5].

The two distinct RNA components, tracrRNA and crRNA was later engineered and combined in to a chimeric sgRNA of about 100 nt long, making the entire system a simple two-component machinery. Manually defining the sgRNA to look for specific sequences repurposes the system to target any desired locus in a genome. When Non Homologous End Joining (NHEJ) tries to repair the double strand break created by Cas9, the error prone repair process introduces indels. When within the coding region of a gene this renders the gene knocked out[2]. NHEJ being the predominant repair mechanism in post-mitotic cells like Neurons knocking out genes with CRISPR cas9 becomes a fairly simpler process. Alternatively, if one wants to edit or replace a particular DNA sequences with a corrected one, Homology Directed Repair (HDR) mediated via a donor sequence supplied externally serves as the template[6]. However, HDR is limited to dividing cells only and gene editing in neurons becomes a bit more complicated process.

The second type of DNA targeting Cas endonuclease is the Type V, Cas12a mentioned previously. The most remarkable of all is its ability to process its own crRNA without a tracrRNA or additional enzymes. This allows for a much shorter sgRNA of about 40 nt as opposed to 100 nt for cas9. This proves particularly useful when multiplexed targeting of genes is required. These marked differences poses cas12a as more advantageous over Cas9. In 2019, Campa and group[7] exploited its properties to target 20 different loci from a single vector containing multiple crRNAs and one cas12a. It recognizes a PAM 5′-TTTN-3′ and generates staggered cuts.

Depending on the target the experimenter wants to achieve different types of CAS proteins, natural or engineered offer its unique assortment of functionalities. The design of sgRNA varies concomitantly. Besides preparations of this "toolkit," the delivery and expression mechanism of the system greatly affects the efficiency of the system often being context specific. In the next section we look into how researchers have tried to perfect and better these mechanisms in attempts to disrupt genes to better understand the nervous system.

2. Delivery and expression mechanisms in applications of CRISPR-Cas9 based disruption of genes in the mammalian CNS

The way to unravel the functions of a gene is to disrupt its expression and look for phenotypical changes. The logistical difficulties associated with the traditional method of gene disruption through germline editing, involved limitations of multi gene targeting, time consumption and generation of prohibitive phenotypes. RNAi based approaches have shown considerable off-target activities and leakages that hinders complete knock-out. CRISPR-Cas9, by its nature, excellently surpasses these limitations and aids the process as discussed before. In 2014 Incontro et al.[8] tested the system by targeting two synaptic proteins, GluN1 subunit of the NMDA receptor and the GluA2 subunit of the AMPA receptor using the CRISPR-Cas9 machinery. They observed complete elimination of GluA2 expression in their hippocampal slice cultures. SpCas9 gene has a large size of about 4.1 kb and hence vectors face the problem of packaging limitations when delivering it with a library of sgRNA. With AAV having a maximum packaging limit of 5 kb, it gets infeasible to incorporate the sgRNA array into the same vector that carries spCas9. One solution is to co-transfect cells with two separate vectors, one having the spcas9 gene and the other containing the sgRNA library. The other way is to develop transgenic mice with Cas9 integrated into the genome.

Swiech et al.[9] in 2015 attempted to interrogate single Mecp2 (methyl CpG binding protein 2) gene as well as multiplexed targets Dnmt1, Dnmt3a and Dnmt3b using an Adeno-associated Virus (AAV) mediated adult intra-cranial delivery using spCas9. They addressed the payload size issue by implementing co-transfection of cells with separate viri carrying spCas9 and the sgRNAs. Characterization of the effects through electrophysiological, genetic, biochemical and behavioral analyses, revealed morphological defects brought on by MeCP2 knock-down. Knock-down of DNMTs in the dentate gyrus of mice brains exhibited context specific freezing deficits post fear conditioning exercises leaving other behavioral parameters like object recognition and open field performances intact.

Earlier in 2014, Platt. et al.[10] addressed the question of developing a reliable delivery vehicle to stably incorporate the spCas9 gene into the genome. Stable genomic integration alleviates loads of hindrances faced while delivery owing to the large size of spcas9. They used the Rosa26 locus to introduce a Cre-dependent spCas9 gene using AAV. The Cre mechanism imparts conditional Cre-recombinase inducible expression of spCas9. Cre-recombinase upon expression flips an oppositely oriented DNA segment into the forward orientation thereby switching it on. They also tried lentiviral and particle based delivery machineries in neurons among other cell types. Besides creating the first transgenic they also established the lack of any toxicity from Cas9 from prolonged expression. They demonstrated high Knock-Out frequencies for the NeuN gene.

The cre-dependent expression regime was also employed by Bäck et al.[11] who used germline modified transgenic rats. Elucidating the Genetics of neuronal functions, development and decay extensively requires use of genetically modified animals. CRISPR-Cas9 mechanisms smartly and greatly reduces the labor and time intensive process developing transgenic animals. It imparts capabilities of simultaneous and multiple site specific gene insertions

(Knock-ins), gene edits and knock-outs with significant ease. Toward neuroscience and engineering, rats are preferred over mice to study behavioral phenotypes. Unfortunately, transgenic rat models are far fewer in variety as well as numbers, when compared to mice models. The CRISPR-Cas9 machinery came as a boon to the neuroscientist as it could rapidly generate a plethora of genetically modified rat models. Rat models that expressed Cre-dependent Cas9 and Cre-dependant Cas9 nickase were created by Bäck and group who reported an improved version of the Cre-recombinase induced (iCre) CRISPR cas9 system that harbored neural cell type specificity in its expression regime.

Bäck et al. tried to work through the disease mechanism and modeling of Parkinson's disease. Let's have a look at the background of their work. Tyrosine hydroxylase (TH) catalyzes the formation of L-dopa, a precursor to the biosynthesis of dopamine. In Parkinson's disease, the progressive degeneration of dopaminergic neurons in the subsantia nigra causes dopamine levels to drop initiating phenotypic implications. L-dopa formation being the rate limiting step in dopamine synthesis (Fig. 1A), knock-down of tyrosine hydroxylase can be expected to mimic a Parkinson's phenotype. Bäck et al. targeted dopaminergic neurons where the Cre recombinase would express under a dopamine transporter promoter that natively controls the rat dopamine transporter that sequesters dopamine to the synaptic cleft (Fig. 1A). The design leads to the expectation that the system should express only in dopaminergic neurons. The rat-line thus generated, DAT-iCre-Cas9, targeted cas9 toward tyrosine hydroxylase (TH) in the dopaminergic neurons in the Substantia Nigra of the brain implicated in Parkinson's disease. AAV mediated infusion resulted in 45% reduction in TH levels as observed through immune-reactivity. 60% reduction was also observed upon infusion into the striatum.

The iCre-Cas9 system's cell lineage specific expression was tested further with sgRNAs that were designed against ManF (Mesencephalic Astrocyte Derived Neurotrophic Factor) and the machinery injected into the midbrain. Situated in the ER and Golgi complexes ManF is important for the survival of Dopaminergic Neurons in the midbrain[12]. Its knock-down is known to induce ER-stress induced cell death[13]. Analysis revealed almost no dopaminergic neuron showing ManF immunoreactivity (only 3%) whereas up to 90% of non-dopaminergic neurons being ManF positive. Animal models expressing such neural lineage specific Cas9 expression, presents a huge benefit to neuroscientific research.

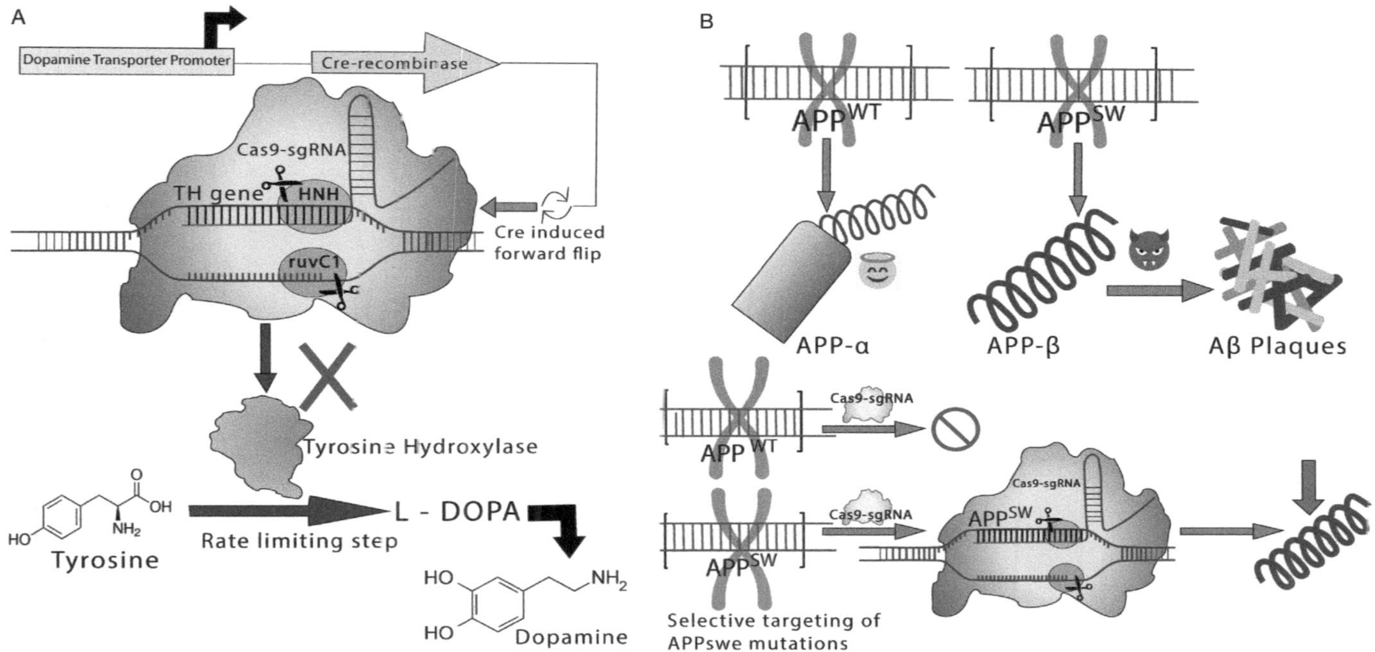

Fig. 1 (A) Cre-dependent CRISPR-cas9 targeting of tyrosine hydroxylase (TH) gene to downregulate dopamine levels. Dopaminergic neurons were targeted in a Cre-recombinase dependent manner where the Cre machinery was being expressed from the Dopamine transporter promoter. Cas9–sgRNA was targeted toward Tyrosine Hydroxylase gene. Tyrosine hydroxylase catalyzes the formation of the dopamine precursor L-Dopa from Tyrosine which is the rate limiting step in Dopamine synthesis. Inhibition of Tyrosine hydroxylase down-regulates Dopamine levels in a bid to mimic a Parkinson's disease model. (B) APP^{SW} represents a mutation of APP^{WT} (wild type) that renders a higher susceptibility to the disease. CRISPR-Cas9 mediated intervention displayed allele-specific targeting APP^{SW} over APP^{WT}. Significant and selective reduction was observed in indel-free APP^{SW} as compared to indel-free APP^{WT}.

Besides Parkinson's disease that stands as one of the most formidable neuropathologies, Alzheimer's disease also continue to remain in the shady areas of human understanding let alone strong therapeutic successes. Besides its sporadic form, Alzheimer's disease is also associated with mutations in the Amyloid precursor protein (APP) genes, presenilin 1 (PSEN1), and presenilin 2 (PSEN2) which triggers beta-amyloid peptide production. György et al.[14] demonstrated a CRISPR mechanism to selectively target the APP^{SW} allele that carries a mutation APPswe (indigenous to Sweden and causing increased Aβ levels) rendering susceptibility to the disease. The group harvested human fibroblasts that carried such mutations and used CRISPR-Cas9 to selectively target APP^{SW} against APP^{WT} in human $APP^{\mathrm{SW/WT}}$ fibroblasts (Fig. 1B). Significant and selective reduction was observed in indel-free APP^{SW} as compared to indel-free APP^{WT}. Further, indels were only observed in APP^{SW} in in vivo observations by the group, while targeting the hippocampus of Tg2576 mice carrying multiple copies of the APPswe mutation, were also in agreement with the in vitro studies. While this is a promising outcome toward mitigation of Alzheimer's, efficiency reported in this work was quite low.

As a novel therapeutic target for Alzheimer's disease, Raikwar et al.[15] targeted Glia Maturation factor (GMF). Earlier it was found by the same group that GMF plays an important part in Alzheimer's pathogenesis by virtue of being a pro-inflammatory molecule. Microglia playing a part in neuro-inflammation and being particularly reactive around Aβ plaques GMF is highly expressed in the AD brain and promotes neurodegeneration. Raikwar et al. down-regulated GMF expression using CRISPR-cas9 based intervention employing saCas9, the smaller ortholog of cas9. AAV mediated transduction of BV2 microglial cell line introducing the constructs showed successful reduction in GMF expression albeit with low efficiencies.

It has long been agreed that understanding one's enemy well is about half the battle won. CRISPR/Cas9 based gene disruptions systems, such as the ones discussed above, goes a long way toward elucidating disease mechanisms in a more efficient, multiplexed and target specific systematic manner. However introducing the system into the battlefield is a prerequisite and the ones we have discussed so far employ viral vectors for the job. Viral transductions come with their own set of problems where one needs to look out for AAV induced immunogenicity, their tropism toward certain cell types and most notably their propensity for off-target activity. In the next section we look into non-viral delivery methods that aim to improve upon some of these issues.

2.1 Non-viral delivery and expression mechanisms

Virus mediated deliveries have exhibited several drawbacks over the years. The most notable of them is Off-target activities. One should note that Lentiviral or Adenoviral deliveries result in prolonged gene expression with AAV showing transgene expression several years after infection. Such prolonged expression may not always be required for example after gene-knock-out has already been effected. CRISPR-Cas9 systems have been observed to show significant off-target activities in prolonged expression which can be explained by the fact that upon exhaustion of target-sites, excess Cas9 complexes have only off-target sites to act upon[16]. Besides the above, viral delivery machineries suffer from other issues like tissue specificity and tropism and in some cases immunogenicity.

Thereafter alternative delivery routes resulting in transient expression was the need of the moment. Staahl and co-workers in 2017[17] addressed this by engineering tissue deliverable nanoparticles to target parts of the brain. The nanoparticles packaged within them Ribonucleoprotein complexes (RNPs)-that are Cas9 proteins already conjugated to the required sgRNA. Earlier in 2016, Chew and his group[18] reported an anti-cas9 immune response in post-natal mice. Reinforcing the claim of safety of the nanoparticle based route of delivery, Staahl et al. reported no significant immunogenicity toward Cas9 in mice arising from dorsal striatum injections of the nanoparticles. These RNPs were linked with a SV40 Nuclear Localization signal that results in enhanced penetrance[19]. The nanoparticle encapsulated package was administered into the V1 primary visual cortex, the hippocampus, dorsal striatum and the S1 primary somatosensory cortex of mice. Expression was also found to be tunable in a RNP-nanoparticle dose dependant manner as reflected by reporter proteins.

In a bid to reduce Aβ plaques to mitigate Alzheimer's disease, CRIPSR delivery was improved further by Park et al. in 2019[20]. They incorporated an amphiphilic peptide R7L10 to the CRISPR RNP complexes (Fig. 2A). The amphiphilic nature of the peptide facilitated the thus generated CRISPR nano-complexes to enter the cell with comparative ease. The RNPs here targeted the β-secretase 1 (Bace1) gene that causes Aβ42 secretion upon differential cleaving of Amyloid precursor protein and ultimately the dreaded Aβ plaques. The CRISPR nano-complexes were injected into the hippocampal CA3 region of 5XFAD mice and achieved a 45% efficiency resulting in a significant reduction in Aβ plaques. CRISPR-Cas9 mediated Bace1 knock-down persisted strongly for about a week post-delivery and slowly

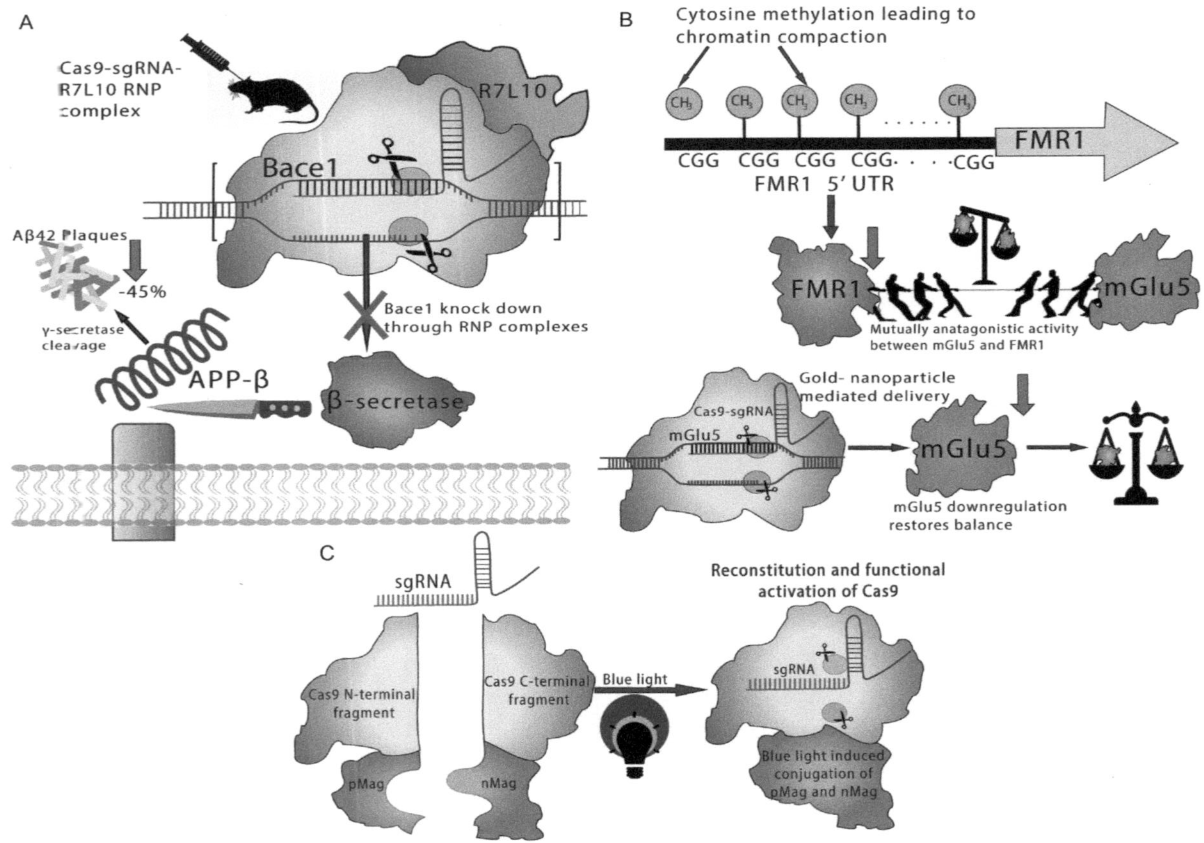

Fig. 2 See figure legend on opposite page.

tapered off in about 3 week time. The work demonstrated that the phenotypical changes including improved contextual and associative memory persists until about 3 months post treatment.

Going beyond Alzheimer's and Parkinson's, Fragile X syndrome is another disease with effectively no prognosis and associated with mental retardation and other physical manifestations especially pronounced in males. It is initiated by epigenetic silencing of the FMR1 gene. In fragile X-syndrome (FXS) the FMR1 gene (Fragile X mental retardation 1) contains an elongated CGG repeat. The FMR1 (situated on the X chromosome) gene contains a CGG repeat at its 5′ UTR which ranges between 5 and 40 usually. In FXS this expands to greater than 200 which attracts excessive methylation at Cytosine residues. This in turn promotes excessive condensation of the chromatin at the locus resulting in shutting down of the gene. FMR1 has been known to antagonistically regulate the Translational activity at the synaptic regions of neurons with mGlu5. With FMR1 shut off, mGlu5 regulation/moderation is lost. In 2018 Lee. et al.[21] tried to restore the balance by targeting mGlu5 through Cas9-RNP mediated knock-down (Fig. 2B). They employed a Gold-nanoparticle based delivery route that resulted in a 50% decrease in protein levels and respectable reduction in FXS behavioral phenotype.

Fig. 2 (A) β-secretase 1 (Bace1) initiates the formation of Aβ plaques by virtue of its differential cleaving of APP. Ribonucleoprotein complexes (sgRNA bound to cas9) were conjugated with R7L10 peptide whose amphiphilic nature encouraged cellular entry. The conjugated RNP complexes were injected into the hippocampal CA3 region of 5XFAD mice and successful cellular entry demonstrated a virus independent delivery mechanism. sgRNAs were targeted to β-secretase 1 (Bace1) that initiates the formation of Aβ plaques by virtue of its differential cleaving of APP. Successful targeting was indicated by 45% reduction of Aβ42 levels. (B) FMR1 and mGlu5 function in a mutually antagonistic manner to regulate translational activity at the synaptic regions of neurons. In Fragile X syndrome, the 5′ UTR region of the FMR1 gene contains excessive CGG repeats that attract methylation of the site resulting in epigenetic silencing of FMR1. With lower leves of FMR1, mGlu5 activity goes off balance and unregulated. In a demonstration of gold nanoparticle mediated delivery of sgRNA-Cas9 ribonucleoprotein complexes, sgRNAs were designed to target mGlu5 in an attempt to reduce mGlu5 levels in the cell and thereby restore balance. (C) Light mediated control of cas9 activity. Cas9 protein was split into two fragments and each fragment was tethered to one of two photosensitive protein components. These photosensitive proteins that had been derived from fungus, had the property to dimerize with each other upon irradiation with Blue light owing to their electrostatic properties. Dimerization of the two proteins called the "Negative" and "Positive" "magnets" (nMag and pMag) would in turn result in a reconstitution of the two fragments of cas9 bringing it to an active form.

Gee et al.[22] in 2020, demonstrated another innovative form of non-viral delivery of CRISPR-Cas components wherein Ribonucleoproteins or RNPs were delivered into iPSC-derived cortical neurons by means of something called NanoMEDIC (nanomembrane-derived extracellular vesicles for the delivery of macromolecular cargo). Extra cellular vesicles are typically composed of the viral envelope and/or structural proteins sans the nucleic acid genetic material of the virus. The corresponding proteins are overexpressed from producer cells to assemble the envelope or vesicle. The target payload gets packaged into the vesicle by dimerization. Gag, the structural polyprotein from retroviruses, has served as the ideal candidate to incorporate the delivery payload. The gag polyprotein also imparts active release mechanisms of extracellular vesicles from cells. The assembled and packaged vesicle buds off from the producer cell membrane and enters the host cell. The payload protein gets released into the host cell by means of pol mediated protease activity. The requirement of pol protease activity however poses the risk of unwanted protease mediated degradation of target proteins. Gee et al. realized an induced dimerization regime where they utilized the high affinity of an FRB (FK506-rapamycin (FR)-binding domain) variant (T2098L) toward a rapamycin analog, AP21967. The addition of a ligand (AP21967) would trigger the specific interaction among the N-terminal fused FRB-SpCas9 and the FKBP12 domains. Within the target cells, the FRB-SpCas9 simply dissociates away effecting its release.

We are now beginning to see the progress of CRISPR-Cas9 in neurobiology from a mechanism to unravel disease pathogenesis to potential therapeutic intervention strategies. The simplicity of its principle remains its strength as novel delivery pathways assist toward a cleaner attack on disease specific problem points. After getting a bit of grasp on the delivery vehicles of CRISPR/Cas9 into neuronal cells, the next issue one needs to address is the control of their expression within the cell. Constitutive expression is often an undesirable property and calls for fine transcriptional or translational control. We now proceed to look into some of the mechanisms to control CRISPR/Cas9 expression temporally within target cells.

2.2 Temporal expression mechanisms in CRISPR-Cas9 systems

CRISPR-Cas9 systems invoke genetic modifications which in cases may be required to be induced temporally. Constitutive systems may pose unwanted phenotypes or developmental aberrations that may be unwanted. Interrogation of several dynamic processes of the cell benefits from temporal regulation of the CRISPR-Cas9 system that could be induced by either external small molecules, intra-cellular cues or physical signals like light.

Back in 2015, the first external small molecule inducible systems that could be induced by a tetracycline dependent promoter were reported by Dow et al.[23] A year later de Solis and group[24] implemented the first Doxycycline inducible CRISPR-Cas9 machinery that was used to interrogate the brain. de Solis and group tried two regimes, one where cas9 expression was inducible from the Dox-dependent modified TRE3G (Kozak removed) promoter and in the other the gRNA expression was inducible from the dox-dependent H1/TO promoter. They targeted the Tet2 genomic locus in mouse neuroblastoma cell line Neuro-2A (N2A) to obliterate an *Eco*RV restriction enzyme site. While the former displayed significant leakage of expression, the inducible sgRNA scheme showed tight regulation. Besides the "On" state of the system showed comparable indel frequencies to that of a constitutively expressing. In vivo infusions in the mouse basolateral amygdala showed indels at target locus 10–14 days post infection. A similar work by Kumar et al.[25,26] demonstrated Dox and cell-specific Cre- inducible system with a smaller orthologue of Cas9, saCas9 (from *Staphylococcus aureus*, 3.2 kb) both in vitro and in vivo. The smaller orthologue allows for a single AAV vector system instead of two.

Response time is an important parameter for scenarios demanding temporal control of CRISPR-Cas9 activity. Besides reducing off-target activities, a shorter response time translates to a finer temporal control of operational window for the rapidly changing intra-cellular world. Translational control of expression usually renders a faster response than the transcriptional regulations discussed above.

Nihongaki et al.[27]. designed a translationally and conditionally regulated Cas9 that could exploit physical cues like Blue light as an activation trigger (Fig. 2C). In the very innovative design by Nihongaki et al. the cas9 protein was split into two fragments and each fragment was tethered to one of two photosensitive protein components. These photosensitive proteins that had been derived from fungus, had the property to dimerize with each other upon irradiation with Blue light owing to their electrostatic properties. Dimerization of the two proteins called the "Negative" and "Positive" "magnets" (nMag and pMag) would in turn result in a reconstitution of the two fragments of cas9 bringing it to an active form (Fig. 2C). This strategy although very convenient had a much slower response than the ones with small molecules besides lesser editing efficiencies.

Davis et al.[16] achieved translational and conditional control of Cas9 by being able to switch it to its active form only in the presence of an externally supplied small molecule 4-hydroxytamoxifen (4-HT) that serves as a ligand for the Human-Estrogen-Receptor binding domain (ERT). The design was implemented by placing the ERT domain within the M. Tuberculosis RecA Intein which was thereby evolved to activate its Intein self-splicing property only in the presence of 4-HT. The newly evolved 4-HT dependent Intein named 37R3-2 was strategically placed upon Cas9 such that its presence would inhibit its catalytic endonuclease domain. Presence of 4-HT triggers self-splicing of 37R3-2 thereby restoring the active form of Cas9. The same molecule 4-HT has also been used to control the intra-cellular localization of Cas9 by conjugating Cas9 to ERT2 domain that facilitates nuclear localization in the presence of 4-HT. Presence of multiple such ERT2 domain on Cas9 efficiently regulates its intra nuclear presence in a reversible manner. Both these systems reported remarkable improvement in specificity and reduction of off-target hits as compared to the wild type variant of Cas9.

3. Gene editing with CRISPR-Cas9 in neurons

Addressing any disease that has its root in an anomaly of the DNA sequence or is associated with a DNA sequence related aggravation, requires us to go a step ahead of just controlled and specific gene silencing. The comparative ease of sequence editing with CRISPR-Cas9 has proven to be the hallmark of its success in biological research. Besides their sporadic forms, several neuropathological conditions have been traced to inherited genetic mutations which increase a person's susceptibility. Others, may be managed with intelligent edits to relevant protein domains. Insertion of protein tags are also of importance in untangling disease mechanisms and following protein localizations within cells. Let's have a look into the advances of CRISPR/Cas9 mediated gene editing strategies in neurobiological research and therapeutics.

3.1 HDR mediated approaches

Gene editing through CRISPR-Cas9 systems is generally made possible through HDR repair pathway. HDR or homology Directed repair usually remains active in the cell during S/G2 phase when DNA replication is active and there is proximity of sister chromatids[28]. Upon double strand break by a sgRNA directed Cas9, the presence of an externally supplied single stranded oligonucleotides (ssODNs) bearing the desired edits and sufficient sequence homology with the flanking regions (donor template), and HDR mediated sequence repair gets invoked that conveniently incorporates the edit. However, such a machinery can only function in mitotic cells and consequently terminally differentiated neurons hit a roadblock here. One way to introduce sequence edits is to introduce them at the neural progenitor stage when HDR is active.

Basal forebrain cholinergic neurons (BFCNs) have been known to be one the first type of cells to be affected during the onset of Alzheimer's.disease. In 2017, Ortiz-Virumbrales et al.[29] derived human BFCNs from hiPSCs harboring PSEN2^{N141I} mutation which triggers Aβ42/40 upregulation and is associated with the pathogenesis of Alzheimer's disease. The group corrected the point mutation through CRISPR/Cas9 intervention through HDR mediated editing in pre-differentiation stages (Fig. 3A). Upon BFCN differentiation, the electrophysiological parameters, which were deficient in the mutant cells, showed signs of return to normalcy post successful gene

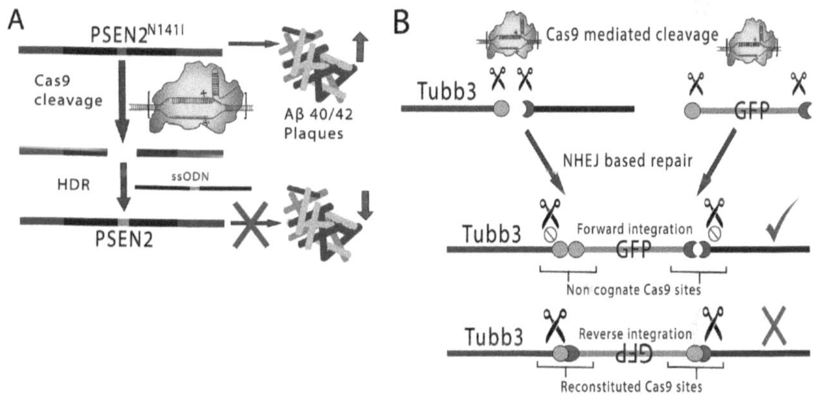

Fig. 3 (A) HDR mediated gene editing using CRISPR-Cas9. PSEN2^{N141I}, a point mutation of the Presinelin2 gene encourages Aβ42/40 formation ultimately causing Alzheimer's disease. ssODN harboring the desired gene edit and flanking homology sequences attempts to correct PSEN2 gene with CRISPR-Cas9 mediated blunt double strand break. HDR in the cell uses ssODN as the template strand incorporating the corrected sequence. (B) Homology-Independent Target Insertion (HITI). The technique frees gene editing techniques of cell cycle stage and availability of HDR. In this CRISPR-Cas9 mediated site specific sequence editing process, GFP was inserted into the 3′ end of *TUBB3* (Tubulin Beta 3 Class III). In HITI, recipient and donor sequences each contain two unique cas9 cleavage sites at the two ends, but are oppositely oriented. sgRNA guided cas9 cleavage generates blunt end breaks in both. NHEJ brings ends of the donor and recipient strands together and integrates them. In case of a correctly oriented NHEJ between ssODN and recipient strand, there is a loss of cognate sgRNA site by virtue of the oppositely placed cas9 sites in the starting material. ssODNs integrated in the reverse direction or rejoining of the original strand may reconstitute the sgRNA cognate sequence and thus result in repetitive cleavage.

correction. Notably they used asymmetric ssODNs with long and short homology arms of 91 and 36 bp respectively. Asymmetrical ssODNs, where the end with a shorter homology arm is oriented toward the area closer to the PAM side, have been shown to generate higher HDR editing efficiencies[30].

Mikuni et al.[31] tried to introduce tags or fluorescent proteins at the N-terminal or C-terminal ends of proteins in day 12 and 15 embryonic neural progenitors. In-utero electroporation into embryonic Swiss Webster mice brains introduced the CRISPR-Cas9 machinery, ssODNs and a piggyback transposase system for stable genomic integration. The sgRNAs targeted N-terminal or C-terminal ends of multiple neural proteins to avoid possible loss of protein function from indels in case of the more prevalent NHEJ based repair. ssODNs provided donor templates that ranged from short tag to larger GFP tags that got successfully integrated through HDR with a 7.5% efficiency (day 12 embryos) which is quite respectable. The process called Single cell Labeling of Endogenous proteins via HDR (SLENDR) opened wide possibilities of robust protein localization and visualization in Neurological research.

While SLENDR was introduced into the brain through in-utero electroporation, the success of the SLENDR technique prompted researchers to execute the process based on an AAV mediated delivery (vSLENDR) that demonstrated gene editing in the mouse mature brain in vivo as well as hippocampal slice cultures in vitro[32] targeting glial cells in addition to post-mitotic neurons and mitotic neuronal progenitors neurons. AAVs have been observed to induce efficient HDR in post-mitotic cells by a hitherto elusive mechanism. Compared to non-viral deliveries, AAVs have been shown to serve as excellent donor templates for Homologous recombination which can be reckoned to be facilitated by AAVs single stranded genome coupled with superior nuclear localization and the presence of inverted terminal repeats. Both the in vivo and in vitro post-mitotic cells displayed up to 30% gene editing efficiencies targeting genes like CaMKIIα as reported by integration of Fluorescent or HA tags.

While success was achieved in these above discussed HDR mediated sequence editing strategies we cannot ignore the fact that neurons are post-mitotic and HDR based DNA edits will never be as efficient in them as mitotic cells. We need to break free from our dependence on HDR in order to expand our possibilities of future neuro-therapeutics based on CRISPR-Cas9. In the following section we explore some of the HDR independent strategies in this context.

3.2 HDR independent approaches

To remove dependence of gene editing solely on HDR and thereby mitotic cell types, other gene editing techniques that are Homology independent or HDR independent were developed. Homology-Independent Target Insertion (HITI) is a tool that utilizes the predominant NHEJ pathway, present in both mitotic and post-mitotic cells, for site specific sequence insertion[33]. In HITI, both recipient and donor sequences contain a cas9 cleavage site but are oppositely oriented (Fig. 3B). sgRNA guided cas9 cleavage generates blunt end breaks in both. In an attempt to repair the breaks NHEJ brings ends of the donor and recipient strands together and integrates them. ssODNs integrated in the reverse direction or rejoining of the original strand may reconstitute the sgRNA cognate sequence and thus result in repeated cleavage. In case of a correctly oriented NHEJ between ssODN and recipient strand, there is loss of cognate sgRNA site by virtue of the oppositely directed sgRNA sites in the starting material (Fig. 3B). This strategy of Suzuki and Group[33] showed remarkable improvement of efficiency of knock-in dividing as well non-dividing cells in post-natal mammalian neurons compared to HDR based mechanisms. They packaged the machinery of HITI into two AAV vectors, with each set in two AAV serotypes 8 and 9. These AAV serotypes are known for safe and efficient infection. To test the HITI system in non-dividing cultured mouse primary neurons, they targeted to insert a GFP at the downstream of the tubb3 gene. Cytoplasmic fluorescence coupled with natural tubb3 localization showed successful integration. EdU labeling differentiated diving cells form non-dividing ones. Tubb3-GFP was also seen in Cre-dependent inducible expression systems of HITI system in the fetal brain after in-utero electroporation.

In another approach called Homology Mediated End Joining (HMEJ), devised by Yao et al. in 2017[34], the donor vector is designed such that the transgene is flanked by about 800 bp pairs of homology arms on both upstream and downstream ends respective to that of a sgRNA directed double strand break site on the host strand. The mechanism somewhat boroughs from the micro-homology mediated end joining (MMEJ) but utilizes longer homology arms instead of just 5–25 bp. This greatly encourages site specific integration and proper orientation. The process achieved high knock-in efficiency of up to 50% in adult mouse neurons with knock-ins higher than HR-, NHEJ- and MMEJ-based strategies. They targeted the neuronal marker dopamine beta-hydroxylase (*Dbh*) gene.

The requirement of unique donor template generation with long homology arms may pose logistical problems for high throughput requirements. To address this question, the third mechanism termed Homology Independent Universal Genome Engineering (HiUGE) was reported by Gao et al. in 2019[35]. This design decouples the donor template cas9 cleavage site from the genomic locus cleavage site. The donor vector cleavage and release is guided by a donor specific sgRNA (DS-sgRNA) while the genomic site is cleaved by an independent genome specific GS-sgRNA. The decoupling allows high throughput multiplexed genetic payload transfer to multiple genomic sites without the need for site specific donor oligonucleotide design. They tested primary neurons from neonates of conditional cas9 mice targeting genes like Tubb3 for an HA-epitope integration. As a demonstration of the Universal nature of the Donor template, the group showed simultaneous integration of donor payload sequence to 12 genes expressed in neurons of Glial cells with highly patterned subcellular localizations.

3.3 Prime editing with Cas9-nickase

Prime editing involves a comparatively newer method to introduce site specific Base edits exploiting the CRISPR-Cas9-nickase machinery. A Cas9-nickase is an engineered Cas9 protein that has lost its ability to inflict a double strand break at the target site but instead creates a single strand nick. These Cas9-nickases are also fused to a Reverse transcriptase domain (Fig. 4). This fused complex targeted to desired sites by means of something called a Prime editing guide RNA or a pegRNA. This pegRNA is a piece of RNA engineering in that it's composed of the sgRNA sequence coupled to an 3′ RNA template addendum containing the desired edit[36]. The two sections are conveniently placed through hairpin loops so that the sgRNA section can interact with the targeted sequence while the edit- containing RNA template can interact with the opposite strand where the nick is inflicted the edit would be introduced (Fig. 4). The sgRNA binding and consequent nick encourages invasion of the edit strand and triggers the action of reverse transcriptase to extend the DNA installing the edit. The pegRNA's RNA template faces the nicked strand in association with the reverse transcriptase domain and consists of a homologous sequence site which anneals to the nicked strand. Reverse transcriptional sequence extension follows from this region and incorporates the desired edit. The 3′ end of the newly synthesized strand ligates to the host strand while the 5′ loose end of the original sequence strand that gets displaced from the nick by the newly synthesized strand, faces

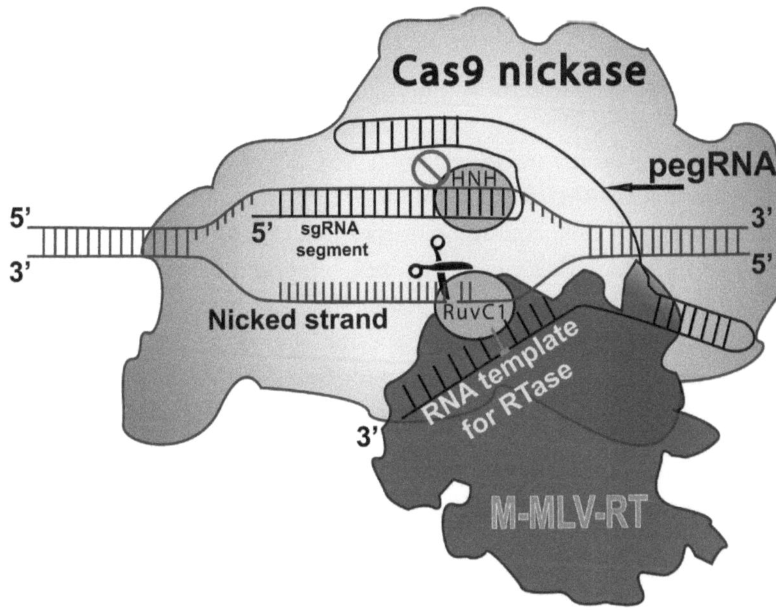

Fig. 4 Prime editing with Cas9 and pegRNA. The Cas9 protein has been rendered partially deactivated with only the RuVC1 domain retaining its catalytic activity thereby generating a single strand nick. Cas9-nickase protein has been tethered to an engineered reverse transcriptase domain (M-MLV-RT) which uses an RNA template to polymerize DNA from 5′ to 3′. This RNA template incorporates the desired base edit in itself and has been added to the sgRNA strand's 3′ prime end, thereby forming the prime editing guide RNA or pegRNA. The pegRNA's RNA template faces the nicked strand in association with the reverse transcriptase domain and consists of a homologous sequence site which anneals to the nicked strand. Reverse transcriptional sequence extension follows from this region and incorporates the desired edit. The 3′ end of the newly synthesized strand ligates to the host strand while the 5′ loose end of the original sequence strand that gets displaced from the nick by the newly synthesized strand, faces 5′ to 3′ exonuclease activity.

5′ to 3′ exonuclease activity. This system has been reported to show lesser indels than HDR mediated editing attempts albeit with a very low editing efficiency of 6–8% in primary neuronal cultures. This system provides for a highly convenient editing technique displaying high fidelity and great potential.

4. The repurposed CRISPR-Cas9: Nuclease deficient-cas9 (dcas9) and neuroscience

Armed as an extremely convenient and customizable genome targeting tool, CRISPR-Cas9 was poised to find applicability in multifarious arenas of research, much beyond those requiring direct interrogation and modification of the DNA sequence. Reaching out to specific target sequences without actually modifying them, becomes a necessity in several research aims. The first step toward was taken by Qi and group in 2013[37], who disabled the nuclease activity of Cas9 through point mutations into its RuvC1 (D10A) and HNH (H840A) domains. The resulting protein, deficient-Cas9 or dCas9, although rendered nuclease deficient, was still fully capable of sgRNA guided target sequence binding with similar specificity and efficiency. dCas9's ability to "seamlessly" conjugate with additional effector domains imparting sequence specific alternative catalytic activities opened up gateways to a plethora of repurposed dcas9 variants to function as synthetic transcription factors, epigenomic editors, imaging "tags" etc.

4.1 DNA methylation modification with Dcas9

Increasing scientific evidences point toward an equal role of epigenetics in cell dynamics, development and interactions as genetics. With the nervous system being no exception, several aspects of its functioning and lineage characteristics involving gene expression regulation is intricately related to the epigenetic landscape. DNA methylation, particularly cytosine methylation status heavily controls neural functionality[38] and therefore implicated in several neuropathological conditions. Fragile X syndrome whose hallmark is an overly condensed chromatin triggered by cytosine methylation on extended CGG repeats near the promoter region of FMR1 is a classic example. Angelman's syndrome and Prader Willi's syndrome are often caused by methylation induced silencing of genetic material on chromosome 15 from one parent while the similar section on the other is defective or absent. Angelman's syndrome results from a defective gene (UBE3A) on the maternal chromosome 15 while methylation causes the paternal counterpart to go silent. Prader Willi's syndrome follows the vice versa where specific genes on the Maternal counterpart is silenced while the paternal section defective or absent. These CpG methylation events occur neurologically in the developing brain resulting in severe intellectual, developmental and physiological defects that have no cure. Rett syndrome is caused by sporadic mutations in the MECP2 gene, a DNA methylation reader, which is said to have a role in shutting off other genes in nerve cells.

Dcas9 protein, when conjugated to epigenetic effectors or modulators, can generate excellent site specific epigenetic editing tools for researches attempting to elucidate epigenome related neuropathologies.

In 2016 Liu and group[39] fused dcas9 to TET1 or DNTM3a, that are involved in DNA demethylation and methylation respectively. Dcas9-TET1 fusion was targeted to BDNF-IV promoter in cultured primary neurons. A sixfold increase in BDNF expression was observed. Significant leakage was also seen from controls with no sgRNA where a twofold increase in expression was observed. This clearly hinted at potential off-target activity. DNTM3a fused to dcas9 has also been employed to regulate expression levels of the SNCA gene. SNCA (encoding Alpha Synuclein) over expression has been associated with Parkinson's disease with PD brains showing SNCA over-expression associated with demethylation of CpGs at *SNCA* intron 1. Kantor et al.[40] in 2018 tried to restore healthy levels of SNCA levels by inducing methylation at the SNCA intron 1 site via dcas9-DNMT3a fusions (Fig. 5A). Transduced dopaminergic neurons derived from hiPSCs from a Parkinson's patient showing SNCA locus triplication exhibited a big success in restoration to normal NCA levels.

To explore therapeutic possibilities to mitigate Fragile X Syndrome (FXS), the hyper-methylated island of the CGG trinucleotide repeat in the 5″ UTR of Fmr1 was targeted for demethylation by dcas9 conjugated to TET1, which resulted in promising results of significant alleviation of the hyper-methylated state and the phenotype of the disease[41] (Fig. 5B). In an in vivo study, grafts of FXS neurons derived from dcas9-TET1treated iPSC, showed enhanced FMR1 levels in the mouse brain.

4.2 Histone modification with Dcas9

Histones are known to allow or disallow access to DNA sequences where heterochromatic states are condensed states offering steric hindrances to DNA acting elements and euchromatic states are permissive toward strand access and gene expression. These landscapes govern cell-fate and function and gene expression by controlling access to their respective cis-regulatory elements[42]. Functional annotations of the DNA regulatory elements can be made possible by means of affecting histones. In 2019, dcas9 fused to histone modifying effectors, such as the catalytic core of the human acetyltransferase p300 (dcas9-p300) or Histone deacetylase 8, HDAC8 (dcas9-HDAC8), were employed[43] to influence the expression of the genes Fos and Npas4. The second enhancer (Enh2) of the neuron immediate early gene, Fos was targeted. While nowhere near as dramatic as the expression changes seen with CRISPRa systems discussed below, dcas9-p300 and dcas9-HDAC8 could weakly increase and decrease Fos activity respectively.

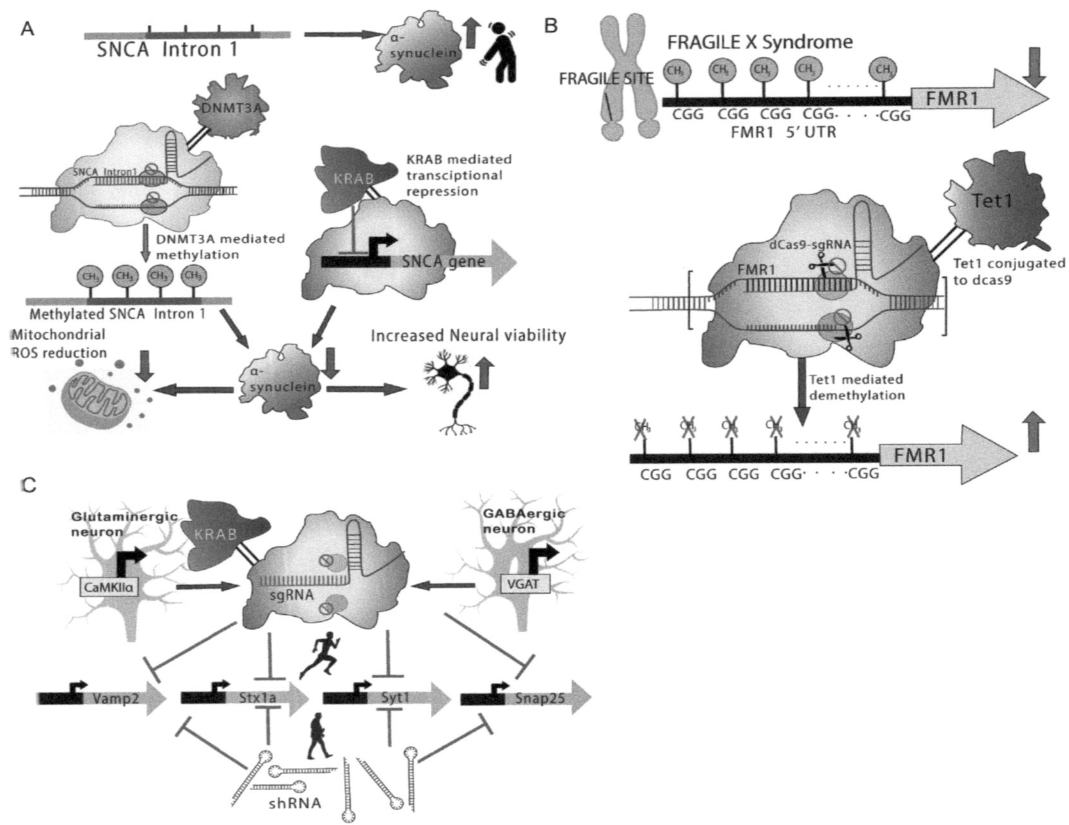

Fig. 5 See figure legend on opposite page.

4.3 CRISPR mediated chimeric transcription factors: CRISPRa & CRISPRi

Artificially rendered transcriptional activation or inactivation finds applications in genetic circuit designs and metabolic engineering and to therapeutic strategies in multiple arenas. The versatility of dcas9 in tethering it to transcription factors was a possibility too good to resist. Synthetic transcriptional activators/silencers from dCas9 were reported by Gilbert et al.[44] in 2013 who fused transcriptional effector domains like the NF-κB trans-activating unit P65 or the tetrameric herpes simplex virus protein 16 (VP64) or the Krüppel-associated box domain of Kox1 (KRAB) to dcas9. While dcas9-VP64 showed efficient transcriptional activation of target gene (CRISPRa), dcas9-KRAB functioned as a targeted transcriptional repressor (CRISPRi). Several groups thereafter went on to interrogate neurodegenerative mechanisms and neurodevelopmental aspects employing these strategies.

Point mutations and multiplications of the SNCA gene encoding the Alpha Synuclein (present on the synaptic tips of neurons) is implicated in the pathogenesis of Parkinson's disease which is associated with severe movement and coordination inabilities. These mutations that are known to cause Alpha synuclein aggregations, have been reported from both

Fig. 5 (A) Over-expression of the SNCA gene leads to higher levels of alpha synuclein, a condition associated with Parkinson's disease. The condition features demethylation of CpGs at intron 1 of SNCA. dCas9, with deactivated nuclease domains is tethered to DNMT3A. sgRNA guides the Cas9=DNMT3A complex to SNCA intron 1 where DNA methyl transferase activity methylates the site and down-regulates alpha synuclein expression from SNCA. Similarly, dcas9 tethered to the transcriptional repressor domain KRAB (Krüppel-associated box domain) inhibits transcription at the SNCA promoter when targeted to the site. A drop in the alpha synuclein levels results in a decrease of mitochondria associated ROS in concerned neural cells along with increased viability. (B) In Fragile X syndrome, the 5′ UTR region of the FMR1 gene contains excessive CGG repeats that attract methylation of the site resulting in epigenetic silencing of FMR1. Tet1 domain, that causes demethylation on DNA sequences, is tethered to dcas9 and targeted to the 5′ UTR of FMR1. The resulting demethylation increases FMR1 levels. (C) Four proteins, known to be actively involved in release of neurotransmitter vesicles, namely Vesicle associated membrane protein II (Vamp2), syntaxin 1A (Stx1a), synaptotagmin I (Syt1) and synaptosome associated protein 25 (Snap25) were targeted through dcas9 tethered to KRAB. CRISPRi repression mediated through KRAB was seen in all 4 genes with the indications of the process being more efficient than corresponding shRNA mediated repression of the same genes. Cell type specificity of CRISPRi activity was observed when the CaMKIIα promoter was employed to drive expression of the system for Glutaminergic neurons and the VGAT promoter for GABAergic neurons.

familial as well as sporadic forms of Parkinson's disease. Besides Alzheimer's, the overabundance of hyper-phosphorylated microtubule associated Tau proteins (encoded by MAPT gene) is also implicated in Parkinson's disease associated with tauopathies and frontotemporal dementia. The overabundance of the Amyloid Precursor protein (APP) is the hallmark of Alzheimer's disease. Huntington's disease pathology has been connected to the expanded CAG repeats in the HTT gene that cause huntingtin protein to show aggregation-prone poly-glutamine tracts. Heman-Ackah. et al.[45] in 2016, tried to elucidate the roles of these four proteins, HTT, MAPT, SNCA and APP and the effects of their repression through CRISPRi mediated intervention on their Transcription Start Sites (TSS). Dcas9-KRAB (Fig. 5A) and dvas9-VPR targeted to SNCA in human iPSC derived neurons showed excellent transcriptional control of Parkinson's associated gene.

In 2015, Frank et al.[46] employed the dcas9-VP64 fusion to elucidate chromatin mechanisms driving neuronal differentiation and maturation. They targeted putative enhancers flanking Grin2c gene in cerebellar granule neurons of the developing mouse. Huang and group[47], in 2017, targeted CRISPRi to the AP-1 binding sequence of the App promoter by injecting carrier AAVs into the newborn mouse cortex. The group attempted to explore how isoforms of ApoE (ApoE4, ApoE3, ApoE2) influence the transcription of amyloid-β precursor protein with ApoE4 causing with the most enhancement of APP levels and ApoE2 showing a protective role. It was observed that addressing each promoter with multiple targeted sgRNAs could yield a better activation through CRISPRa. As the number of Transcription factors recruiting at the locus and the cooperativities between them seemed to have a positive influence of the efficiency and intensity of gene activation. Multimeric recruitment of these transcription factors therefore should generate a more efficient activation system. Tanenbaum et al.[48] in 2014, fused an array of tandemly repeating GCN4 epitopes to the dcas9 protein. On the other hand, they fused VP64 with short chain variable fragments (scFv) that are known to have high affinity toward the GCN4 epitopes. The resulting combination displayed multimeric recruitment of VP64 on the GVN4 array and thereby a much stronger transcriptional influence on the targeted DNA sequence. This tag of GCN4 sites on dcas9 was termed as the Supernova Tag or SunTag.

In a clever move of sgRNA engineering, Konnermann et al.[49] in 2015 improved the sgRNA sequence by replacing non interacting stem loops of the sgRNA chain with the MS2 aptamers that could recruit MS2 coat proteins (MCPs). These MCPs have the potential to be fused to a transcriptional effector like p65 or the human heat shock factor 1 (HSF1). It was observed that the transcriptional activation of the ASCL1 and MYOD1 genes was further enhanced when dcas9 mediates recruitment of transcriptional activators at the site was heterogeneous. This meant that a recruitment of a combination of MS2 bound p65 working with a dcas9-VP64 was more potent than a MS2 bound VP64 attached to a dcas9-VP64. This hints at some sort of synergy in heterogeneous recruitments. MS2 could be fused to two different transcriptional effectors in tandem like a MS2-p65-HSF1 which proved to be one of the most effective combinations in activating transcription.

Moving ahead with these findings, Chavez et al.[50] in 2015 went on to screen putative transcriptional activators to map their activation efficiency. While three activators, VP64, p65 and Epstein-Barr-Virus trans-activator (Rta) emerged as the most potent, the fusions of p65 and Rta to dcas9 had lower activity than that of VP64. Further improvements in efficiency were realized by fusing P65 and Rta to the original dcas9-VP64 (dcas9-VP64-P65-Rta, or dcas9-VPR). Up to 320-fold enhancement of gene activation in the triple-effector fusion highlighted the synergistic effects among them.

Improved fused chimeric activator complexes of the likes of the above, were further screened for maximal transcription from NeuroD1 and Asc1 genes revealing the aptamer based technique used by Konermann et al. as the most consistent. Similar in design, dcas9 based gene repression systems have also been developed and screened for gene repression efficiency. Yeo et al. in 2018 reported a dcas9-KRAB-MeCP2 fusion complex as the most potent and robust among dcas9 based inhibitor systems.

In a comprehensive study of the dcas9-VP64-P65-Rta (VPR) efficacy on neuronal cells, Savell et al.[51] in 2019 lentivirally delivered it to rodent neurons in vitro and in vivo reporting their robust, tunable and modular gene activation with multiplexing capabilities. The targeted the unique promoters of the multi-transcript gene brain-derived neurotrophic factor (BdnF) (that is implicated in memory formation and learning and synaptic plasticity) in primary rat cortical, hippocampal, and striatal neuron cultures as well as diverse brain structures in vivo including the adult rat prefrontal cortex (PFC), hippocampus, and nucleus accumbens. They optimized the system for neurons by placing the CRISPR machinery under neurons specific Synapsin promoter which showed better activity than several ubiquitous promoters. Targeting multiple sgRNAs per promoter yielded better activation re-iterating previous reports. Also the system showed multiple gene targeting efficiencies in a coordinated manner.

In vivo interrogations of the neurons largely benefit from these CRISPRa systems and tried to improve upon as was seen in a work by Zhou et al.[52] in 2018. The team combined components of multiple systems discussed above to render a dCas9-SunTag-p65-HSF1 Transcriptional activator by replacing the VP64 tetramers of the SunTag system with p65-HSF1 domains. They accomplished multiple gene activation in the mouse brain in vivo and reported a stronger CRISPRa effect than what was exhibited by the previous SunTag, dcas9-VPR or the RNA aptamer based design implemented by Konermann et al. (discussed above). Viral delivery of these large number of components was surpassed by generating a transgenic mouse line the incorporated a Cre-dependent expression of the dCas9-SunTag-p65-HSF1 system. To impart neural cell-type specificity for in vivo operations, the Cre system was placed under the control of hSyn or CamKIIα that are specific to cells in the brain and the sgRNAs were targeted for 8 protein coding genes and 2 lncRNAs. With the successful establishment of this in vivo multiplexed gene targeting system operating in the brain, the transgenic mouse line proved to be particularly useful.

Looking at CRISPRi based interrogations in vivo in the mammalian brain, a lentiviral mediated targeting of neuron specific genes like vesicle associated membrane protein II (Vamp2), syntaxin 1A (Stx1a), synaptotagmin I (Syt1) and synaptosome associated protein 25 (Snap25)-all four known to be actively involved in release of neurotransmitter vesicles- was attempted by Zheng et al.[53] in 2018 upon hippocampal neurons (Fig. 5C). A Dcas9-KRAB repressor system, the group compared the efficacies of gene down-regulation as compared to shRNA driven knock-downs (Fig. 5C). A clear winner was found in CRISPRi systems that exhibited up to 90% down-regulation with shRNA showing much lower efficiency. Cell-type specificity was imparted to the system by driving dcas9-KRAB expression from CaMKIIα promoter for glutaminergic neurons and VGAT promoter for GABAergic neurons (Fig. 5C). As the system was introduced into the dentate gyrus of the brain, signal transduction decreased to 20% and was lim ited to the granular layer of the cerebellum. Almost complete knock-down of Syt1 was observed in a cell type specific manner. Patch clamp studies demonstrated concomitant decrease in excitatory postsynaptic current (EPSC) from CRISPRi disrupted neurotransmission. Phenotypes involving memory and learning were observed in mice behavioral studies that were in accordance with cell type specific down-regulation of Syt1 causing alteration of the Inhibitory-Excitatory (I-E) ratio with the hippocampus. With Dcas9-KRAB affecting glutaminergic neurons, a shift was observed toward inhibition as indicated by reduction in performance of spatial memory in solving mazes. Fear response was also damped in such mice. Mice receiving GABAergic neuron specific CRISPRi showed heightened fear responses indicating I-E shift toward excitation. Further the off-target activity of the dcas9-KRAB fusion was assayed by targeting the Syt1 gene with a library of sgRNAs that contained subtle mismatches to the actual nucleotide sequence. It was observed that the subtle mismatches in sgRNAs resulted in largely unperturbed expression levels of Syt1 strongly indicating against off targeting activities of the system.

5. RNA-targeting with CRISPR in the neuron

So far we have been discussing cas9 mediated targeting of DNA sequences in a neurological perspective. However, as alluded to previously in this chapter, there is a group of Class II- Type VI RNA targeting Cas13 proteins that can be similarly used to target RNA oligos. Targeting RNA through CRISPR finds wide application in scenarios demanding RNA visualization, post-transcriptional interceptions including Knock-down and base-editing or isoform modifications. Cas13 protein family differs from its cousin Cas9 in more ways than one. This single effector protein, is comprised of two higher eukaryotes and prokaryotes nucleotide (HEPN) Ribonuclease domains that function independent of each other[54]. One of these domains is known to process the pre-crRNA by identifying them for their stem loop structures thereby eliminating the need for a separate tracr-RNA to act as adapter. They also eliminate the need for specific PAM sequences for self nonself differentiation. It instead hunts for Protospacer flanking sequence (PFS) for LshCas13a, situated at the 3′ end of the spacer sequence and known to be comprised of a single A, U, or C base pair.

Cas13a tolerates single mismatches between the crRNA and target sequence, however, cutting efficiency of Cas13a is reduced when two mismatches are present. The protospacer flanking sequence (PFS) for LshCas13a, which is analogous to the PAM sequence for Cas9, is located at the 3′ end of the spacer sequence and consists of a single A, U, or C base pair. This is while saying that certain variants of Cas13 do not even require a PFS for successful RNA cleavage[55]. Cas13 variants have been found to have very little sequence conservation between save for HEPN domain. Cas13a, a variant of cas13 has been found to tolerate up to one base mismatch. Beyond that cleavage efficiency takes a plunge. A marked difference between cas13a and cas9 is that cas13a retains its active form after just one successful crRNA guided RNA cleavage whereas cas9 goes into an inactive state. Cas9 requires repeated sgRNA sequence matches and PAM availability for each individual strand but cas13a continues RNA nuclease activity irrespective of sequence match or a PFS. The continued nuclease activity maybe an evolutionary tactic to cause cell death or cell dormancy of a phage infected cell before viral load in it increases and the infection spreads. Studies by several groups have established the superiority of cas13 mediated RNA interceptions compared to shRNA mediated RNAi techniques.

A fairly comprehensive study by Konermann et al. 2018[56] comparing the CRISPRi mediated transcriptional inhibition as effected by dcas9-KRAB against the post-transcriptional inhibition accomplished by a cas13d effector from *Ruminococcus flavefaciens* (CasRx) when targeted toward the genes B4GALNT1, ANXA4, and HOTTIP in HEK293FT revealed marked superiority of the later with median knock-down efficiency reaching 96% as opposed to 53% for CRISPRi. shRNA mediated knock-downs also yielded 53% efficiency with significant off-target activity which was practically absent in CasRx systems. Reports from other studies put beside this, suggest CasRx being more efficient than Cas13 variants than Cas13a or cas13b. The latter two showed 80% and 66% knock-down efficiency as opposed to 97% from CasRx seen in here. Moreover, CasRx also exhibited capacity of multiplexed RNA targeting as was tested with 14sgRNAs against including coding and non-coding transcripts. An added advantage of CasRx its small size of about 2.9 kb as opposed to the large sizes of Cas13 variants. This greatly facilitates viral delivery mechanisms where packaging size limitations often become a hindrance.

In a newer study in 2020[57], CasRx found a beautiful use in conversion of Müller glia into retinal ganglion cells (RGCs) as well as astrocytes into Dopaminergic neurons. The elegant approach saw a post transcriptional downregulation of a single RNA-binding protein, polypyrimidine tract-binding protein 1 (Ptbp1) by targeting its mRNA through AAV borne CasRx and two gRNAs. It resulted in the highly efficient conversion of Müller glia into retinal ganglion cells (RGCs). The conversion displayed appearance of several morphological characters of RGCs on converted cells and a partial return of vision in mice inflicted with drug induced retinal damage. The strategy was also employed to locally knockdown ptbp1 in the striatum which resulted in local astrocytes becoming dopaminergic. This diminished symptoms of motor dysfunction in a Parkinson's disease mouse model. This technique shows a strong promise of future disease management for Parkinson's as astrocytes are quite abundant at the site of dopaminergic neuronal loss seen in PD.

RNA targeting deficient-CasRx (dCasRx) systems provide an excellent platform for exon skipping strategies which form the basis of therapeutic interventions of several diseases. Tau protein related diseases like Frontotemporal dementia with parkinsonism linked to chromosome 17 (FTDP-17), for example, stems from a point mutation in the exon 10 of MAPT gene which dictates the Tau protein isoform that would be expressed[58]. As seen from the disease pathologies of Duchenne muscular dystrophy (DMD) and its mitigation approaches, skipping an aberrant exon along a few flanking others exons during post transcriptional processing, results in a restoration of the reading frame with a slightly shorter protein but with rescued functionality[59]. Pre-mRNA splicing is an extensive coordination event between *cis*-acting elements, *trans*-acting spliceosomes and several heterogeneous nuclear RNPs (hnRNPs) that direct alternative splicing routes through effecting exon exclusion[60,61]. DcasRx was fused to hnRNPa1-CTD and was targeted to multiple putative splicing elements inducing exon skipping events in a fluorescent splice reporter.[56] The design was then targeted to the MAPT gene in cortical neurons from patient derived iPSCs, to try and attempt skipping of the aberrant exon 10 hoping to diminish levels of the Tau pathogenic isoform 4R. Isoform 4R and 3R ratios were restored to healthy ranges (about 50%) after dCasRx-hnRNPa1 and three versions of the exon 10 targeting sgRNA were delivered through AAV routes.

6. Perspectives and concluding remarks

We have taken a brisk walk through the CRISPR-Cas systems applied to Neuroscience and neuropathological therapeutics. It is beyond doubt that possibilities are limitless with this excellent machinery of selectively targeting nucleotide sequences. Besides DNA, CRISPR based systems have been used on RNA targets with their effector domains ranging from their plain endonuclease activity to introduce indels or edits to prime editing with nickase, or histone modifiers to DNA methylation/demethylation systems or fluorescence tags. While each of these system showed enhanced efficiency compared to traditional methods, CRISPR-Cas systems also face their own set of challenges. In the context of neuroscience the most crucial of these is the Blood Brain Barrier. Methods to overcome this usually revolve around nanoparticle engineering or modified viral coat proteins that can cross the blood brain barrier. Modified capsids were developed for AAVs by Chan et al.[62] in 2017 reporting success in crossing the BBB. They utilized a method called Cre REcombinase-based AAV Targeted Evolution (CREATE)[63] to develop an improved version of the AAV viral capsid AAV-PHP·B (called AAV-PHP.eB) that can more efficiently transduce cells in the brain and spinal cord. However these systems need to be further improved for an efficient delivery process as direct cranial injection based methods have extremely limited reach to deeper tissue.

Tissue specificity and viral vector tropism is also a matter that should not be ignored. Murlidharan et al.[64] addressed the question of restricting AAV transduction to the CNS and off-target transduction such as the liver by developing a chimeric AAV2g9 which brings together desirable CNS specific attributes from its parents AAV2 and AAV9. AAV2g9 containing CRISPR-Cas9 targeting the schizophrenia risk gene MIR137, displayed preferential neuronal transduction with marked decrease in off-target glial tropism in the brain. Injection of AAV2g9 with sgRNA payload, through Cerebrospinal fluid of SPR/Cas9 knock-in mice showed minimal off-target organ gene silencing activities such as the liver.

The third parameter that can make or break the efficacy of a CRISPR-Cas system is its immunogenicity. More often than not, humans have inherent immunogenicity to host microbes of the Cas protein primarily *S. pyogenes* and *Staphylococcus aureus* as these are rampantly present on our skin. A study by Charlesworth et al.[65] in 2019 reported SaCas9 triggered humoral response and specific antigen T-cells in human subjects. Orthologs of Cas proteins from other non-immunogenic sources could be an answer but concerns of efficiency and off-target activities of these need to be comprehensively verified. Specificity and efficacy of targeting is an important parameter that can have devastating consequences in the therapeutic context if deleterious off-target activities are present. Low efficiencies as observed in HDR based mechanisms are not desirable. Translational medicine in future is poised to increasingly rely on CRISPR mechanisms and ailments of the Nervous system that are yet to see a feasible curative strategy from modern medicine, look forward to these betterments in the CRISPR-Cas system.

References

1. Ishino Y, Shinagawa H, Makino K, Amemura M, Nakatura A. Nucleotide sequence of the iap gene, responsible for alkaline phosphatase isoenzyme conversion in Escherichia coli, and identification of the gene product. *J Bacteriol.* 1987;169:5429–5433.
2. Jinek M, et al. A programmable dual-RNA-guided DNA endonuclease in adaptive bacterial immunity. *Science.* 2012;337:816–821.
3. Shamkov S, Smargon A, Scott D, et al. Diversity and evolution of class 2 CRISPR–Cas systems. *Nat Rev Microbiol.* 2017;15(3):169–182. https://doi.org/10.1038/nrmicro.2016.184.
4. Xie H, et al. Erratum to: SaCas9 requires 5′-NNGRRT-3′ PAM for sufficient cleavage and possesses higher cleavage activity than SpCas9 or FnCpf1 in human cells. *Biotechnol J.* 2018;13:1–21. https://doi.org/10.1002/biot.201700561 [1700561].
5. Lee CM, Cradick TJ, Bao G. The neisseria meningitidis CRISPR-Cas9 system enables specific genome editing in mammalian cells. *Mol Ther.* 2016;24:645–654.
6. Mali P, et al. RNA-guided human genome engineering via Cas9. *Science.* 2013;339:823–826.
7. Campa CC, Weisbach NR, Santinha AJ, Incarnato D, Platt RJ. Multiplexed genome engineering by Cas12a and CRISPR arrays encoded on single transcripts. *Nat Methods.* 2019;16:887–893.
8. Incontro S, Asensio CS, Edwards RH, Nicoll RA. Efficient, complete deletion of synaptic proteins using CRISPR. *Neuron.* 2014;83:1051–1057. https://doi.org/10.1016/j.neuron.2014.07.043.
9. Swiech L, et al. In vivo interrogation of gene function in the mammalian brain using CRISPR-Cas9. *Nat Biotechnol.* 2015;33:102–106. https://doi.org/10.1038/nbt.3055.
10. Platt RJ, et al. CRISPR-Cas9 knockin mice for genome editing and cancer modeling. *Cell.* 2014;159:440–455. https://doi.org/10.1016/j.cell.2014.09.014.
11. Bäck S, et al. Neuron-specific genomic modification in the adult rat brain using CRISPR-Cas9 transgenic rats. *Neuron.* 2019;102:105–119.e8.
12. Petrova PS, et al. MANF: a new mesencephalic, astrocyte-derived neurotrophic factor with selectivity for dopaminergic neurons. *J Mol Neurosci.* 2003;20:173–188. https://doi.org/10.1385/JMN:20:2:173.
13. Apostolou A, Shen Y, Liang Y, Luo J, Fang S. Armet, a UPR-upregulated protein, inhibits cell proliferation and ER stress-induced cell death. *Exp Cell Res.* 2008;314:2454–2467. https://doi.org/10.1016/j.yexcr.2008.05.001.
14. György B, et al. CRISPR/Cas9 mediated disruption of the Swedish APP allele as a therapeutic approach for early-onset Alzheimer's disease. *Mol Ther–Nucleic Acids.* 2018;11:429–440.
15. Raikwar SP, et al. Targeted gene editing of glia maturation factor in microglia: a novel Alzheimer's disease therapeutic target. *Mol Neurobiol.* 2019;56:378–393. https://doi.org/10.1007/s12035-018-1068-y.
16. Davis KM, Pattanayak V, Thompson DB, Zuris JA, Liu DR. Small molecule-triggered Cas9 protein with improved genome-editing specificity. *Nat Chem Biol.* 2015;11:316–318. https://doi.org/10.1038/nchembio.1793.
17. Staahl BT, et al. Efficient genome editing in the mouse brain by local delivery of engineered Cas9 ribonucleoprotein complexes. *Nat Biotechnol.* 2017;35:431–434.
18. Chew WL, et al. A multifunctional AAV-CRISPR-Cas9 and its host response. *Nat Methods.* 2016;13:868–874. https://doi.org/10.1038/nmeth.3993.
19. Liu J, Gaj T, Wallen MC, Barbas CF. Improved cell-penetrating zinc-finger nuclease proteins for precision genome engineering. *Mol Ther–Nucleic Acids.* 2015;4:e232.
20. Park H, et al. In vivo neuronal gene editing via CRISPR–Cas9 amphiphilic nanocomplexes alleviates deficits in mouse models of Alzheimer's disease. *Nat Neurosci.* 2019;22:524–528.

21. Lee B, et al. Nanoparticle delivery of CRISPR into the brain rescues a mouse model of fragile X syndrome from exaggerated repetitive behaviours. *Nat Biomed Eng.* 2018;2:497–507. https://doi.org/10.1038/s41551-018-0252-8.
22. Gee P, et al. Extracellular nanovesicles for packaging of CRISPR-Cas9 protein and sgRNA to induce therapeutic exon skipping. *Nat Commun.* 2020;11:1334. https://doi.org/10.1038/s41467-020-14957-y.
23. Dow LE, et al. Inducible in vivo genome editing with CRISPR-Cas9. *Nat Biotechnol.* 2015;33:390–394. https://doi.org/10.1038/nbt.3155.
24. de Solis CA, Ho A, Holehonnur R, Ploski JE. The development of a viral mediated CRISPR/Cas9 system with doxycycline dependent gRNA expression for inducible In vitro and In vivo genome editing. *Front Mol Neurosci.* 2016;9:1–14.
25. Kumar N, et al. The development of an AAV-based crispr sacas9 genome editing system that can be delivered to neurons in vivo and regulated via doxycycline and cre-recombinase. *Front Mol Neurosci.* 2018;11:1–14.
26. Zhou XX, et al. A single-chain photoswitchable CRISPR-Cas9 architecture for light-inducible gene editing and transcription. *ACS Chem Biol.* 2018;13:443–448.
27. Nihongaki Y, Kawano F, Nakajima T, Sato M. Photoactivatable CRISPR-Cas9 for optogenetic genome editing. *Nat Biotechnol.* 2015;33:755–760.
28. Saleh-Gohari N, Helleday T. Conservative homologous recombination preferentially repairs DNA double-strand breaks in the S phase of the cell cycle un human cells. *Nucleic Acids Res.* 2004;32:3683–3688.
29. Ortiz-Virumbrales M, et al. CRISPR/Cas9-Correctable mutation-related molecular and physiological phenotypes in iPSC-derived Alzheimer's PSEN2N141I neurons. *Acta Neuropathol Commun.* 2017;5:1–20.
30. Chu VT, et al. Increasing the efficiency of homology-directed repair for CRISPR-Cas9-induced precise gene editing in mammalian cells. *Nat Biotechnol.* 2015;33:543–548. https://doi.org/10.1038/nbt.3198.
31. Mikuni T, Nishiyama J, Sun Y, Kamasawa N, Yasuda R. High-throughput, high-resolution mapping of protein localization in mammalian brain by in vivo genome editing. *Cell.* 2016;165:1803–1817. https://doi.org/10.1016/j.cell.2016.04.044.
32. Nishiyama J, Mikuni T, Yasuda R. Virus-mediated genome editing via homology-directed repair in mitotic and postmitotic cells in mammalian brain. *Neuron.* 2017;96:755–768.e5. https://doi.org/10.1016/j.neuron.2017.10.004.
33. Suzuki K, et al. In vivo genome editing via CRISPR/Cas9 mediated homology-independent targeted integration. *Nature.* 2016;540:144–149.
34. Yao X, et al. Homology-mediated end joining-based targeted integration using CRISPR/Cas9. *Cell Res.* 2017;27:801–814.
35. Gao Y, et al. Plug-and-play protein modification using homology-independent universal genome engineering. *Neuron.* 2019;103:583–597.e8.
36. Anzalone AV, Randolph PB, Davis JR, et al. Search-and-replace genome editing without double-strand breaks or donor DNA. *Nature.* 2019;576:149–157.
37. Qi LS, et al. Repurposing CRISPR as an RNA-γuided platform for sequence-specific control of gene expression. *Cell.* 2013;152:1173–1183.
38. Smith ZD, Meissner A. DNA methylation: roles in mammalian development. *Nat Rev Genet.* 2013;14:204–220.
39. Liu KI, et al. A chemical-inducible CRISPR-Cas9 system for rapid control of genome editing. *Nat Chem Biol.* 2016;12:980–987. https://doi.org/10.1038/nchembio.2179.
40. Kantor B, et al. Downregulation of SNCA expression by targeted editing of DNA methylation: a potential strategy for precision therapy in PD. *Mol Ther.* 2018;26:2638–2649. https://doi.org/10.1016/j.ymthe.2018.08.019.
41. Liu XS, et al. Rescue of fragile X syndrome neurons by DNA methylation editing of the FMR1 gene. *Cell.* 2018;172:979–992. https://doi.org/10.1016/j.cell.2018.01.012.

42. Yarrington RM, Verma S, Schwartz S, Trautman JK, Carroll D. Nucleosomes inhibit target cleavage by CRISPR-Cas9 in vivo. *Proc Natl Acad Sci U S A*. 2018;115:9351–9358.
43. Chen L, Lin YT, Gallegos DA, et al. Enhancer histone acetylation modulates transcriptional bursting dynamics of neuronal activity-inducible genes. *Cell Rep*. 2019;26:1174–1188.
44. Gilbert LA, Larson MH, Morsut L, et al. CRISPR-mediated modular RNA-guided regulation of transcription in eukaryotes. *Cell*. 2013;154:442–451.
45. Heman-Ackah SM, Bassett AR, Wood MJA. Precision modulation of neurodegenerative disease-related gene expression in human iPSC-derived neurons. *Sci Rep*. 2016;6:1–12.
46. Frank CL, Liu F, Wijayatunge R, et al. Regulation of chromatin accessibility and Zic binding at enhancers in the developing cerebellum. *Nat Neurosci*. 2015;18:647–656.
47. Huang YWA, Zhou B, Wernig M, Südhof TC. ApoE2, ApoE3, and ApoE4 differentially stimulate APP transcription and Aβ secretion. *Cell*. 2017;168:427–441. https://doi.org/10.1016/j.cell.2016.12.044.
48. Tanenbaum ME, Gilbert LA, Qi LS, Weissman JS, Vale RD. A protein-tagging system for signal amplification in gene expression and fluorescence imaging. *Cell*. 2014;159:635–646.
49. Konermann S, et al. CRISPR-Cas9 complex. *Nature*. 2015;517:61422–61427.
50. Chavez A, et al. Highly-efficient Cas9-mediated transcriptional programming. *bioRxiv*. 2014;12:012880.
51. Savell KE, et al. A neuron-optimized CRISPR/dCas9 activation system for robust and specific gene regulation. *eNeuro*. 2019;6:1–17.
52. Zhou H, et al. In vivo simultaneous transcriptional activation of multiple genes in the brain using CRISPR-dCas9-activator transgenic mice. *Nat Neurosci*. 2018;21:440–446.
53. Zheng Y, et al. Author Correction: CRISPR interference-based specific and efficient gene inactivation in the brain. *Nat Neurosci*. 2018;21:894. https://doi.org/10.1038/s41593-018-0077-5.
54. Abudayyeh OO, et al. C2c2 is a single-component programmable RNA-guided RNA-targeting CRISPR effector. *Science*. 2016;353. https://doi.org/10.1126/science.aaf5573.
55. Cox DBT, Gootenberg JS, Abudayyeh OO, et al. RNA editing with CRISPR-Cas13. *Science*. 2018;358:1019–1027.
56. Konermann S, et al. Transcriptome engineering with RNA-targeting type VI-D CRISPR effectors. *Cell*. 2018;173:665–676.e14. https://doi.org/10.1016/j.cell.2018.02.033.
57. Zhou H, et al. Glia-to-neuron conversion by CRISPR-CasRx alleviates symptoms of neurological disease in mice. *Cell*. 2020;181:590–603. https://doi.org/10.1016/j.cell.2020.03.024.
58. Boeve BF, Hutton M. Refining frontotemporal dementia with parkinsonism linked to chromosome 17: introducing FTDP-17 (MAPT) and FTDP-17 (PGRN). *Arch Neurol*. 2008;65:460–464.
59. Nelson CE, Wu Y, Gemberling MP, et al. Long-term evaluation of AAV-CRISPR genome editing for Duchenne muscular dystrophy. *Nat Med*. 2019;25:427–432.
60. Matera AG, Wang Z. Erratum: a day in the life of the spliceosome. *Nat Rev Mol Cell Biol*. 2014;15:294. 108-122.
61. WANG Y, et al. Mechanism of alternative splicing and its regulation. *Biomed Rep*. 2015;3:152–158.
62. Chan KY, et al. Engineered AAVs for efficient noninvasive gene delivery to the central and peripheral nervous systems. *Nat Neurosci*. 2017;20:1172–1179. https://doi.org/10.1038/nn.4593.

63. Deverman BE, et al. Cre-dependent selection yields AAV variants for widespread gene transfer to the adult brain. *Nat Biotechnol.* 2016;34:204–209. https://doi.org/10.1038/nbt.3440.
64. Murlidharan G, et al. CNS-restricted transduction and CRISPR/Cas9-mediated gene deletion with an engineered AAV vector. *Mol Ther–Nucleic Acids.* 2016;5:e338.
65. Charlesworth CT, et al. Identification of preexisting adaptive immunity to Cas9 proteins in humans. *Nat Med.* 2019;25:249–254. https://doi.org/10.1038/s41591-018-0326-x.

Index

Note: Page numbers followed by "*f*" indicate figures and "*t*" indicate tables.

D

E

S

T

U

V

W

X

Y

Z

Papier | Fördert gute Waldnutzung
FSC® C083411

Kolibri 360 GmbH
Ferdinand-Jühlke-Straße 7
99095 Erfurt, Deutschland
produktsicherheit@kolibri360.de